Impedance Spectroscopy and its Application in Biological Detection

This book includes basics of impedance spectroscopy technology, substrate compatibility issues, integration capabilities, and several applications in the detection of different analytes. It helps explore the importance of this technique in biological detection, related micro/nanofabricated platforms and respective integration, biological synthesis schemes to carry out the detection, associated challenges, and related future directions. The various qualitative/quantitative findings of several modules are summarized in the form of the detailed descriptions, schematics, and tables.

Features:

- Serves as a single source for exploring underlying fundamental principles and the various biological applications through impedance spectroscopy.
- Includes chapters based on nonbiological applications of impedance spectroscopy and IoT-enabled impedance spectroscopy-based methods for detection.
- Discusses derivations, substrates, applications, and several integrations.
- Describes micro/nanofabrication of impedance-based biological sensors.
- Reviews updated integrations like digital manufacturing and IoT.

This book is aimed at researchers and graduate students in material science, impedance spectroscopy, and biosensing.

Emerging Materials and Technologies

Series Editor: Boris I. Kharissov

The *Emerging Materials and Technologies* series is devoted to highlighting publications centered on emerging advanced materials and novel technologies. Attention is paid to those newly discovered or applied materials with potential to solve pressing societal problems and improve quality of life, corresponding to environmental protection, medicine, communications, energy, transportation, advanced manufacturing, and related areas.

The series takes into account that, under present strong demands for energy, material, and cost savings, as well as heavy contamination problems and worldwide pandemic conditions, the area of emerging materials and related scalable technologies is a highly interdisciplinary field, with the need for researchers, professionals, and academics across the spectrum of engineering and technological disciplines. The main objective of this book series is to attract more attention to these materials and technologies and invite conversation among the international R&D community.

Polymer Processing
Design, Printing and Applications of Multi-Dimensional Techniques
Abhijit Bandyopadhyay and Rahul Chatterjee

Nanomaterials for Energy Applications
Edited by L. Syam Sundar, Shaik Feroz, and Faramarz Djavanroodi

Wastewater Treatment with the Fenton Process
Principles and Applications
Dominika Bury, Piotr Marcinowski, Jan Bogacki, Michal Jakubczak, and Agnieszka Jastrzebska

Mechanical Behavior of Advanced Materials: Modeling and Simulation
Edited by Jia Li and Qihong Fang

Shape Memory Polymer Composites
Characterization and Modeling
Nilesh Tiwari and Kanif M. Markad

Impedance Spectroscopy and its Application in Biological Detection
Edited by Geeta Bhatt, Manoj Bhatt and Shantanu Bhattacharya

Nanofillers for Sustainable Applications
Edited by N.M Nurazzi, E. Bayraktar, M.N.F. Norrrahim, H.A. Aisyah, N. Abdullah, and M.R.M. Asyraf

For more information about this series, please visit: www.routledge.com/Emerging-Materials-and-Technologies/book-series/CRCEMT

Impedance Spectroscopy and its Application in Biological Detection

Edited by
Geeta Bhatt, Manoj Bhatt and
Shantanu Bhattacharya

CRC Press
Taylor & Francis Group
Boca Raton London New York

CRC Press is an imprint of the
Taylor & Francis Group, an **informa** business

Designed cover image: © Geeta Bhatt, Manoj Bhatt and Shantanu Bhattacharya

First edition published 2024
by CRC Press
6000 Broken Sound Parkway NW, Suite 300, Boca Raton, FL 33487-2742

and by CRC Press
4 Park Square, Milton Park, Abingdon, Oxon, OX14 4RN

CRC Press is an imprint of Taylor & Francis Group, LLC

ISBN: 978-1-032-41442-3 (hbk)
ISBN: 978-1-032-41443-0 (pbk)
ISBN: 978-1-003-35809-1 (ebk)

DOI: 10.1201/9781003358091

Typeset in Times
by codeMantra

Contents

PART I *Fundamentals of impedance spectroscopy*

PART II *Applications of impedance-based biological sensors*

PART III   Technological integrations and developments

Preface

Biological/biomolecular detection is one of the most influential MEMS applications as it is a socially relevant field that observes several real-world applications such as bedside diagnostics, food/water screening, and animal/plant procreation. Most of the MEMS devices present their applications directly/indirectly for the detection of a range of biomolecules, namely, proteins, carbohydrates, lipids, DNA, RNA, cells, viruses, drugs, and also several derived entities that directly/indirectly affect human health in several ways. To take appropriate measures or to monitor their effects properly, we should efficiently detect these molecules. There exist a range of biomolecular detection techniques such as colorimetric/optical, mass-based sensing, and electrochemical/electrical, out of which electrochemical impedance spectroscopy-based detection has gained popularity due to its highly electronic integrable nature, fast detection speed, easy response readability, and low detection cost. The research in this field is growing in a fast manner and is further expected to rise in the coming times due to the high utility of impedance spectroscopy-based biological detection. Also, the advent of the digital manufacturing era will further enhance the possibilities of making this type of detection more cost-effective and user-friendly, as lab-on-phone and IoT integration are very attractive options to integrate with. This further enhances the value of this research field immensely and hence, keeping these aspects in mind, the need for a thorough book on the topic that would present all the detailed aspects in one place and help the several researchers, technologists, and academicians to move ahead in this research field with all the background knowledge at one place is a must.

Keeping the current and future requirements of impedance spectroscopy-based biological detection, this book provides a detailed discussion on several related topics and presents a collection of articles with the fundamentals of impedance spectroscopy technology, various substrate compatibility issues, several integration capabilities, and several applications in the detection of different analytes. This kind of detection is an interdisciplinary field, including material science, physics, electrochemistry, electronic engineering, mechanical engineering, curve fitting and analysis techniques, and biological methods.

Readers of this book will be able to explore the basics of impedance spectroscopy, its importance in biological detection, various related micro/nano-fabricated platforms and respective integration, various biological synthesis schemes to carry out the detection, associated challenges, and related future directions. Through the inclusion of the most updated experimental techniques, synthesis protocols, integration modules, and technology development, this book is expected to educate graduate students, technologists, postdoctoral fellows, and professionals on diversified fronts, and hence, serving as an in-hand knowledge material as well as handbook.

Each chapter of this book is contributed by leading experts in the field of material science, physics, electrochemistry, electronic engineering, and mechanical engineering to ensure that readers of the book are facilitated with the most updated and worthy information through detailed critical reviews and perspectives of various modules. The various qualitative and quantitative findings of several modules will be summarized in the form of detailed descriptions, attractive schematics, and lucid tables.

About the editors

Dr. Geeta Bhatt is presently employed as an Assistant Professor in the Department of Mechanical Engineering at BITS Pilani, Pilani, Rajasthan. She completed her Ph.D. in design and fabrication of gene amplification/transformation devices from the Indian Institute of Technology, Kanpur, India in 2020, M.E. from Thapar University, Patiala, India in 2013, and B.Tech. from the Dehradun Institute of Technology, Dehradun, India in 2011. She has published two patents, 11 peer-reviewed research articles, and eight book chapters along with some international conference proceedings. Her area of expertise is MEMS fabrication, microdevices and bacteria/DNA detection, microfluidics, micro-manufacturing, and surface modification.

Dr. Manoj Bhatt is presently employed as an Assistant Professor in the Department of Electronics and Communication at Govind Ballabh Pant University of Agriculture & Technology, Pant Nagar. He has done his Ph.D. in behavioural modeling and analysis of digital predistorter for HPA using neural committee machine from the Govind Ballabh Pant University of Agriculture and Technology, Pant Nagar, Uttarakhand, India in 2021, M.Tech. in Digital Signal Processing from the Govind Ballabh Pant Institute of Engineering and Technology, Pauri Garhwal, Uttarakhand, India in 2012, and B.Tech. from DBIT, Dehradun, India in 2010. He has published four research papers in peer-reviewed journals, one conference proceeding, and one book chapter. His research interests include communication systems, signal processing, propagation modeling for both uniform and nonuniform propagation scenarios, predistortion techniques, and regression analysis through neural networks.

Prof. Shantanu Bhattacharya (Ph.D.) is FRSC, Abdul Kalam National Fellow, GVMM Chair, and Professor of Mechanical Engineering at the Indian Institute of Technology Kanpur. He served as Head of Design Interdisciplinary program between 2017 and 2020 at IIT Kanpur. Prior to this, he completed his M.S. in Mechanical Engineering from Texas Tech University, Lubbock, Texas, and a Ph.D. in Bioengineering from the University of Missouri, Columbia, USA. He also completed postdoctoral training at the Birck Nanotechnology Center, Purdue University. His main research interests are design and development of micro- and nano-sensors and actuation platforms, nano-energetic materials, micro- and nano-fabrication technologies, water remediation using visible light photocatalysis, and product design and development. He has many awards and accolades to his credit which includes the Institution of Engineers Young Engineer Award, the Institute for Smart Structures and Systems Young Scientist Award, the Best Mechanical Engineering Design Award (National Design Research Forum, IEI), fellowship from the Royal Society of Chemistry, UK (FRSC), The Institution of Engineers (India) (FIE), and Institution of Electronics and Telecommunication Engineering (FIETE), NASI Reliance Platinum

Jubilee Award from the National Academy of Sciences, Dr. R.S. Khandpur Award of IETE, and Er. M.P. Baya National Award of the IEI. He is a senior member of IEEE and has been bestowed with the prestigious Abdul Kalam Technological Innovation National Fellowship from the Indian National Academy of Engineering. He has guided many Ph.D. and master's students and has many international journal publications, patents, books, and conference proceedings to his name.

Contributors

Manoj Singh Adhikari,
LPU Phagwara, Jalandhar 144001,
  Punjab, India

Adreeja Basu,
Biological Science, St. John's
  University, Jamaica, New York, NY
  11420, USA

Aviru Kumar Basu,
Quantum Material and Device Unit,
  Institute of Nanoscience and
  Technology, Mohali 140306, Punjab,
  India

Geeta Bhatt,
Department of Mechanical Engineering,
  BITS Pilani 333031, Rajasthan, India

Manoj Bhatt,
Department of Electronics &
  Communication Engineering,
  GBPUAT, Pantnagar 263153,
  Uttarakhand, India

Shantanu Bhattacharya,
Department of Mechanical Engineering,
  Indian Institute of Technology
  Kanpur 208016, UP, India

Jasdeep Bhinder,
École de technologie supérieure
  Montréal, Quebec H3C 1K3, Canada

Sagnik Sarma Choudhury,
Department of Mechanical Engineering,
  Indian Institute of Technology
  Kanpur 208016, UP, India

Vibhas Chugh,
Quantum Material and Device Unit,
  Institute of Nanoscience and
  Technology, Mohali 140306, Punjab,
  India

Manshu Dhillon,
Quantum Material and Device Unit,
  Institute of Nanoscience and
  Technology, Mohali 140306, Punjab,
  India

Ankur Gupta,
Department of Mechanical Engineering,
  Indian Institute of Technology
  Jodhpur-342037, Rajasthan, India

Rishi Kant,
Department of Mechanical Engineering,
  Harcourt Butler Technical
  University, Kanpur 208002, UP,
  India

Nitish Katiyar,
Department of Mechanical Engineering,
  Indian Institute of Technology
  Kanpur 208016, UP, India

Vinay Kishnani,
Department of Mechanical Engineering,
  Indian Institute of Technology
  Jodhpur 342037, Rajasthan, India

Kapil Manoharan,
Department of Mechanical Engineering,
  Indian Institute of Technology
  Kanpur 208016, UP, India

Sanjay Mathur,
Department of Electronics &
  Communication Engineering,
  GBPUAT, Pantnagar 263153,
  Uttarakhand, India

Keerti Mishra,
National University of Singapore,
  Singapore 119077

Abhishek Naskar,
Quantum Material and Device Unit,
  Institute of Nanoscience and
  Technology, Mohali 140306, Punjab,
  India

Mohit Pandey,
Department of Mechanical Engineering,
  Indian Institute of Technology
  Kanpur 208016, UP, India

Ramesh N. Pudake,
Amity Institute of Nanotechnology,
  Amity University UP, Noida 201301,
  UP, India

Mayank Punetha,
BTKIT Dwarahat, Almora-263653,
  Uttarakhand, India

Mohammed Rashiku,
Department of Mechanical
  Engineering, Indian Institute of
  Technology Kanpur 208016, UP,
  India

Rajkumar Saha,
Mathematics, City University of
  New York, New York, NY
  10017, USA

Ranamay Saha,
Department of Mechanical
  Engineering, Indian Institute
  of Technology Kanpur 208016,
  UP, India

Gaganpreet Singh,
Department of Physics, Stockholm
  University, 11419 Stockholm,
  Sweden

Mitesh Upreti,
Department of Electronics &
  Communication
  Engineering, GBPUAT,
  Pantnagar-263153, Uttarakhand,
  India

Part I

Fundamentals of impedance spectroscopy

1 A historical perspective on impedance spectroscopy and its application in biological detection

Geeta Bhatt, Manoj Bhatt, and
Shantanu Bhattacharya

CONTENTS

1.1   INTRODUCTION: HISTORICAL PERSPECTIVE

The history of EIS dates back to the 1880s when Oliver Heaviside introduced impedance in the electrical engineering domain (Macdonald 1992). The implementation represented that the EIS data greatly involves complex terms in it; hence, the
research associated with data simplification/interpretation started. His initial experiments were very soon elaborated by C.P. Steinmetz and A.E. Kennelly by proposing
a complex representation of the data and corresponding vector diagrams. The several incremental enhancements in this field were majorly supported by the dielectric
response field through the Cole–Cole plot in 1941, which is commonly used today
also for the 2D representation of impedance response. This was the first basic representation scheme adapted for common use. With progression in research, the 3D
representation plots were finally introduced in 1981 (Ross Macdonald, Schoonman,
and Lehnen 1981), which enhanced the modeling capability of EIS.

In EIS, the impedance data is usually fit to an equivalent circuit for the parameter extraction procedure. Hence, significant studies and advancements have also
been done in this domain to express the significance of various resistance, capacitance, inductance, and constant phase elements. The various circuit components are
significant under specific conditions and are representative of particular processes

DOI: 10.1201/9781003358091-2

happening in the system. Macdonald and Garber in the year 1977 proposed a very acceptable analysis method in mathematical modeling or equivalent circuit domain, termed a complex nonlinear least square fitting method (Macdonald and Garber 1977).

The primary experimental study in EIS is dominated by the double-layer formulation/response. A double layer consisting of a narrow band of positive and negative ions is commonly formulated when a solution comes in contact with the metal electrode (Liu et al. 2008). The inner Helmholtz plane (IHP) and outer Helmholtz plane (OHP) constitute the layers formulated by the condensed solvent molecular layer and diffusion layer, respectively. Corresponding to the formulated double layer, different components of the equivalent circuit arise because of different phenomena that occur in the system, viz. solution resistance, polarization resistance, electron transfer across IHP/OHP or medium resistance, the non-faradaic capacitance of the double layer, and the mass transport of reactants/products across the double layer. Hence, the characterization of the double layer is a very important aspect of EIS measurement/interpretation.

As EIS measurements are primarily mathematical data-driven, EIS studies with respect to the mathematical formulation, electrical equivalent circuit fitting, and experiments are extensively being explored. These studies have been further refined through the usage of the latest, most accurate, and fast-measuring equipment that has been available from various manufacturers.

1.2   UTILIZATION OF EIS-BASED MEASUREMENTS IN BIOLOGY

EIS is a great characterization tool that acquires the electric phenomenon at the electrode–analyte interface. Its application regime ranges from biological to nonbiological analytes in solid, liquid, or gas mode. EIS can either directly (non-faradaic mode—no redox probes) or indirectly (faradaic mode—redox probes incorporation) measure the analyte response. As compared with other electrical/electrochemical measurement techniques, viz. amperometric, voltammetric, and conductimetric, EIS offers a prominent advantage of using low input measurement signals (generally AC sinusoidal voltage), which reduces the possibilities of sample contamination and hence can be efficiently used in various scenarios without spoiling the analyte sample.

EIS can be conveniently applied to any biological regime from whole cell mode (Gómez, Bashir, and Bhunia 2002) to individual nucleic acid/protein (Bhatt et al. 2021). The wide area of EIS measurements utilizes various detection routes, viz. adsorption (Kant et al. 2017), intercalator probe binding (Defever et al. 2011), polymerase chain reaction (PCR) (Bhattacharya et al. 2008), or antibody/antigen interaction (Taylor, Marenchic, and Spencer 1991) to monitor various biological systems like cells/bacteria (Varshney et al. 2005), DNA/RNA (Bhatt et al. 2019), proteins (Rodriguez, Kawde, and Wang 2005), drugs (Li, Liu, and Luong 2005), aptamers (Rodriguez, Kawde, and Wang 2005), antibody/antigens (Suehiro et al. 2006), and ions (Day et al. 2018). Except for biological analytes, EIS is extensively used for nonbiological applications also, viz. characterizing batteries/fuel cells (Tröltzsch,

Kanoun, and Tränkler 2006), paint/lubricants (Smiechowski and Lvovich 2002), and electroactive thin films (Inzelt 2007).

Although there are various detection routes, such as optical, mechanical, chemical, and electrical/electrochemical ones, EIS-based devices have been gaining high importance and replacing the other detection routes due to the several advantages it offers. It has also been observed that it has replaced market-sold optical/chemical detection techniques. To elaborate, among various *in vitro* DNA detection schemes, qPCR (quantitative PCR) (Nayak et al. 2013) carried out through fluorescence-based detection is the most promising technique that produces accurate and reproducible results. However, qPCR is an expensive technique, and, hence, this detection trend is slowly shifting toward electronic detection where impedance-based detection has evolved as one of the most emerging techniques that provide fast detection at a low cost due to its easy electronic integrability, miniaturizability, and lesser requirement of additional entities. The impedance-based detection lacks specificity in the case when electronic labels have been missing, which increases the cost of detection. Hence, cost reduction and specificity enhancement are the primary forte of the present EIS research regime.

1.3   BOOK ORGANIZATION

Because of several crises observed in the recent past, the concept of the medical industry has been redefined to integrate the emerging need for point-of-care/bedside diagnostics. It is now readily adaptable as per the changing need of the hour and, hence, introduces flexibility. In this respect also, EIS has been coming up with many possibilities due to its highly integrable, miniaturized, and rapid nature. This high level of involvement of EIS in the biological field drives us to compile the various literature available in this field, and, hence, in this direction, this book has been extensively compiled to include several aspects of EIS. This book has been subdivided into three major sections, namely, fundamentals of impedance spectroscopy, applications of impedance-based biological sensors, and technological integrations and developments. These subsections discretely discuss fundamental mathematical modeling, equivalent circuit fitting, microfluidic integration, various biological/environmental applications of EIS, noninvasive EIS techniques/devices, recent trends in biomedical devices, and the importance of EIS in this field, nonbiological applications of EIS, and various technological integrations, such as CMOS, polymers, and aptamer-based smart sensors in EIS measurements. This book finally concludes on the challenges that have been observed in EIS measurements and how they can be addressed.

In this line, specific titles covered in this book are:

1   Basic principles of impedance spectroscopy
2   Integrating microfluidics and sensing for capability enhancement
3   Characterization of bioanalytical applications using impedance spectroscopy
4   Impedance spectroscopy and environmental monitoring
5   Application of noninvasive impedance spectroscopy techniques
6   Progress in biomedical devices and importance of impedance spectroscopy

REFERENCES

Bhatt, Geeta, Swati Gupta, Gurunath Ramanathan, and Shantanu Bhattacharya. 2021. "Integrated DEP Assisted Detection of PCR Products with Metallic Nanoparticle Labels Through Impedance Spectroscopy." *IEEE Transactions on Nanobioscience* 21 (4): 502–10.

Bhatt, Geeta, Keerti Mishra, Gurunath Ramanathan, and Shantanu Bhattacharya. 2019. "Dielectrophoresis Assisted Impedance Spectroscopy for Detection of Gold-Conjugated Amplified DNA Samples." *Sensors and Actuators, B: Chemical* 288: 442–53. https://doi.org/10.1016/j.snb.2019.02.081.

Bhattacharya, Shantanu, Shuaib Salamat, Dallas Morisette, Padmapriya Banada, Demir Akin, Yi-Shao Liu, Arun K. Bhunia, Michael Ladisch, and Rashid Bashir. 2008. "PCR-Based Detection in a Micro-Fabricated Platform." *Lab on a Chip* 8 (7): 1130. https://doi.org/10.1039/b802227e.

Day, C., S. Søpstad, H. Ma, C. Jiang, A. Nathan, S. R. Elliott, F. E. Karet Frankl, and T. Hutter. 2018. "Impedance-Based Sensor for Potassium Ions." *Analytica Chimica Acta* 1034: 39–45. https://doi.org/10.1016/j.aca.2018.06.044.

Defever, Thibaut, Michel Druet, David Evrard, Damien Marchal, and Benoit Limoges. 2011. "Real-Time Electrochemical PCR with a DNA Intercalating Redox Probe." *Analytical Chemistry* 83 (5): 1815–21. https://doi.org/10.1021/ac1033374.

Gómez, Rafael, Rashid Bashir, and Arun K. Bhunia. 2002. "Microscale Electronic Detection of Bacterial Metabolism." *Sensors and Actuators, B: Chemical* 86 (2–3): 198–208. https://doi.org/10.1016/S0925-4005(02)00175-2.

Inzelt, G. 2007. "Charge Transport in Conducting Polymer Film Electrodes." *Chemical and Biochemical Engineering Quarterly* 21 (1): 1–14.

Kant, Rishi, Geeta Bhatt, Poonam Sundriyal, and Shantanu Bhattacharya. 2017. "Relevance of Adhesion in Fabrication of Microarrays in Clinical Diagnostics." In *Adhesion in Pharmaceutical, Biomedical and Dental Fields*, edited by K. L. Mittal, and F. M. Etzler, 257–98. Scrivener Publishing LLC.

Li, Chen Zhong, Yali Liu, and John H. T. Luong. 2005. "Impedance Sensing of DNA Binding Drugs Using Gold Substrates Modified with Gold Nanoparticles." *Analytical Chemistry* 77 (2): 478–85. https://doi.org/10.1021/ac048672l.

Liu, Yi Shao, Padmapriya P. Banada, Shantanu Bhattacharya, Arun K. Bhunia, and Rashid Bashir. 2008. "Electrical Characterization of DNA Molecules in Solution Using Impedance Measurements." *Applied Physics Letters* 92 (14): 143902. https://doi.org/10.1063/1.2908203.

Macdonald, J. Ross. 1992. "Impedance Spectroscopy." *Annals of Biomedical Engineering* 20 (3): 289–305. https://doi.org/10.1007/BF02368532.

Macdonald, J. Ross, and J. A. Garber. 1977. "Analysis of Impedance and Admittance Data for Solids and Liquids." *Journal of The Electrochemical Society* 124 (7): 1022–30. https://doi.org/10.1149/1.2133473.

Nayak, Monalisha, Deepak Singh, Himanshu Singh, Rishi Kant, Ankur Gupta, Shashank Shekhar Pandey, Swarnasri Mandal, Gurunath Ramanathan, and Shantanu Bhattacharya. 2013. "Integrated Sorting, Concentration and Real Time PCR Based Detection System for Sensitive Detection of Microorganisms." *Scientific Reports* 3: 3266. https://doi.org/10.1038/srep03266.

Rodriguez, Marcela C., Abdel Nasser Kawde, and Joseph Wang. 2005. "Aptamer Biosensor for Label-Free Impedance Spectroscopy Detection of Proteins Based on Recognition-Induced Switching of the Surface Charge." *Chemical Communications* no. 34: 4267–69. https://doi.org/10.1039/b506571b.

Ross Macdonald, J., J. Schoonman, and A. P. Lehnen. 1981. "Three Dimensional Perspective Plotting and Fitting of Immittance Data." *Solid State Ionics* 5 (C): 137–40. https://doi.org/10.1016/0167-2738(81)90211-3.

Smiechowski, Matthew F., and Vadim F. Lvovich. 2002. "Electrochemical Monitoring of Water-Surfactant Interactions in Industrial Lubricants." *Journal of Electroanalytical Chemistry* 534 (2): 171–80. https://doi.org/10.1016/S0022-0728(02)01106-3.

Suehiro, Junya, Akio Ohtsubo, Tetsuji Hatano, and Masanori Hara. 2006. "Selective Detection of Bacteria by a Dielectrophoretic Impedance Measurement Method Using an Antibody-Immobilized Electrode Chip." *Sensors and Actuators, B: Chemical* 119 (1): 319–26. https://doi.org/10.1016/j.snb.2005.12.027.

Taylor, Richard F., Ingrid G. Marenchic, and Richard H. Spencer. 1991. "Antibody- and Receptor-Based Biosensors for Detection and Process Control." *Analytica Chimica Acta* 249 (1): 67–70. https://doi.org/10.1016/0003-2670(91)87009-V.

Tröltzsch, Uwe, Olfa Kanoun, and Hans Rolf Tränkler. 2006. "Characterizing Aging Effects of Lithium Ion Batteries by Impedance Spectroscopy." *Electrochimica Acta* 51: 1664–72. https://doi.org/10.1016/j.electacta.2005.02.148.

Varshney, Madhukar, Liju Yang, Xiao Li Su, and Yanbin Li. 2005. "Magnetic Nanoparticle-Antibody Conjugates for the Separation of Escherichia Coli O157:H7 in Ground Beef." *Journal of Food Protection* 68 (9): 1804–11. https://doi.org/10.4315/0362-028X-68.9.1804.

2 Basic principles of impedance spectroscopy

Geeta Bhatt, Gaganpreet Singh,
and Shantanu Bhattacharya

CONTENTS

2.1   INTRODUCTION

Point-of-care (POC) diagnostics is a critical need of the medical industry, and several techniques have evolved in the past few decades to address this need. Various methods of detection, such as rapid assays, lateral flow assays (Kumar, Bhushan, and Bhattacharya 2016), enzyme-linked immunosorbent assays, polymerase chain reaction (PCR) (Bhattacharya et al. 2008; Nayak et al. 2013), and electrophoresis, are continually refined and miniaturized into devices to meet the requirements of rapid, reliable, and sensitive detection with minimum possible human intervention. Many POC diagnostics devices utilize electrokinetic phenomena, such as dielectrophoresis (DEP) (Bhatt et al. 2017), electrophoresis (Ghosh et al. 2011), electroosmosis, and streaming potential, for carrying out the detection of the targets. The extensive use of electrokinetic phenomena is mainly attributed to the charged nature of the

DOI: 10.1201/9781003358091-3

analytes (nucleic acids or proteins), helping in carrying out the diagnostics. The biological system characterization or most commonly diagnostics is achieved through various characterization techniques including colorimetric/optical (Bhatt, Kant, and Bhattacharya 2019), electrochemical/electrical (Bhatt et al. 2019), and mass-based methods (Basu, Basu, and Bhattacharya 2020), in a standalone manner or by integrating them with additional techniques. The various methods offer several benefits along with different disadvantages, where colorimetric detection, which is a constituents/concentration-based detection technique is prone to exposure to environmental conditions. Optical detection schemes are quite sophisticated in nature and are quite dependent on the system conditions and mechanical or mass-based detections are fabrication-dependent, which is a very crucial task. Due to these shortcomings, electrochemical detection is presently being studied quite a lot, as it observes a highly flexible and integrating nature. This technique is quite capable of being easily integrated into several other complementary techniques involving optical, mechanical, magnetic, and various electrokinetic modules.

There are various electrochemical detection schemes available for monitoring/analyzing the biological systems, viz., cyclic voltammetry (CV) (Ye et al. 2003), differential pulse voltammetry (Yeung, Lee, and Hsing 2006), squarewave voltammetry (SWV) (Meric et al. 2002), potentiometric stripping analysis (Marrazza et al. 2000), and electrochemical impedance spectroscopy (EIS) (Liu et al. 2008). Among the stated techniques, most utilize the intercalating molecule to make the analyte detectable, while EIS can be efficiently performed with or without any intercalating molecule, which makes detection technology rather simpler. EIS measures and elaborates various system conditions through the obtained values of impedance across a wide range of frequency sweeps. In detail, EIS detects an analyte by observing the change in impedance due to a change in the constituents of the analyte solution. Impedance of the analyte solution is mostly influenced by the double layer, which is formed at the interface of the electrode and the analyte. As we talk about POC devices, it becomes essential to find out new strategies to carry out miniaturization, to provide a quantifiable estimate of the target, and also to machine-read the signals generated for the quantification and provide user convenience. The impedance-based detection method for diagnostics is easily integrable and compatible with electronic detection platforms, and it can be easily mapped into the overall scheme of POC devices. Further impedance-based detection also facilitates easy quantification of the analyte sample where a good process of curve fits the impedance profile through various circuit models, thus making the quantitative estimation accurate and easy. The only step that may be needed as an add-on to the impedance-based detection process is a specificity enhancement step and this is being accomplished through various means by different researchers.

The simplest system to describe impedance measurement is liquid electrochemistry. The measurement electrodes can be of any configuration like two electrodes (Bhatt et al. 2019, 2021), three electrodes (Yang, Ruan, and Li 2003), or four electrodes (Wei, Wang, and Zhang 2013) and also of different sizes/shapes ranging from macroscale to microscale or parallel plate electrode to ring-shaped or some modified shapes of electrodes (Nakano et al. 2015) and chemistries (Park and Park 2009) on several substrates. These days, the electrodes printed on paper are also used for carrying out an impedance measurement, which efficiently reduces the electrode size and cost, making the system more economical (Lu et al. 2012). Keeping the utility of EIS analysis in mind, this chapter explains the fundamental principles and associated parameters/applications of EIS. The various sections of the chapter detail the fundamental principles of impedance spectroscopy and various related aspects, like, electrical analogies to various processes (conductivity and diffusion in electrolytes, adsorption, charge transfer at the electrode–electrolyte interface, polarization), impedance measurements in a basic electrochemical system and corresponding representation, mass and charge transfer modes, fluid flow and associated microfluidics; equivalent circuit diagrams; various impedance spectroscopy modules; and common applications of impedance spectroscopy.

2.2 FUNDAMENTAL PRINCIPLES OF IMPEDANCE SPECTROSCOPY

Electrical impedance was conceptualized in the 1880s by Oliver Heaviside and the researchers soon translated its representation in terms of vectors and complex (Orazem and Tribollet 2008). It is a more generalized term than pure resistor/capacitor to represent the circuit's ability to resist the current flow. If $V(t) = V_A sin\omega t$ is the applied sinusoidal voltage signal (function of radial frequency ω and time t) and $I(t) = I_A \sin(\omega t + \phi)$ is the response current signal in a circuit (shifted by phase angle ϕ), the complex impedance ($Z(\omega)$) is given by Eq. (2.1).

$$Z(\omega) = \frac{\tilde{V}(\omega)}{\tilde{I}(\omega)} \tag{2.1}$$

where $\tilde{V}(\omega)$ and $\tilde{I}(\omega)$ are the Fourier transforms of $V(t)$ and $I(t)$. When an applied voltage and current are plotted on the x and y axis, respectively, the Lissajous figure (oval in shape) appears (Figure 2.1a), which was used as the means to measure impedance in initial days. By expressing $V(t)$ and $I(t)$ in terms of complex function, impedance $(Z(\omega))$ (Z_A being the absolute value of impedance [magnitude]) can be expressed as a combination of in-phase or real (Z_{REAL}) and out-of-phase or imaginary (Z_{IM}) parts, as:

$$Z(\omega) = Z_A (cos\phi + jsin\phi) = Z_{REAL} + jZ_{IM} \tag{2.2}$$

Figure 2.1b shows the various components of impedance when plotted in terms of Z_{REAL} vs $-Z_{IM}$ called as Nyquist plot. It can be observed from the figure that this plot is corresponding to ω ranging from ∞ to 0. Starting from this basic brief, the impedance analysis covers various dimensions related to system characteristics, viz. electrical analogies related to various system processes, basic electrochemical system representation, and various mass/charge transfer modes in biological systems and associated fluid dynamics, which are discussed below.

2.2.1 ELECTRICAL ANALOGIES TO VARIOUS PROCESSES

When impedance spectroscopy is utilized for characterizing any physical/chemical process/system, it is modeled in terms of several electrical parameters that represent

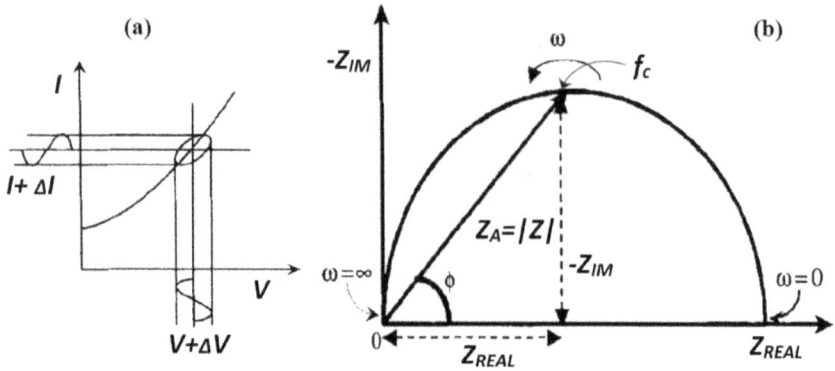

FIGURE 2.1 Representation of complex impedance data: (a) Lissajous figure; (b) complex impedance plot. [Reprinted with permission from Lvovich (2012). Copyright (2012) Wiley].

discrete processes/behaviors occurring in the system. These approximations tend to develop unique equivalent circuits for various processes, and these circuits can be simple or complex depending on the state of the system. While assigning electrical analogs to the processes, two general shortcomings are observed: one is the geometrical dependence of current distribution in the system and the other is the inconsistent behavior of bulk and interface parameters as a function of frequency (Barsoukov and Macdonald 2005). Although the small signal is generally considered in impedance spectroscopy (which implies linear behavior) as opposed to CV, where the information deconvolution is difficult, there are still several phenomena that take place in impedance measurements. The most important parameters that are taken into account while characterizing biological samples are sample conductivity, diffusion, and interfacial phenomenon (adsorption and charge transfer at the electrode–electrolyte interface).

2.2.1.1   Conductivity and diffusion in electrolytes (Newman 1973; Ibl 1983)

The Poisson equation is commonly used to relate electric field gradient and charge density in a medium with a uniform dielectric constant and it is given by Eq. (2.3).

$$\nabla^2 \varnothing = \frac{-\rho}{\varepsilon_s \varepsilon_0} \qquad (2.3)$$

where $\varnothing$ is the electric potential, ρ is the charge density given by $\rho = F \sum z_i c_i$ (c_i is the local concentration of a species with z_i charge), ε_0 is free space permittivity, and ε_s is a static dielectric constant given by $\varepsilon_s = 1 + \dfrac{P_s}{E\varepsilon_0}$ (P_s is total polarization and E is an electric field). The constant magnitude F / ε_0 in the system results in the generation of very high electric fields due to minute changes in electroneutrality. Hence, for bulk electrolyte, it is convenient to consider $\sum z_i c_i = 0$ for systems with considerable separation, which follows that the bulk electrolyte is electrically neutral.

It follows that the Laplace equation $\nabla^2\varnothing = 0$ can approximately model the condition of the bulk of the electrolyte system.

Additionally, mass conservation in the system is given by,

$$\frac{\partial c_i}{\partial t} = -\nabla.j_t + R_i \tag{2.4}$$

which shows that the accumulation rate of i species in an element of volume is given by the summation of negative flux divergence $(-\nabla.j_t)$ and any term that represents an increase or decrease of i species due to a chemical reaction (R_i). Further electric charge density can be related to charged species flux through the following equation.

$$i = F\sum z_i j_i \tag{2.5}$$

where F is the Faraday constant and the generalized flux, j_i is given as

$$j_i = -c_i u_i \left[\nabla\eta_i + \sum_j \alpha_{ij}\nabla\eta_i \right] \tag{2.6}$$

where u_i denotes the characteristic mobility, η_i denotes electrochemical potential, α_{ij} denotes the coefficients representing the effect of driving potential (given as $\nabla\eta_i$) on i. And in terms of diffusion coefficient, this equation formulates as,

$$j_i = -\frac{D_k c_i \nabla\mu}{RT} - z_i F c_i u_i \nabla\varnothing \tag{2.7}$$

where R is the ideal gas constant, T is the thermodynamic temperature, and D_k is the component diffusion coefficient. This relation maintains Nernst–Einstein proportionality between mobility and diffusion coefficient, even after relaxing ideality condition.

2.2.1.2 Diffusion in electrolyte

The absence of an electric field and additional R_i term reduces the mass conservation of the system to,

$$\frac{\partial c_i}{\partial t} = -\nabla.j_i = \nabla.(D_i \nabla c_i) \tag{2.8}$$

For constant D_i, this equation turns to,

$$\frac{\partial c_i}{\partial t} = D_i \frac{\partial^2 c_i}{\partial x^2} \tag{2.9}$$

This equation justifies the assumption of diffusion current or flux is proportional to the concentration gradient and also justifies the continuity equation. Taking Laplace of the above equation and solving it further for the electrode–electrolyte interface $(x=0)$ in case of semi-infinite diffusion into the electrode, the solution is obtained as follows:

$$\{\Delta c\}_{x=0} = \frac{\{\Delta i\}}{zF\sqrt{pD}} \tag{2.10}$$

where $\Delta i = -zFDd\Delta c / dx$ is the AC and $p = \sigma + j\omega$ is the complex frequency variable. For small modifications around equilibrium, $\{\Delta c\}_{x=0}$ and Δv (AC component of voltage) can be related as $\dfrac{\Delta v}{\Delta c} = \left(\dfrac{dE}{dc}\right)$, where $\left(\dfrac{dE}{dc}\right) = \dfrac{RT}{zFc}$ represent electrode potential change with respect to concentration. Further solving it and limiting it for the condition $\sigma = 0$,

$$Z(j\omega) = \frac{\left(\dfrac{dE}{dc}\right)\left(\dfrac{1}{\sqrt{\omega}} - \dfrac{j}{\sqrt{\omega}}\right)}{zF\sqrt{2D}} \tag{2.11}$$

Hence it can be observed that complex impedance through diffusion is inversely proportional to the square root of frequency, and in the Nyquist plot, it is represented as an inclined line, at 45° to the real impedance axis.

2.2.1.3 Adsorption

At the interface, mobile ions and solvent layers are present, which are affected by the electrostatic force acting on it. The electrostatic forces help orient the dipoles of the solvent and observe the ion distribution with respect to the distance at the interface. It is also possible that the ions may chemically react with the electrode and break through the layer of the solvent. This condition is called specific adsorption. In the case of a common electrochemical system, the locus of the specifically adsorbed molecule's center is known as the inner Helmholtz plane. The neutral molecules may further be adsorbed at the interface, which may alter the faradaic current.

For drawing an electrical response through the adsorbed species, an equation relating the surface concentration of the adsorbed molecules, Γ and electrolyte concentration just outside the double layer, Γ_0 is developed. For predicting this phenomenon, various authors have proposed several theories for predicting the effect of adsorbed species on the derived capacitance. The simplest theory in this regard has been proposed by Langmuir, which assumes that the adsorption-free energy remains the same throughout the surface and there is no interaction among the adsorbed molecules. In this condition, the Γ and Γ_0 are related by the expression,

$$\frac{\Gamma}{\Gamma_0 - \Gamma} = \frac{\theta}{1-\theta} = a_i^b \exp\left(\frac{-\Delta G_i^0}{RT}\right)\exp\left(\frac{-\emptyset z_i F}{RT}\right) \tag{2.12}$$

where θ is a constant leading to frequency and time-dependent conductivity and diffusion coefficient, a_i^b is the bulk activity of i species, and ΔG_i^0 is the free energy of adsorption. The adsorption capacitance can be calculated through adsorbed species charge using the following equation,

$$C = \frac{dq}{d\emptyset} = \frac{d(\theta q_i)}{d\emptyset} = q_i\left(\frac{z_i F}{RT}\right)\theta(1-\theta) \tag{2.13}$$

2.2.1.4 Charge transfer at the electrode–electrolyte interface

The rate of charge transfer in an oxidation–reduction reaction at the interface is given by,

$$-i_F = nF\left[k_f c_o - k_b c_R\right] \tag{2.14}$$

where i_F is faradaic current density, c_o and c_R are concentrations of reactant and products, and k_f and k_b are forward and backward reaction rate constants. The current includes the DC or the steady-state part (measured through dc potential E and dc concentration c_o and c_R) and ac part (measured through changing voltage ΔE and concentrations Δc_i). The faradaic impedance is the ratio of the transformation of the ac components of voltage and current.

$$Z_F = \frac{\{\Delta E\}}{\{\Delta i_F\}} \tag{2.15}$$

After elaborating,

$$Z_F = \frac{1}{(\partial i_F / \partial E)}\left[1 - \sum\left(\frac{\partial i_F}{\partial c_i}\right)\frac{\{\Delta c_i\}}{\{\Delta i_F\}}\right] \tag{2.16}$$

where the first term is charge transfer resistance, R_{ct}. At the equilibrium potential, E_r, the net current is zero, and, in this circumstance, R_{ct} can be expressed as

$$R_{ct} = \frac{RT}{nFi_o} \tag{2.17}$$

where i_o is exchange current density. For semi-infinite diffusion to the planar interface, Z_F is expressed as

$$Z_F = R_{ct} + \frac{\sigma}{\sqrt{\omega}}(1-j) \tag{2.18}$$

In this equation, the second term corresponds to Warburg impedance, where σ is the Warburg constant given as

$$\sigma = \frac{RT}{\sqrt{2}n^2 F^2 A}\left(\frac{1}{D_o^{\frac{1}{2}}C_o(x,t)} + \frac{1}{D_R^{\frac{1}{2}}C_R(x,t)}\right) \tag{2.19}$$

where A is the electrode area, n is valency, and D and C are diffusion coefficient and concentration, respectively, of oxidized (subscript O) and reduced (subscript R) species.

2.2.1.5 Polarization

Normally, the small signal consideration usually removes the electrode polarization issue, but some specified voltage conditions may observe the electrodes as ideally polarizable. Depending on the Fermi level state in the electrolyte and electrode, some charge may flow in some direction with a field created on the electrolyte side where the mobile charges from the electrolyte distribute themselves. Hence, in this system, metal charge density is limited to the electrode surface. Boltzmann equation, Eq. (2.20) demonstrates any excess charge density at any point in the system.

$$\rho(x) = \sum z_i F c_i = \sum z_i F c_i^0 \exp\left(\frac{-z_i F \varnothing}{RT}\right) \tag{2.20}$$

Further, the Poisson equation, represented by Eq. (2.21) defines the relation between charge and potential.

$$\frac{d^2 \varnothing}{dx^2} = \frac{-1}{\varepsilon \varepsilon_0} \sum z_i F c_i^0 \exp\left(\frac{-z_i F \varnothing}{RT}\right) \tag{2.21}$$

This equation can be further solved to obtain diffuse double layer (space charge) capacitance, C_d of the system (which behaves as a parallel plate capacitor) and for the symmetrical electrolyte ($z_+ = z_-$), it is observed as,

$$C_d = \sqrt{\frac{2z^2 F^2 \varepsilon \varepsilon_0 c_i^0}{RT}} \cosh\left(\frac{2F \varnothing_0}{2RT}\right) \tag{2.22}$$

where $\varnothing_0$ is the value of the potential at $x = 0$ ($\varnothing = 0$ at $x = \infty$). The electric field promoted concentration variation causing a double layer can be observed up to the thickness of the order of Debye length, denoted by the term L_D.

$$L_D = \sqrt{\frac{RT \varepsilon \varepsilon_0}{2z^2 F^2 c_i^0}} \tag{2.23}$$

As the concentration of the electrolyte increases, the thickness of the diffuse layer reduces. In a polar solvent solution (e.g. water), a net dipole near the interface is present and a charge monolayer in the electrode vicinity exists, which makes the system as two capacitor systems (connected in series), where double layer (Gouy–Chapman) capacitance extends beyond this inner (Stern) layer capacitance. The inner layer capacitance, C_i (capacitance per unit area) is given by Eq. (2.24).

$$C_i = \frac{\varepsilon \varepsilon_0}{d} \tag{2.24}$$

2.2.2   IMPEDANCE MEASUREMENTS IN A BASIC ELECTROCHEMICAL SYSTEM AND CORRESPONDING REPRESENTATION

Along with a low AC voltage input and wide frequency range, impedance spectroscopy utilizes an offset voltage (usually a DC voltage) to study the behavior of electrochemical processes and it primarily focuses on charge and material-exchange kinetics study at the interface of the sample and electrode. Charge carrier shift at the electrode–electrolyte interface is governed by an electrochemical reaction known as electrolysis (elaborated in Section 2.1). The electrochemical characteristics of various species (capacity to release or accept the electrons) define the limit of their electrochemical discharge at a particular potential of the polarized electrodes. The opposition encountered by the system toward the combined discharge current across the Helmholtz plane near the electrode is given by charge transfer resistance, R_{ct}. R_{ct} is often considered polarization resistance for low potential values. Because electrolysis monitors the limit of accumulation/depletion of charge/mass at the electrode–electrolyte interface, there is a development of the concentration gradient in the bulk solution and the electrode. When the electrochemical reaction leads to the generation of a local concentration gradient around the electrode, the abundant species from the solution try to refill the depleting species at the electrode by diffusing toward the interface. At the same time, the released species diffuses toward the solution, where their concentration is low. This diffusion-based mass transport due to the concentration gradient that occurs in the diffusion layer can be represented as a complex diffusion impedance element (Warburg impedance), Z_w.

There are several species accumulated on the interface that do not participate in the electrolysis (diffusing or discharging) at the interface, but their charges are countered by the oppositely charged species on the interface forming a capacitor with an overall double layer capacitance, C_d. This capacitor in the system is formed by the species except the ones involved in charging/diffusing; hence it behaves as a parallel electrical pathway (energy storage component) and is placed parallel to R_{ct} and Z_w. An uncompensated minimum solution resistance in the system is denoted by R_s. Charge transport at the electrode–electrolyte interface and the corresponding equivalent Randle electrical circuit comprising all these parameters have been depicted in Figure 2.2.

From the equivalent circuit, the overall impedance can be observed as:

$$Z(\omega) = R_s + \frac{R_p + Z_w}{1 + j\omega C_d (R_{ct} + Z_w)} \tag{2.25}$$

The Warburg impedance, Z_w can be given as

$$Z_w = \left(\frac{2}{\omega}\right)^{1/2} \sigma \tag{2.26}$$

For analyzing the behavior of a system, all these impedance parameters are extracted from the recorded impedance data through various fitting procedures and these basics of impedance spectroscopy are utilized for different systems to draw conclusions on their behavior. Impedance changes in a system are usually monitored through a set of electrodes covered with the analyte solution (stationary/flowing). Primarily, various aspects associated with the impedance-based measurement of a system include ways of mass and charge transfer modes and physics, such as fluid flow and microfluidics.

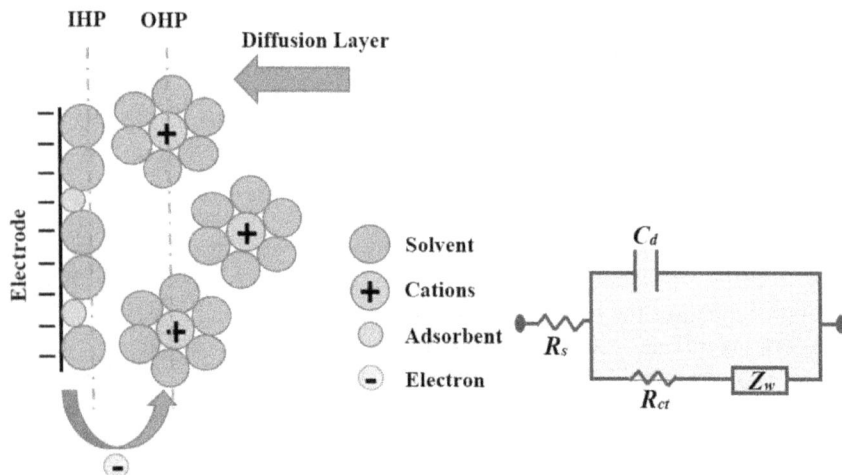

FIGURE 2.2 (a) Charge transport at the electrode–electrolyte interface (IHP: inner Helmholtz plane, OHP: outer Helmholtz plane); (b) equivalent Randle electrical circuit. [Reprinted with permission from Bhatt and Bhattacharya (2018). Copyright (2018) Springer].

2.2.3 MASS AND CHARGE TRANSFER MODES

The mass/charge transfer in electrochemical systems, which monitors the electro-chemical impedance measurements of the solution is completely dependent on the solution constituents. Generally, the solution can be chemical/biological depending on the requirement. In chemical systems, either the constituents of the solution remain constant or modify depending on the kind of reaction taking place in the system, which can completely vary from system to system, but in biological measurements, the means through which the solution constituents change mostly remains the same for a particular system and are only interpreted discretely depending upon the measurement need.

From a biological perspective, the impedance measurements can be obtained using either solution comprising bacteria or DNA. If the solution contains bacteria, energy metabolism (catabolism) and ionic exchange monitor the release of ion content and the corresponding solution conductivity (Owicki and Parce 1992). Catabolism in bacteria is the process of consumption of carbohydrates (sugar) and oxygen to produce organic acids and carbon dioxide (Figure 2.3a). Figure 2.3a also shows that the production of organic acids can be directly through catabolism procedure, for example, lactic acid or through indirect mode where released carbon dioxide further reacts with water molecules to produce carbonic acid. The ionic content produced in the solution is directly proportional to the concentration of the bacteria, which in turn modifies the conductivity of the solution. Along with catabolism, ion exchange through the cell membrane also modifies ion content in the solution. The cell membrane that monitors the membrane potential and the osmotic pressure of the cell facilitates the active transport of sodium and potassium ions, which do not primarily yet change the solution's electrical characteristics (Figure 2.3a).

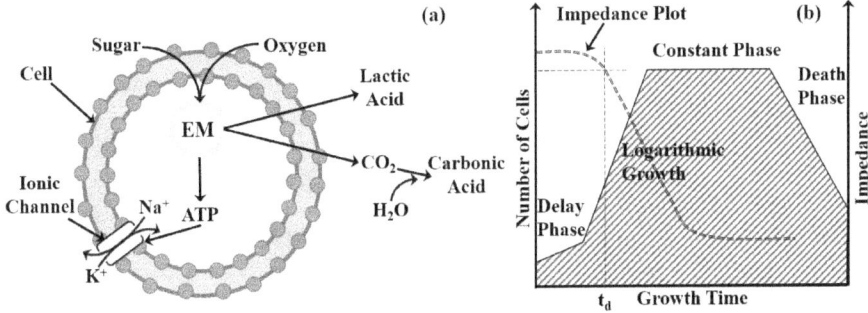

FIGURE 2.3 (a) Ion release mechanism in bacteria through energy metabolism and ion exchange through ionic channel; (b) effect of bacterial growth on impedance plot in solution.

The impedance measurement in case of bacterial presence can be direct or indirect (Ramírez et al. 1989). In direct measurements, the effect of these processes is directly measured through a set of electrodes containing bacterial solution over it. In indirect measurement, the measuring electrodes do not measure the electric response of cells directly, rather the produced ionic content is absorbed in a different solution, and normally potassium hydroxide shows a reduction in conductance of the solution for increasing ionic concentration. The impedance value directly depends on the concentration of the cells/bacteria and the way bacterial growth continues. Figure 2.3b shows an impedance vs growth time plot, which shows that initially with respect to a particular cell concentration, the impedance is constant and slowly impedance reduces and crosses the threshold value at detection time t_d. At this time, most of the growth content of the solution has been metabolized and converted into the final product. Beyond this point, impedance reduces constant, which is corresponding to the replication phase (logarithm growth) of the bacteria, and lastly, impedance value saturates for the constant bacterial concentration.

The impedance measurements in biological systems primarily comprise contribution through two impedance components, electrode–solution interface and medium (Gomez, Bashir, and Bhunia 2002). It is observed in the literature that prominent impedance changes are recorded around 10 Hz to 10 kHz frequency, and at the low-frequency domain, <10 Hz, the overall impedance mostly comprises the capacitance component (double layer capacitance), while the capacitance effect diminishes beyond 10 kHz. At higher frequencies beyond 10 kHz, impedance becomes frequency independent, and it is mostly dominated by medium resistance.

DNA carries an overall negative charge due to the presence of a phosphate backbone in it, and whenever DNA is present in the solution, counterion clouds are generated. The generated counterion clouds lead to the formation of dipoles in the solution, and the dipole formation/relaxation under the effect of the electric field changes the overall electrical behavior of the solution (Liu et al. 2008). The counterion clouds can be diffused/condensed in nature, as can be observed in Figure 2.4a, and it is also pertinent to mention that the behavior of DNA in the electric field is enormously dependent on the frequency of the electric field (Baker-Jarvis, Jones, and Riddle 1998).

It is well characterized in the literature that DNA exhibits three relaxation frequency domains, i.e., the α-relaxation in the frequency range of 1–100 Hz due to the end-to-end rotation of DNA (longitudinal polarization of diffused counterions), the β-relaxation in the frequency range of 100 kHz–1 MHz due to the axial rotation of DNA (through condensed counterions), and the γ-relaxation state above 1 GHz due to the polarizability of water molecules surrounding the DNA in solution (Baker-Jarvis, Jones, and Riddle 1998; Figure 2.4b).

An increase in the concentration of the DNA or template size reduces the overall impedance of the system as charge transport enhances due to an increase in the concentration/template length (Liu et al. 2008), which directly increases the DNA-counterion cloud-based dipole numbers. This enhanced charge transport is majorly observed due to increased capacitance/reduced impedance in the system. It has also been observed that double-stranded DNA exhibits more dielectric relaxation as compared to the single-stranded denatured DNA, primarily due to the coiling effect (Takashima 2007).

Also, the EIS measurements can be direct or indirect in nature. In direct measurements, the effect of solution constituents on electrical parameters is directly measured, while in the case of indirect measurements, the various constituents of the solution are labeled with the intercalators, for example, ferrocene, and the interaction of these labels with the electrode interprets the solution characteristics.

2.2.4 Fluid Flow and Associated Microfluidics

For processing a particular analyte on the fabricated detection platform (set of electrodes in a specific pattern), an efficient microfluidic integration on the sensors is

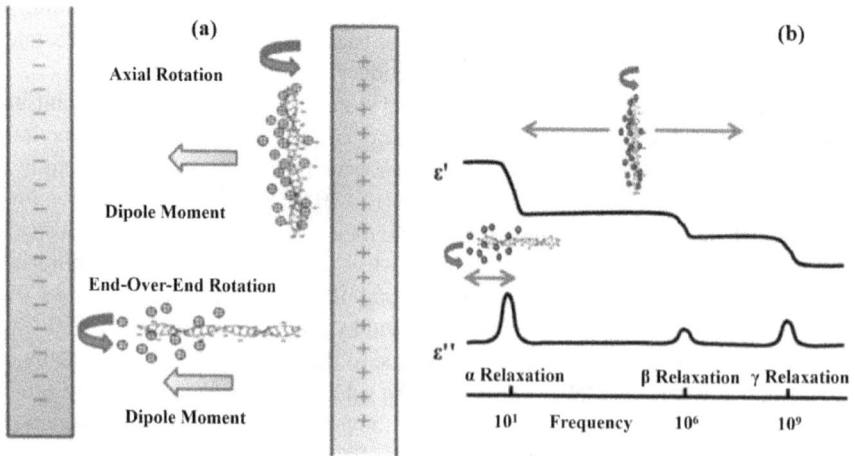

FIGURE 2.4   (a) Frequency-dependent behavior of DNA in an electric field; (b) three dielectric relaxation points in DNA die to end-over-end rotation (at low frequencies), axial rotation (at high frequencies), and water polarizability showing α, β, and γ relaxation, respectively. [Reprinted with permission from Salm et al. (2011). Copyright (2011) Springer].

further accomplished. This integration is a very important aspect as flow pattern/ kinetics are the most important factor for carrying out effective detection. As flow can be categorized in terms of laminar or turbulent flow and for proper measurements, laminar flow in the system is an essential requirement. Hence there is always a need to include an accurate combination of flow regimes through an adequate microfluidic system. Microfluidic (commonly microchannels) integration (Kant, Singh, and Bhattacharya 2017) (Kant et al. 2013) to microdevices can be carried out through various processes using several materials. The various materials that were commonly used for fabricating microchannels were silicon and glass in the initial days, which has now broadened to various other materials like polymers, composites, paper, and metals. The various processes that are used for fabricating microchannels are micromachining, photolithography, embossing, replica molding, injection molding, etc. While choosing a particular material/process for microchannel fabrication/ integration, the most important parameter that is observed is the application of the device as the analyte/detection process compatibility is the most important thing, and then the fabrication cost, reliability, ease of integration, etc. Polydimethylsiloxane (PDMS) is one of the most common polymer materials that is used in microfluidics these days due to its extensive moldability, which gives the advantage of fabricating complex features easily, biocompatibility with biological analytes (due to its inert nature), cost, and availability.

In terms of principle, microfluidics is a broad term that studies fluid behavior/flow at the microscale, typically of the size scale of 100 nm to 500 µm. At the microscale, volumetric forces are dominated by the surface forces and hence the theories applicable to macroscale fluid flow are not applicable at this scale. At this scale, surface tension and fluid resistance start dominating the system and the associated fluid flow dynamics/pattern is governed by these effects. As the sample volume requirement at the microscale is very less, this field observes numerous applications in the biomedical field, where the sample prices are an important aspect. It is extensively used in lab-on-chip technology, DNA microchips, drug delivery, micro-propulsion, etc. Microfluidic systems are most commonly used for mixing, separating, and delivering fluid with very less energy consumption.

In microfluidic systems, Navier–Stokes equations govern the fluid flow, where the velocity profile of the fluid (Doering and Gibbon 1995) is derived as

$$\begin{cases} \dfrac{\partial u}{\partial t} + (u.\nabla)u = -\dfrac{1}{\rho}\nabla p + v\nabla^2 u + f \\ \\ \qquad\qquad \nabla.u = 0 \end{cases} \tag{2.27}$$

where u is the fluid velocity field, p is the pressure field, ρ is the density, v is kinematic viscosity, and f is the external acceleration field caused due to the electrostatic field or gravity. These equations are derived through mass and momentum conservation laws assuming that the fluid has incompressible (constant density over time and space) and Newtonian (uniform viscosity) qualities.

Setting dimensionless parameters as $V^* = LV$, $u^* = u/U$, $t^* = t/(L/U)$, $p^* = pL/\rho vU$, and $f^* = f/f_0$ (L is the characteristic length of geometry, U is the typical velocity of fluid, and f_0 is the acceleration field's typical intensity), Navier–Stokes equation is reduced to (Tritton 1976),

$$Re\left(\frac{\partial u^*}{\partial t} + \left(u^*.\nabla^*\right)u^*\right) = -\nabla^* p^* + \nabla^{*2} u^* + \frac{f_0 L}{U^2} f^* \qquad (2.28)$$

where Re is the Reynolds number given as $Re = \dfrac{UL}{v}$. The dimensionless Reynolds number, which is named after Osborne Reynolds, is most commonly used to define the fluid flow pattern in a system and it is given by the ratio of inertial force to viscous force in the fluid.

$$Re = \frac{Inertial\ force}{Viscous\ force} \qquad (2.29)$$

In microfluidic systems, Re is low (<1) and the flow is laminar in nature. It can be concluded from Eq. (2.28) that at low Re, the right side of the equation dominates, and in this condition, flow is governed by Stokes equation [derived from Eq. (2.27)],

$$v\nabla^2 u = -\frac{1}{\rho}\nabla p - f \qquad (2.30)$$

This equation presents parallel streamlines of the flow and in this case, inertial force is dominated by viscous force. The other important dimensionless number that is used to determine whether the flow is diffusion or convection based in microfluidic devices is the Peclet number. It is given as

$$Pe = \frac{TR_{ad}}{TR_d} = \frac{uL}{D} \qquad (2.31)$$

where TR_{ad} is advective transport rate, TR_d is diffusive transport rate, u is the flow velocity, L is the characteristic length, and D is the diffusion constant. If the value of Pe is greater than 1 or lesser than 1, diffusion or convection dominates, respectively.

2.3 EQUIVALENT CIRCUIT DIAGRAMS

The electrical circuit modeling of impedance data in terms of Randles and Warburg was initiated after the evolution of the double layer theory by Frumkin and Grahame. It is the simplest model that represents the complete impedance spectroscopy behavior in terms of resistance and capacitance. The presence of various circuit elements in the equivalent circuit is interpreted through impedance data plots. The impedance data is commonly plotted in the form of a Bode plot and an Nyquist plot. A Bode plot is drawn for phase (x axis) and logarithmic impedance data points (x axis) against the frequency sweep (y axis), while an Nyquist plot is drawn between the real (x

axis) and imaginary (y axis) part of the impedance. The Bode plot represents the overall behavior of obtained data and the slope of impedance change expresses the possibility of the presence of discretely separated time constants. The specific values of various components of impedance (R_s, C_d, and R_{ct}), which is the basis of drawing equivalent circuits, are usually extracted through the Nyquist plot. The impedance of the Nyquist plot can be represented as a vector of magnitude |Z| making an angle ϕ, phase angle with the x axis, as shown in Figure 2.1b. There is a term called the time constant, given by $\tau = R_{ct}C_d$ corresponding to an equivalent circuit and a "critical" or "characteristic" relaxation frequency, $f_c = 1/2\pi\tau$ associated with it.

2.3.1 GRAPHICAL REPRESENTATION AND PARAMETER EXTRACTION

The behavior of impedance plots can be explicitly presented in terms of the equivalent circuit. Depending upon the impedance data, the plot, their fitting, and corresponding equivalent circuit changes. Table 2.1 presents three common types of equivalent circuits and corresponding Nyquist and Bode plots associated with them. It is worth mentioning that the Nyquist and Bode plots presented in the table are just representative and do not have specific extracted parameters associated with them.

As can be observed from the table, the first equivalent circuit represents the case, when only C_d and R_{ct} are in the system (connected in parallel). In this case, the Nyquist plot is presented as the complete semicircular plot without any additional curve. The presence of a single semicircle points toward the possibility of the existence of a single charge transfer phenomenon. It is observed that at very high frequencies, $\omega = \infty$, the system comprises only uncompensated solution resistance, R_s and at low frequency, $\omega = 0$, the system comprises total resistance, $R_s + R_{ct}$, as depicted in Table 2.1. The time constant term, τ is further utilized to calculate C_d of the system. The Bode plot represents the presence of a single slope in the plotted impedance, while it represents a single peak in the phase plot.

As the impedance through capacitance is frequency-dependent ($Z(\omega) = \dfrac{1}{j\omega C_d}$, when the circuit comprises only capacitance), the current flow in such systems is quite dependent on the f_c value, when the impedance through the capacitive and resistive components is similar. At $f > f_c$, the capacitive element serves as the least impedance path for current flow, while impedance is likely primary capacitive with a larger phase angle value. However, at $f < f_c$, impedance through capacitance becomes high and the resistive path serves as the least impedance path with a phase angle approaching zero degrees.

In the second case, the system comprises C_d and R_{ct} along with the Warburg impedance, Z_w (connected in series). The Warburg component presents the condition of mass transport in the system. As can be observed from the table, the presence of Z_w can be confirmed by observing an inclined line in the Nyquist plot. The slope of the observed inclined line is ≈ 1. The effect of Z_w is prominent in the lower frequency and can be quoted as $Z_w = \infty$ at $\omega = 0$. This effect keeps reducing with frequency and is observed to present no effect at a higher frequency. The presence of Z_w in the Bode plot can be seen by reducing impedance and phase at lower frequency instead of constant values in the first case.

TABLE 2.1

Common equivalent circuits and corresponding Nyquist and Bode plots associated with them

Equivalent circuit	Nyquist plot	Bode plot

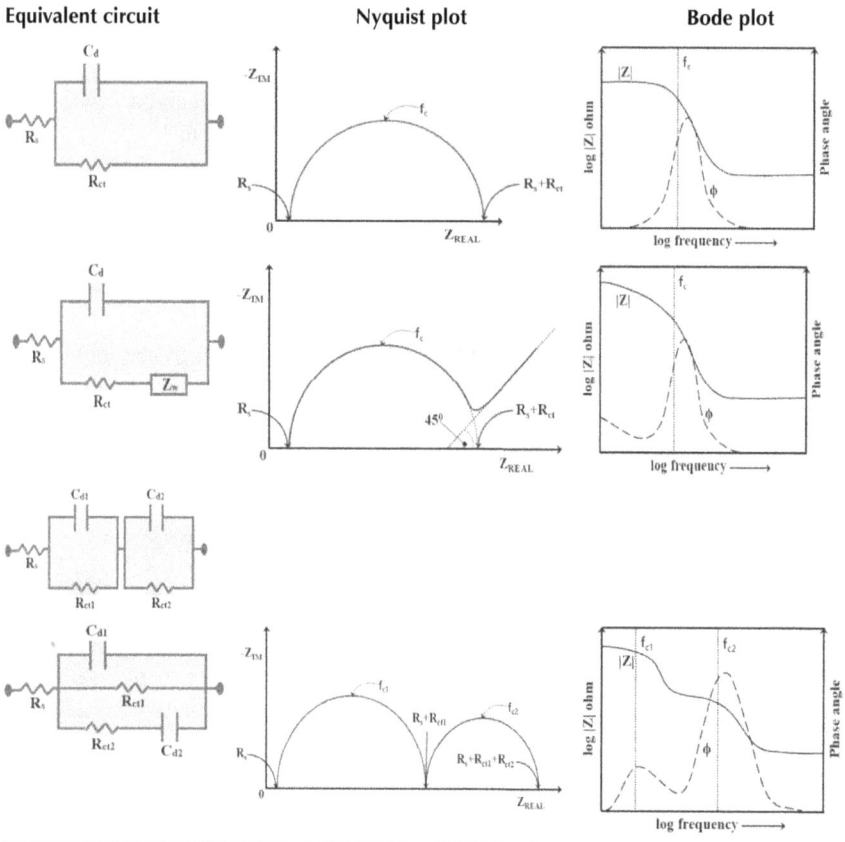

The third case of the equivalent circuit can be seen as the presence of two C_d and R_{ct} values and the Nyquist system comprising two semicircles. This kind of Nyquist plot exists when there is a heterogeneous phenomenon (two different types of charge transport phenomenon at varied time scales) occurring in the system. Bode plots in such systems can be seen as the presence of two distinct impedance reduction slopes and two peaks in the phase plots.

2.4   VARIOUS IMPEDANCE SPECTROSCOPY MODULES

Many times, the conventional EIS technique alone is not enough to interpret all the characteristic features of the electrochemical system under investigation. In this case, various characterization techniques are simultaneously used to study the reaction kinetics (Piro et al. 2005; Ding et al. 2007). Hence, to make EIS more application-driven, along with EIS, several modified modules concerning impedance measurement are quantified. These are AC voltammetry, Fourier transform impedance spectroscopy (FTEIS), nonlinear higher-harmonics impedance analysis, and scanning photo-induced impedance spectroscopy (SPIM).

AC voltammetry is the extension of CV (Bard and Faulkner 2001). In this, a DC potential along with an AC signal is superimposed and applied to the working electrode and the corresponding electrode response is measured. The simplified formulation is obtained for high perturbation AC frequency, where AC and DC responses can be effectively differentiated. The average surface concentration of electroactive species remains the same in this case as well when compared to the case with no AC perturbation. However, the overall surface concentration here is the result of long-term diffusion (due to DC potential) and rapid diffusional fluctuations (due to AC potential) caused by the overall signal, which is similar to bulk concentration through AC potential. Because of these characteristics, AC voltammetry can differentiate process phenomena existing at varied time scales and is capable of exploring process kinetics.

The faradaic impedance in this case comprising both charge transfer resistance and diffusion is given by:

$$Z = \sqrt{\left(R_{ct} + \frac{\sigma_D}{\sqrt{\omega}}\right)^2 + \left(\frac{\sigma_D}{\sqrt{\omega}}\right)^2} \qquad (2.32)$$

Peak current in this case is recorded as:

$$I_p = \frac{z^2 F^2 A C_{ox}^* \sqrt{\omega D}}{4 R_G T} V_A \qquad (2.33)$$

where C_{ox}^* is bulk concentration and V_A is AC potential. As peak current is proportional to the various quantities, AC voltammetry can be efficiently utilized for analytical electrochemistry-based analysis of the system.

FTEIS (Pettit et al. 2006) and potentiodynamic EIS (PDEIS) (Ragoisha et al. 2004) combine EIS with CV analysis to benefit from the advantage of both methods. As EIS solely does not provide time-dependent analysis of the systems and particularly for low-frequency analysis (needing longer sampling time), voltage-dependent analysis is required for studying reaction kinetics. Also, CV analysis solely is not capable of double layer analysis, and, hence, a merger of these two processes is an attractive option when a multidimensional process investigation is required. The formulation in this case is usually obtained using Fourier transform and the recorded impedance at every frequency, in this case, is given by:

$$Z(j\omega) = K_F \frac{1}{I_G} F\{V(t)\} \qquad (2.34)$$

where K_F is the constant that depends on the implementation of the Fourier transform, I_G is the generic term representing signal demonstrating transient, and $F\{V(t)\}$ is the Fourier transform of the continuous-time function of potential $V(t)$ in the time interval $[0, t]$ and is given by,

$$F\{V(t)\} = \Delta t \sum_{i}^{N-1} V_i \exp(-j2\pi f t_i) \qquad (2.35)$$

where $t_i = i\Delta t$, $V_i = V(i\Delta t)$, $i = 0,1, 2,...N$ for the frequency interval $[0, f_0]$ ($f_0 = 1/2\Delta t$). This formulation does not consider the overall impedance response in terms of continuous function but a discrete set of data points. In FTEIS, several impedance plots (Nyquist or Bode) are measured in parallel to CV measurements, but due to their complex nature, the measurements are limited up to 10 mV excitation signal. While this complexity is a bit resolved in the case of PDEIS, where 3D spectra comprising complex impedance characterization with respect to frequency as well as electrode potential are recorded along with the CV plots. The limitation of this method is the limited recorded points for the process.

Nonlinear higher-harmonics impedance analysis deals with high perturbation signals. In EIS, a small AC perturbation is usually provided to linearly approximate the nonlinear electrochemical system. The measurements are usually independent of the perturbation signal amplitude and no higher-harmonic responses are present there. But in highly resistive materials, higher perturbation of the order of 1 V is provided for smooth current passage, which in turn causes prominent harmonics and nonlinearities, which is utilized in nonlinear EIS as a factor for pattern recognition. Bulk electrolyte follows Ohm's law and hence remains linear, but such high perturbation usually makes system response nonlinear by changing the number of ions per unit volume, viscosity, and bulk-solution response at high frequency. In nonlinear electrochemical systems, higher (second order and up) harmonic signals are free of linear contributions in the impedance measurements, like solution resistance and charging currents of double layer. As higher harmonics arise due to nonlinear heterogeneous kinetic processes like adsorption, diffusion, and charge transfer, these nonlinear EIS processes are helpful in the study of mass and charge transport at the interface without any interference, which is not feasible in linear EIS. Another advantage from nonlinear EIS is that the kinetic parameter extraction can be carried out using a single DC offset potential. The response current signal in this modification is obtained by the following expression:

$$I = I_{DC} + I_0 \sin(\omega t + \theta) + I_1 \sin(2\omega t + \theta_1) + I_2 \sin(3\omega t + \theta_2) + ... + Noise \qquad (2.36)$$

where I_{DC} current's DC component, I_1, I_2, I_3,... are the current's harmonic distortion, and θ, θ_1, θ_2 are phase components. Hence overall, nonlinear EIS deals with the application of perturbation signal, measurement of current and voltage, Fourier transform of input and output signal (time dependent), calculation of real and imaginary parts of the output for higher harmonics, back conversion into the frequency domain, and the harmonic coefficient's calculation. This analysis is commonly utilized in observing complicated several-stage kinetic problems (Wilson, Schwartz, and Adler 2006), for example, adsorption impedance (Lvovich and Smiechowski 2008), corrosion rates (Darowicki 1995), and diffusion effects (Elkin et al. 2001).

SPIM (Moritz et al. 2000) is another modification of EIS, where impedance characterization of electrochemical systems or thin films (laterally resolved) is accomplished through the measurement of photocurrent at field-effect structures, which

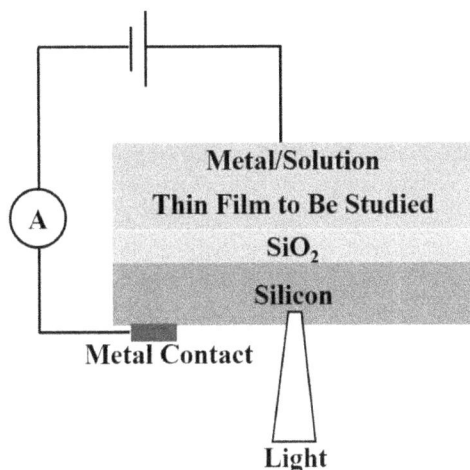

FIGURE 2.5 Experimental setup of SPIM (Moritz et al. 2000).

consists of a gate electrode (generally a metal film) and thin insulator along with a semiconductor substrate. Figure 2.5 shows a schematic of the setup utilized in SPIM measurements.

The figure shows that a DC potential is applied between silicon (semiconductor substrate) and the gate electrode; however, because of the presence of an insulator (SiO_2), it does not produce DC currents, but it helps in controlling the space charge region around the semiconductor–insulator interface. The light beam focused on silicon produces AC currents in the system by producing electron-hole pairs, which are separated in the space charge region in an inversion-biased state. These charges collect on both sides of the insulator and charge the capacitor, which is discharged as the illumination stops. In the case of measurements, if a thin film is to be studied, it reduces the developed photocurrent. In the case of impedance measurement in the electrolyte solution state, the electrode is placed inside the gate electrode (solution format). This measurement does not need any type of probe/microelectrode placement over the sample.

Recently, the modified EIS approach has also been proposed, which is a multi-time-scale fractional order model to study long and short time-scaled battery dynamics, both in the time and frequency domains (Ruan et al. 2021).

2.5 COMMON APPLICATIONS OF IMPEDANCE SPECTROSCOPY

EIS is a broadly studied electrochemical process, which is used to quantify various systems and reaction kinetics. These are biological systems (cell suspensions, protein adsorption, DNA amplification, and biomedical devices), electroactive polymer thin films, industrial colloids and lubricants, insulating films, paints, batteries, and fuel cells.

Due to enormously offered advantages, electrochemical measurements of biological systems are gaining popularity over several high-quality techniques, such as fluorescence-assisted, magnetic-assisted, plate-culture methods and clinical laboratory methods. Biological sensors/biosensors have been widely using EIS as a detection alternative to various options, such as optical/colorimetric, mechanical, ultrasonic, and magnetic methods. The most important advantage EIS offers in biosensors is the possibility of detection in a very small time, like in a few minutes, with an adequate level of sensitivity. EIS-based biosensing is quite versatile, flexible, cost-effective, and easily capable of electronically integrating into MEMS designs. The application of EIS in biological systems was established in 1950, by Schwan through dielectric spectroscopy of biological cell dispersions (Schwan 1959). Since then, this field grew in multiple dimensions to express its utility in biological suspensions characterization (Asami et al. 1996; Jones 2003), Coulter counter and cytometer (capacitive) for biological particles detection (Sohn et al. 2000; Gagnon et al. 2008), glucose monitoring through enzyme-based biosensors (Caduff et al. 2006), electrophoretic/dielectrophoretic separation of ions, drugs, cells/bacteria, and DNA/RNA (Brett and Brett 1993; Gomez, Bashir, and Bhunia 2002; Varshney et al. 2005; Bhatt et al. 2019; Bhatt and Bhattacharya 2019). The current research focus is primarily to obtain the detection with extremely high sensitivity and specificity. Hence, the researchers are currently using various modes/means to enhance the sensitivity. By following various detection routes, antibody/antigen interaction, intercalator binding, PCR, or simply adsorption, EIS is capable of monitoring the extremely diverse fields of biological systems, such as DNA/RNA (Bhatt et al. 2019), aptamer (Rodriguez, Kawde, and Wang 2005), drugs (Li, Liu, and Luong 2005), proteins (Rodriguez, Kawde, and Wang 2005), antibody/antigens (Suehiro et al. 2006), and cells/bacteria (Varshney et al. 2005), as stated in Table 2.2. EIS can be efficiently utilized to devise POC diagnostic systems and health monitoring systems (by examining several biological parameters).

Electroactive or electrochemically active polymer films offer various advantages, such as great stability, blendability with regular polymers, and awesome optical and electrical properties. These polymers express excellent capability of reversibly modifying electrical conductors from insulators to excellent conductors through a partial oxidation/reduction process (doping process). The oxidation/reduction and corresponding charge transport in these polymers occur either due to concentration gradient-driven electron exchange among redox sites if segmental motion is allowed in the polymer, or due to delocalized electrons capable of moving in the conjugated systems. Table 2.2 lists various polymers studied as electroactive thin films. Among the various polymers, PANI and PPY are two commonly utilized ones in this domain, due to easy synthesis, rich doping-related chemistries, high environmental stability, and excellent electrical conductivity (Inzelt 2007). The synthesis/characterization of these thin films via electrochemical means can be accomplished in a standalone format (by keeping it between two solutions/metal conductors) or also in as-deposited thin film over conducting surface. These polymers are capable of presenting discrete characteristics through composition modification/application conditions.

The utilization of lubricants in transportation and industrial applications is always admired due to their high-utility usage. Lubricants are primarily used to reduce friction, wash contaminant particles, protect contact surfaces, heat dissipation, and balance corrosive acids generated through combustion procedures. Lubricants commonly comprise a mineral/synthetic oil base (continuous phase) with some additives (polar molecules observed at the discontinuous phase) with low electrical conductivity.

TABLE 2.2
Application of EIS in various fields

No.	Application area	Key features	Applications
1.	Biological systems • DNA/RNA (Bhatt et al. 2019) • Aptamer (Rodriguez, Kawde, and Wang 2005) • Drugs (Li, Liu, and Luong 2005) • Protein (Rodriguez, Kawde, and Wang 2005) • Antibody/antigen interaction (Suehiro et al. 2006) • Cells/bacteria (Gomez, Bashir, and Bhunia 2002) • Biological fluids and various interactions	Impedance behavior of various biological analytes is discrete, depending upon the constituents of the analyte, which vary from one system to another	• Point-of-care diagnostics (Bhatt et al. 2019) • Biological system monitoring/response (Caduff et al. 2006)
2.	Electroactive polymer thin films (Inzelt 2007) (Inzelt and Láng 1994) • Polyaniline (PANI) (Musiani 1990) • Nafion • Polypyrrole (PPY) • Quinone-based electroactive polymers (Piro et al. 2005)	Electro-oxidation potential plays a key role in these studies. E.g., At low potential, PANI changes from nonconductive leucoemeraldine to highly conductive emeraldine, whereas at high potential, it changes to an insulating grainline	• Sensors (Flueckiger, Ko, and Cheung 2010) • Actuators (Hong, Almomani, and Montazami 2014)
3.	Industrial colloids/lubricants, Insulating films/paints • Lubricants' performance characterization for various additives • Monitoring oil degradation through various contaminants like water (Smiechowski and Lvovich 2002), glycol (Wang and Lee 1997), and soot • Paints/insulating thin films	Linear as well as nonlinear impedance characterization is used (Lvovich 2012). This film degradation usually represents two semicircle-based Nyquist plots	• Oil characterization (Wang and Lee 1994) • Lubricant degradation study (Guan et al. 2011) • Electrochemical monitoring device to study lubricant performance (Smiechowski and Lvovich 2003) (Smiechowski and Lvovich 2002)

(Continued)

TABLE 2.2 (Continued)

No.	Application area	Key features	Applications
4.	Batteries and fuel cells (Tröltzsch, Kanoun, and Tränkler 2006) (Yuan et al. 2010) • Polymer electrolyte membrane fuel cell (Springer et al. 1996) • Direct methanol fuel cell (Mueller and Urban 1998) • Proton exchange membrane fuel cell (Yuan et al. 2010) • Reversible fuel cell (Hu et al. 2014)	The diverse operational conditions for batteries/fuel cells constantly manipulate the conditions of their components and make the EIS pattern complex	Fuel cell components and their performance analysis

The electrochemical analysis of lubricants is widely studied as it represents excellent adsorption and charge/mass transfer features at the lubricant–electrode interface. Hence, it is commonly utilized to study the effectiveness of the lubricant through lubricant characterization, lubricant performance, and its degradation during usage in engines. EIS is commonly used for engine oil characterization (Wang and Lee 1994), lubricant oxidation study (Lvovich and Smiechowski 2006), oil degradation monitoring (Guan et al. 2011), and building electrochemical monitoring devices (Smiechowski and Lvovich 2003) as stated in Table 2.2. Along with lubricants, the insulation coating of thin films can also be characterized through the EIS measurements. Every coating has a characteristic impedance pattern (mostly capacitive) and as most of the coatings degrade during the continuous usage period, the impedance behavior becomes complex. This resulting behavior is mostly dependent on the usage environment, for example, if the environment is corrosive in nature, it will form pores in the coating due to penetration through the coating and further resulting in the formation of discretely patterned liquid–metal interface, which is further prone to corrosion (Gamry Instruments 2011; Lalić and Martinez 2019). The layer-by-layer dissolution of these films forms altogether a specific interface (with reduced film thickness), which present discrete (increased capacitance) behavior with time. The degraded coatings usually present a two-circle Nyquist plot as shown in Table 2.1.

The other application area of EIS measurements is the analysis of batteries/fuel cells (Tröltzsch, Kanoun, and Tränkler 2006; Yuan et al. 2010). These are stationary/portable power sources exposed to various operating conditions, which makes their analysis for quality-testing purposes a must. EIS is capable of providing important information about batteries/fuel cell components and their performance. EIS can be efficiently used to analyze various types of batteries (Tröltzsch, Kanoun, and Tränkler 2006) and fuel cells, such as polymer electrolyte membrane fuel cells

(Springer et al. 1996), direct methanol fuel cells (Mueller and Urban 1998), proton exchange membrane fuel cells (Yuan et al. 2010), and reversible fuel cells (Hu et al. 2014). EIS is also integrated with several optical/electrical/mechanical modules to express utility in complex systems. It is also utilized for the dielectric analysis of highly resistive composite materials (having particle conduction) (Barsoukov and Macdonald 2005) and ionic colloidal suspensions, dielectrophoretic spectroscopy of particles (using AC electrokinetics) (Bhatt et al. 2019).

2.6 CONCLUSION AND FUTURE PERSPECTIVES

This chapter explains various fundamental aspects of EIS measurement, including the basic impedance formulation, electrical analogy formulation, their implementation in equivalent circuit representation, impedance data presentation and its interpretation, associated microfluidics, and its common application areas. The applications observe the diverse nature of EIS measurements and their capability to quantify a variety of systems. It is generally observed that this method is an easy-to-use and easily integrable method and can be easily combined with various other methods/modifications for increased measurement dimensions.

REFERENCES

Asami, K., T. Yonezawa, H. Wakamatsu, and N. Koyanagi. 1996. "Dielectric Spectroscopy of Biological Cells." *Bioelectrochemistry and Bioenergetics* 40: 141–45. https://doi.org/10.1016/0302-4598(96)05067-2.

Baker-Jarvis, James, Chriss A. Jones, and Bill Riddle. 1998. "Electrical Properties and Dielectric Relaxation of DNA in Solution." *Technical Note NIST TN,* National Institute of Standards and Technology, Gaithersburg, MD no. 1509.

Bard, Allen J., and Larry R. Faulkner. 2001. *Electrochemical Methods: Fundamentals and Application.* Vol. 38. Wiley.

Barsoukov, Evgenij, and J. Ross Macdonald. 2005. *Impedance Spectroscopy: Theory, Experiment, and Applications.* Wiley.

Basu, Aviru Kumar, Adreeja Basu, and Shantanu Bhattacharya. 2020. "Micro/Nano Fabricated Cantilever Based Biosensor Platform: A Review and Recent Progress." *Enzyme and Microbial Technology* 139: 109558. https://doi.org/10.1016/j.enzmictec.2020.109558.

Bhatt, Geeta, and Shantanu Bhattacharya. 2018. "DNA-Based Sensors." In *Environmental, Chemical and Medical Sensors*, edited by Shantanu Bhattacharya, Avinash Kumar Agarwal, Nripen Chanda, Asok Pandey, and Ashis Kumar Sen, 343–70. Springer. https://doi.org/10.1007/978-981-10-7751-7_15.

———. 2019. "Biosensors on Chip: A Critical Review from an Aspect of Micro/Nanoscales." *Journal of Micromanufacturing* 2 (2): 198–219. https://doi.org/10.1177/2516598419847913.

Bhatt, Geeta, Swati Gupta, Gurunath Ramanathan, and Shantanu Bhattacharya. 2021. "Integrated DEP Assisted Detection of PCR Products with Metallic Nanoparticle Labels through Impedance Spectroscopy." *IEEE Transactions on Nanobioscience* 21 (4): 502–10.

Bhatt, Geeta, Rishi Kant, Keerti Mishra, Kuldeep Yadav, Deepak Singh, Ramanathan Gurunath, and Shantanu Bhattacharya. 2017. "Impact of Surface Roughness on Dielectrophoretically Assisted Concentration of Microorganisms over PCB Based Platforms." *Biomedical Microdevices* 19 (2): 1–11. https://doi.org/10.1007/s10544-017-0172-5.

Bhatt, Geeta, Rishi Kant, and Shantanu Bhattacharya. 2019. "Enhanced Fluorescence-Based Detection of Vibrio Cells over Nanoporous Silica Substrate." In *Lecture Notes in Mechanical Engineering*, 1–9. Pleiades Publishing. https://doi.org/10.1007/978-981-13-6412-9_1.

Bhatt, Geeta, Keerti Mishra, Gurunath Ramanathan, and Shantanu Bhattacharya. 2019. "Dielectrophoresis Assisted Impedance Spectroscopy for Detection of Gold-Conjugated Amplified DNA Samples." *Sensors and Actuators, B: Chemical* 288: 442–53. https://doi.org/10.1016/j.snb.2019.02.081.

Bhattacharya, Shantanu, Shuaib Salamat, Dallas Morisette, Padmapriya Banada, Demir Akin, Yi-Shao Liu, Arun K. Bhunia, Michael Ladisch, and Rashid Bashir. 2008. "PCR-Based Detection in a Micro-Fabricated Platform." *Lab on a Chip* 8 (7): 1130. https://doi.org/10.1039/b802227e.

Brett, M. A. C., and A. M. O. Brett. 1993. *Electrochemistry, Principles, Methods, and Applications*. Oxford University Press.

Caduff, A., F. Dewarrat, M. Talary, G. Stalder, L. Heinemann, and Yu Feldman. 2006. "Non-Invasive Glucose Monitoring in Patients with Diabetes: A Novel System Based on Impedance Spectroscopy." *Biosensors and Bioelectronics* 22 (5): 598–604. https://doi.org/10.1016/j.bios.2006.01.031.

Darowicki, K. 1995. "Corrosion Rate Measurements by Non-Linear Electrochemical Impedance Spectroscopy." *Corrosion Science* 37 (6): 913–25. https://doi.org/10.1016/0010-938X(95)00004-4.

Ding, Xiaoqin, Meng Yang, Jingbo Hu, Qilong Li, and Angus McDougall. 2007. "Study of the Adsorption of Cytochrome c on a Gold Nanoparticle - Modified Gold Electrode by Using Cyclic Voltammetry, Electrochemical Impedance Spectroscopy and Chronopotentiometry." *Microchimica Acta* 158 (1–2): 65–71. https://doi.org/10.1007/s00604-006-0683-x.

Doering, Charles R., and J. D. Gibbon. 1995. *Applied Analysis of the Navier-Stokes Equations*. *Applied Analysis of the Navier-Stokes Equations*. Cambridge University Press. https://doi.org/10.1017/cbo9780511608803.

Elkin, V. V., V. Ya Mishuk, V. N. Alekseev, and B. M. Grafov. 2001. "Polarization Diagram of the Electrochemical Impedance of Second Order: A Theory That Accounts for the Diffusion of Reactants." *Russian Journal of Electrochemistry* 37 (4): 399–408. https://doi.org/10.1023/A:1016682124341.

Flueckiger, Jonas, Frank K. Ko, and Karen C. Cheung. 2010. "Electrospun Electroactive Polymer and Metal Oxide Nanofibers for Chemical Sensor Applications." In *ASME International Mechanical Engineering Congress and Exposition, Proceedings (IMECE)* 10: 583–89. https://doi.org/10.1115/IMECE2010-39220.

Gagnon, Zachary, Jason Gordon, Sharamik Sengupta, and Hsueh Chia Chang. 2008. "Bovine Red Blood Cell Starvation Age Discrimination Through a Glutaraldehyde-Amplified Dielectrophoretic Approach with Buffer Selection and Membrane Cross-Linking." *Electrophoresis* 29 (11): 2272–79. https://doi.org/10.1002/elps.200700604.

Gamry Instruments. 2011. "EIS of Organic Coatings and Paints." *Application Note*: 1–7.

Ghosh, Arnab, Tarak K. Patra, Rishi Kant, Rajeev Kr Singh, Jayant K. Singh, and Shantanu Bhattacharya. 2011. "Surface Electrophoresis of Ds-DNA Across Orthogonal Pair of Surfaces." *Applied Physics Letters* 98 (16). https://doi.org/10.1063/1.3565238.

Gomez, Rafael, Rashid Bashir, and Arun K. Bhunia. 2002. "Microscale Electronic Detection of Bacterial Metabolism." *Sensors and Actuators, B: Chemical* 86 (2–3): 198–208. https://doi.org/10.1016/S0925-4005(02)00175-2.

Guan, L., X. L. Feng, G. Xiong, and J. A. Xie. 2011. "Application of Dielectric Spectroscopy for Engine Lubricating Oil Degradation Monitoring." *Sensors and Actuators, A: Physical* 168 (1): 22–29. https://doi.org/10.1016/j.sna.2011.03.033.

Hong, Wangyujue, Abdallah Almomani, and Reza Montazami. 2014. "Influence of Ionic Liquid Concentration on the Electromechanical Performance of Ionic Electroactive Polymer Actuators." *Organic Electronics* 15 (11): 2982–87. https://doi.org/10.1016/j.orgel.2014.08.036.

Hu, Lan, Ivan Rexed, Göran Lindbergh, and Carina Lagergren. 2014. "Electrochemical Performance of Reversible Molten Carbonate Fuel Cells." *International Journal of Hydrogen Energy* 39: 12323–29. https://doi.org/10.1016/j.ijhydene.2014.02.144.

Ibl, N. 1983. "Fundamentals of Transport Phenomena in Electrolytic Systems." *Comprehensive Treatise of Electrochemistry*. https://doi.org/10.1007/978-1-4615-6690-8_1.

Inzelt, G. 2007. "Charge Transport in Conducting Polymer Film Electrodes." *Chemical and Biochemical Engineering Quarterly* 21 (1): 1–14.

Inzelt, G., and G. Láng. 1994. "Model Dependence and Reliability of the Electrochemical Quantities Derived from the Measured Impedance Spectra of Polymer Modified Electrodes." *Journal of Electroanalytical Chemistry* 378 (1–2): 39–49. https://doi.org/10.1016/0022-0728(94)87055-1.

Jones, Thomas B. 2003. "Basic Theory of Dielectrophoresis and Electrorotation." *IEEE Engineering in Medicine and Biology Magazine* 22 (6): 33–42. https://doi.org/10.1109/MEMB.2003.1304999.

Kant, Rishi, Deepak Singh, and Shantanu Bhattacharya. 2017. "Digitally Controlled Portable Micropump for Transport of Live Micro-Organisms." *Sensors and Actuators, A: Physical* 265: 138–51. https://doi.org/10.1016/j.sna.2017.05.016.

Kant, Rishi, Himanshu Singh, Monalisha Nayak, and Shantanu Bhattacharya. 2013. "Optimization of Design and Characterization of a Novel Micro-Pumping System with Peristaltic Motion." *Microsystem Technologies* 19 (4): 563–75. https://doi.org/10.1007/s00542-012-1658-y.

Kumar, Sanjay, Pulak Bhushan, and Shantanu Bhattacharya. 2016. "Development of a Paper-Based Analytical Device for Colorimetric Detection of Uric Acid Using Gold Nanoparticles-Graphene Oxide (AuNPs-GO) Conjugates." *Analytical Methods* 8 (38): 6965–73. https://doi.org/10.1039/c6ay01926a.

Lalić, Marina Mrđa, and Sanja Martinez. 2019. "A Novel Application of EIS for Quantitative Coating Quality Assessment During Neutral Salt Spray Testing of High-Durability Coatings." *Acta Chimica Slovenica* 66 (4): 513–22. https://doi.org/10.17344/acsi.2019.5113.

Li, Chen Zhong, Yali Liu, and John H T Luong. 2005. "Impedance Sensing of DNA Binding Drugs Using Gold Substrates Modified with Gold Nanoparticles." *Analytical Chemistry* 77 (2): 478–85. https://doi.org/10.1021/ac048672l.

Liu, Yi Shao, Padmapriya P. Banada, Shantanu Bhattacharya, Arun K. Bhunia, and Rashid Bashir. 2008. "Electrical Characterization of DNA Molecules in Solution Using Impedance Measurements." *Applied Physics Letters* 92 (14): 143902. https://doi.org/10.1063/1.2908203.

Lu, Juanjuan, Shenguang Ge, Lei Ge, Mei Yan, and Jinghua Yu. 2012. "Electrochimica Acta Electrochemical DNA Sensor Based on Three-Dimensional Folding Paper Device for Specific and Sensitive Point-of-Care Testing." *Electrochimica Acta* 80: 334–41. https://doi.org/10.1016/j.electacta.2012.07.024.

Lvovich, Vadim F. 2012. *Impedance Spectroscopy-Applications to Electrochemical and Dielectric Phenomena*. Wiley.

Lvovich, Vadim F., and Matthew F. Smiechowski. 2006. "Impedance Characterization of Industrial Lubricants." *Electrochimica Acta* 51: 1487–96. https://doi.org/10.1016/j.electacta.2005.02.135.

———. 2008. "Non-Linear Impedance Analysis of Industrial Lubricants." *Electrochimica Acta* 53 (25): 7375–85. https://doi.org/10.1016/j.electacta.2007.12.014.

Marrazza, Giovanna, Giacomo Chiti, Marco Mascini, and Mario Anichini. 2000. "Detection of Human Apolipoprotein E Genotypes by DNA Electrochemical Biosensor Coupled with PCR." *Clinical Chemistry* 46 (1): 31–37. https://doi.org/10.1016/S0009-8981(02)00413-8.

Meric, Burcu, Kagan Kerman, Dilsat Ozkan, Pinar Kara, Selda Erensoy, Ulus Salih Akarca, Marco Mascini, and Mehmet Ozsoz. 2002. "Electrochemical DNA Biosensor for the Detection of TT and Hepatitis B Virus from PCR Amplified Real Samples by Using." *Talanta* 56 (5): 837–46. https://doi.org/10.1016/S0039-9140(01)00650-6.

Moritz, W., I. Gerhardt, D. Roden, M. Xu, and S. Krause. 2000. "Photocurrent Measurements for Laterally Resolved Interface Characterization." *Fresenius' Journal of Analytical Chemistry* 367 (4): 329–33. https://doi.org/10.1007/s002160000409.

Mueller, Jens T., and Peter M. Urban. 1998. "Characterization of Direct Methanol Fuel Cells by Ac Impedance Spectroscopy." *Journal of Power Sources* 75 (1): 139–43. https://doi.org/10.1016/S0378-7753(98)00109-8.

Musiani, Marco M. 1990. "Characterization of Electroactive Polymer Layers by Electrochemical Impedance Spectroscopy (EIS)." *Electrochimica Acta* 35 (10): 1665–70. https://doi.org/10.1016/0013-4686(90)80023-H.

Nakano, M., Z. Ding, H. Kasahara, and J. Suehiro. 2015. "Rapid Size Determination of PCR Amplified DNA by Beads-Based Dielectrophoretic Impedance Spectroscopy." In *Transducers-2015 18th Internation Conference on Solid-State Sensors, Actuators and Microsystems (TRANSDUCERS)*, 1530–32.

Nayak, Monalisha, Deepak Singh, Himanshu Singh, Rishi Kant, Ankur Gupta, Shashank Shekhar Pandey, Swarnasri Mandal, Gurunath Ramanathan, and Shantanu Bhattacharya. 2013. "Integrated Sorting, Concentration and Real Time PCR Based Detection System for Sensitive Detection of Microorganisms." *Scientific Reports* 3: 3266. https://doi.org/10.1038/srep03266.

Newman, John S. 1973. *Electrochemical Systems*. Prentice-Hall, Inc., Englewood Cliffs, New Jersey. https://doi.org/10.1002/aic.690190440.

Orazem, Mark E., and Bernard Tribollet. 2008. *Electrochemical Impedance Spectroscoppy*. Wiley.

Owicki, John C., and J W Parce. 1992. "Biosensors Based on the Energy Metabolism of Living Cells: The Physical Chemistry and Cell Biology of Extracellular Acidification." *Biosensors & Bioelectronics* 7 (4): 255–72. https://doi.org/10.1016/0956-5663(92)87004-9.

Park, Jin Young, and Su Moon Park. 2009. "DNA Hybridization Sensors Based on Electrochemical Impedance Spectroscopy as a Detection Tool." *Sensors* 9 (12): 9513–32. https://doi.org/10.3390/s91209513.

Pettit, C. M., P. C. Goonetilleke, C. M. Sulyma, and D. Roy. 2006. "Combining Impedance Spectroscopy with Cyclic Voltammetry: Measurement and Analysis of Kinetic Parameters for Faradaic and Nonfaradaic Reactions on Thin-Film Gold." *Analytical Chemistry* 78 (11): 3723–29. https://doi.org/10.1021/ac052157l.

Piro, B., J. Haccoun, M. C. Pham, L. D. Tran, A.rubin, H. Perrot, and C. Gabrielli. 2005. "Study of the DNA Hybridization Transduction Behavior of a Quinone-Containing Electroactive Polymer by Cyclic Voltammetry and Electrochemical Impedance Spectroscopy." *Journal of Electroanalytical Chemistry* 577 (1): 155–65. https://doi.org/10.1016/j.jelechem.2004.12.002.

Ragoisha, G. A., A. S. Bondarenko, N. P. Osipovich, and E. A. Streltsov. 2004. "Potentiodynamic Electrochemical Impedance Spectroscopy: Lead Underpotential Deposition on Tellurium." *Journal of Electroanalytical Chemistry* 565 (2): 227–34. https://doi.org/10.1016/j.jelechem.2003.10.014.

Ramírez, Nardo, Angel Regueiro, Olimpia Arias, and Rolando Contreras. 1989. "Electrochemical Impedance Spectroscopy: An Effective Tool for a Fast Microbiological Diagnosis." *Foundations* 31 (2): 72–78.

Rodriguez, Marcela C., Abdel Nasser Kawde, and Joseph Wang. 2005. "Aptamer Biosensor for Label-Free Impedance Spectroscopy Detection of Proteins Based on Recognition-Induced Switching of the Surface Charge." *Chemical Communications* 34: 4267–69. https://doi.org/10.1039/b506571b.

Ruan, Haijun, Bingxiang Sun, Jiuchun Jiang, Weige Zhang, Xitian He, Xiaojia Su, Jingji Bian, and Wenzhong Gao. 2021. "A Modified-Electrochemical Impedance Spectroscopy-Based Multi-Time-Scale Fractional-Order Model for Lithium-Ion Batteries." *Electrochimica Acta* 394: 139066. https://doi.org/10.1016/j.electacta.2021.139066.

Salm, Eric, Yi Shao Liu, Daniel Marchwiany, Dallas Morisette, Yiping He, Arun K. Bhunia, and Rashid Bashir. 2011. "Electrical Detection of DsDNA and Polymerase Chain Reaction Amplification." *Biomedical Microdevices* 13 (6): 973–82. https://doi.org/10.1007/s10544-011-9567-x.

Schwan, H. P. 1959. "Alternating Current Spectroscopy of Biological Substances." *Proceedings of the IRE* 47 (11): 1841–55. https://doi.org/10.1109/JRPROC.1959.287155.

Smiechowski, Matthew F., and Vadim F. Lvovich. 2002. "Electrochemical Monitoring of Water-Surfactant Interactions in Industrial Lubricants." *Journal of Electroanalytical Chemistry* 534 (2): 171–80. https://doi.org/10.1016/S0022-0728(02)01106-3.

———. 2003. "Iridium Oxide Sensors for Acidity and Basicity Detection in Industrial Lubricants." *Sensors and Actuators, B: Chemical* 96 (1–2): 261–67. https://doi.org/10.1016/S0925-4005(03)00542-2.

Sohn, L. L., O. A. Saleh, G. R. Facer, A. J. Beavis, R. S. Allan, and D. A. Notterman. 2000. "Capacitance Cytometry: Measuring Biological Cells One by One." *Proceedings of the National Academy of Sciences of the United States of America* 97 (20): 10687–90. https://doi.org/10.1073/pnas.200361297.

Springer, T. E., T. A. Zawodzinski, M. S. Wilson, and S. Gottesfeld. 1996. "Characterization of Polymer Electrolyte Fuel Cells Using AC Impedance Spectroscopy." *Journal of The Electrochemical Society* 143 (2): 587–99. https://doi.org/10.1149/1.1836485.

Suehiro, Junya, Akio Ohtsubo, Tetsuji Hatano, and Masanori Hara. 2006. "Selective Detection of Bacteria by a Dielectrophoretic Impedance Measurement Method Using an Antibody-Immobilized Electrode Chip." *Sensors and Actuators, B: Chemical* 119 (1): 319–26. https://doi.org/10.1016/j.snb.2005.12.027.

Takashima, Shiro. 2007. "Dielectric Dispersion of Deoxyribonucleic Acid." *The Journal of Physical Chemistry* 70 (5): 1372–80. https://doi.org/10.1021/j100877a006.

Tritton, D. J. 1976. *Physical Fluid Dynamics. Clarenton Press*. Clarenton Press. https://doi.org/10.1088/0031-9112/28/9/048.

Tröltzsch, Uwe, Olfa Kanoun, and Hans Rolf Tränkler. 2006. "Characterizing Aging Effects of Lithium Ion Batteries by Impedance Spectroscopy." *Electrochimica Acta* 51: 1664–72. https://doi.org/10.1016/j.electacta.2005.02.148.

Varshney, Madhukar, Liju Yang, Xiao Li Su, and Yanbin Li. 2005. "Magnetic Nanoparticle-Antibody Conjugates for the Separation of Escherichia Coli O157:H7 in Ground Beef." *Journal of Food Protection* 68 (9): 1804–11. https://doi.org/10.4315/0362-028X-68.9.1804.

Wang, Simon S., and Han S. Lee. 1994. "The Development of in Situ Electrochemical Oil-Condition Sensors." *Sensors and Actuators: B. Chemical* 17 (3): 179–85. https://doi.org/10.1016/0925-4005(93)00867-X.

Wang, Simon S., and Han Sheng Lee. 1997. "The Application of a.c. Impedance Technique for Detecting Glycol Contamination in Engine Oil." *Sensors and Actuators, B: Chemical* 40 (2–3): 193–97. https://doi.org/10.1016/S0925-4005(97)80261-4.

Wei, Zhenbo, Jun Wang, and Xi Zhang. 2013. "Monitoring of Quality and Storage Time of Unsealed Pasteurized Milk by Voltammetric Electronic Tongue." *Electrochimica Acta* 88: 231–39. https://doi.org/10.1016/j.electacta.2012.10.042.

Wilson, J. R., D. T. Schwartz, and S. B. Adler. 2006. "Nonlinear Electrochemical Impedance Spectroscopy for Solid Oxide Fuel Cell Cathode Materials." *Electrochimica Acta* 51: 1389–1402. https://doi.org/10.1016/j.electacta.2005.02.109.

Yang, Liju, Chuanmin Ruan, and Yanbin Li. 2003. "Detection of Viable Salmonella Typhimurium by Impedance Measurement of Electrode Capacitance and Medium Resistance." *Biosensors and Bioelectronics* 19 (5): 495–502. https://doi.org/10.1016/S0956-5663(03)00229-X.

Ye, Y. K., J. H. Zhao, F. Yan, Y. L. Zhu, and H. X. Ju. 2003. "Electrochemical Behavior and Detection of Hepatitis B Virus DNA PCR Production at Gold Electrode." *Biosensors and Bioelectronics* 18 (12): 1501–08. https://doi.org/10.1016/S0956-5663(03)00121-0.

Yeung, S. S. W., T. M. H. Lee, and I. Ming Hsing. 2006. "Electrochemical Real-Time Polymerase Chain Reaction." *Journal of the American Chemical Society* 128 (41): 13374–75. https://doi.org/10.1021/ja065733j.

Yuan, Xiao Zi, Chaojie Song, Haijiang Wang, and Jiujun Zhang. 2010. *Electrochemical Impedance Spectroscopy in PEM Fuel Cells: Fundamentals and Applications*. Springer. https://doi.org/10.1007/978-1-84882-846-9.

3 Integrating microfluidics and sensing for capability enhancement

Mohammed Rashiku, Mohit Pandey, and Shantanu Bhattacharya

CONTENTS

3.1 INTRODUCTION

Biosensors combine a biological component for detection and a physicochemical component for conversion into a measurable signal. These devices can detect the presence and nature of different biological entities, such as viruses, bacteria, DNA, proteins, and cells, qualitatively as well as quantitatively (Dixit and Kaushik 2016; Basu et al. 2019; Bhatt and Bhattacharya 2019; Bhatt, Manoharan et al. 2019; Rashiku and Bhattacharya 2019). The classification of these biosensors can vary depending upon different parameters but are generally classified based on the transduction method employed in electrochemical, optical, thermal, and piezoelectric biosensors.

Microfluidics can be considered the daughter of electrophoresis, as most early microfluidic devices were fabricated for electrophoretic separation. It deals with the handling of fluids using microchannels with dimensions of tens of micrometers. The range of volumes usually varies from picoliters to microliters. It enables biological and chemical assessments to be performed in a very compact space with much lower volumes of reagents and analyte solutions (Chauhan, Pandey, and Bhattacharya 2019; Pandey, Srivastava et al. 2019). It finds applications in immunoassays, cell separation (Bhatt et al. 2017), and DNA amplification (Bhatt, Mishra et al. 2019; Bhatt et al.

DOI: 10.1201/9781003358091-4

2021). Some of the significant advantages of microfluidic devices include laminar flow, which makes mathematical modeling of flow patterns and concentrations possible, with a necessary substantial reduction in the amount of reagents, an increase in the speed of detection and the ability to integrate sensors at a low fabrication cost. Microfluidics forms the foundation for micro total analysis systems (µTAS) or lab-on-a-chip (LOC). This chapter will deal with biosensors based on electrochemical impedance for sensing operations in the microfluidic domain through the integration of microfluidics.

In the LOC arrangement, the microfluidic channels are integrated with an array of components like filters, microvalves, micromixers, and microreactors. LOCs are miniaturized devices capable of integrating all the processes for performing biological or chemical experiments typically done in a diagnostic lab. A single LOC can carry out various tasks. LOCs find applications in clinical diagnostics, drug delivery, life science research, and environmental monitoring. Digital microfluidics (DMF) is a new fluid manipulation technique that allows liquid droplets in the picoliters to microliters range to be handled automatically in individually addressable droplets. DMF works on the electrowetting-on-dielectric (EWOD) principle. DMF is highly compatible with electrical-based detection techniques and thus can be easily incorporated with electrochemical impedance-based detection strategies.

This chapter will be dealing with biosensors integrating microfluidic principles and electrochemical impedance-based sensing. In electrical impedance, the total opposition offered by a circuit to current is measured. It includes both resistance and reactance. Here resistance is a measure of the restriction provided by an electrical circuit element to electricity, while reactance is a measure of obstruction offered by an electrical circuit element to electricity. There are two types of impedance-based measurement techniques: non-faradaic and faradaic impedance. Non-faradaic impedance measurement does not require a redox probe. Cell growth on the electrode surface induces a shift in impedance due to the insulating effects of cell membranes. The insulation effects are impacted by the number, size, and morphological behavior of the cells growing on the electrode surface. Faradaic EIS is an impedance measurement method carried out in the presence of a redox probe. In EIS, a sinusoidal potential is applied, and the impedance is established by evaluating the current generated by the redox species at the electrode surface. EIS can sense the formation of various complexes (antigen–antibody complexes, biotin–avidin complexes, etc.) on the electrode surfaces. EIS enables the examination of the electrode's electron charge transfer kinetics.

In a system of bacterial cells/pathogens, measurement of impedance change is carried out using metal electrodes immersed in the medium where cells are cultured. The medium's impedance changes as ionic metabolites are released into it. The ionic metabolites are released either due to energy metabolism or ion exchange (like K^+ and Na^+) through the cell membrane. Energy metabolism consumes oxygen and sugars and generates carbon dioxide, lactic acid, and acetic acid. The hydration of carbon dioxide to form carbonic acid increases the ionic release. Ion exchange through the cell membrane plays only a tiny part in releasing ionic species. A prevalent detection method is monitoring the impedance of a pair of electrodes dipped in the medium. A positive detection is reported if the impedance changes beyond a particular value. The time taken for pathogen detection in conventional impedance-based methods rises with the decline in the initial concentration of pathogens. So, the impedance techniques may take quite a long time if the initial pathogen concentration is negligible. The issues arising from low initial concentration can be circumvented by trapping a few cells into a tiny volume, effectively increasing the concentration. Such cases call for the integration of microfluidics with impedance-based detection. For example, by trapping 1,000 cells to a 1 μL volume, a concentration level of 10^6 cells/mL can be achieved. Integration of microfluidics with impedance-based techniques dramatically reduces the detection time. Such integration can be beneficial in detecting pathogens in food, water, or bodily fluids, where the pathogen concentration may be deficient.

3.2   MATERIALS USED FOR FABRICATION OF BIOMEMS DEVICES

There are three categories of material generally used for BioMEMS-related devices. They are (i) silicon, glass, and other microelectronics-related materials, (ii) polymeric materials like (poly)dimethylsiloxane (PDMS), polymethyl methacrylate (PMMA), and (iii) biological materials like cells, tissues, and proteins.

Silicon is the backbone of commercial semiconductor electronics and is frequently used in microfabrication. Silicon is widely used to fabricate cantilevers, accelerometers, beams, etc. Many MEMS-based devices, sensors, and actuators are heavily dependent on silicon. Thus, silicon directly and indirectly serves many industries like automobile, aerospace, and healthcare (Bhattacharya et al. 2019; Pandey, Tatiya et al. 2019a, 2019b; Tatiya et al. 2019). Numerous wet/dry etching techniques can be used to make microstructures in silicon. Layers can be added to silicon using chemical or physical vapor deposition. Optical detection-based devices are easy to fabricate with glass due to optical transparency. Glass can be used to manufacture airtight devices due to its strong covalent bond formation. Disadvantages of glass processing include high cost, large time consumption, and process complexity. Since glass is amorphous, wet etching can be used to form only isotropic structures.

PMMA is a transparent thermoplastic polymer commonly known as acrylic. It is used as a structural material for device fabrication and a masking material for patterning. Hot embossing, laser ablation, electron beam lithography, and X-ray lithography are some techniques used for the patterning of PMMA. PMMA has attractive features like a low friction coefficient, the ability to bond at low temperatures using solvents, high chemical resistance, low fabrication cost, biocompatibility, and good electrical insulation. Despite its many advantages, there are many disadvantages to PMMA that limit its use. PMMA is hydrophobic, leading to the adsorption of biomolecules. PMMA is unsuitable for organic solvents and high-temperature processes. Autofluorescence of PMMA restricts the use of PMMA for optical detection. Another commonly used material in this category is polydimethylsiloxane (PDMS), a low-cost viscoelastic polymer belonging to the class of silicones. A pre-polymer and a curing agent are combined to create PDMS. Advantages of PDMS include the low cost of fabrication, high reproducibility of features, optical transparency, biocompatibility, and self-sealing capability. The low interfacial energy of PDMS makes it highly hydrophobic. But, this interfacial energy can be modified using oxygen plasma, and this makes PDMS ideal for the fabrication of closed microfluidic chambers and channels. Most of the limitations of PDMS are similar to those of PMMA, such as hydrophobicity, autofluorescence, and incompatibility with organic solvents and high temperatures. These disadvantages sometimes become very crucial for the selection of the right material. However, advanced fabrication techniques, such as laser processing, plasma processing, and additive manufacturing, are often integrated to overcome these hurdles and widen the use of these polymers. Bulk modification and surface modifications are also commonly used to increase the usage of these polymers (Kumar et al. 2019; Sundriyal, Pandey, and Bhattacharya 2020; Choudhury, Pandey, and Bhattacharya 2021; Pandey, Rashiku, and Bhattacharya 2021; Pandey et al. 2022).

The use of biological materials in micro-device fabrication is relatively new and unexplored. Possible applications of biological materials include the construction of biological structures, such as making artificial organs and developing cell-based arrays. Many nanomaterials were also explored recently to amalgamate these structures for increased functionalization and improved selectivity of the sensor (Dubey et al. 2018; Jangir et al. 2018; Tatiya, Pandey, and Bhattacharya 2020). Microfluidic paper-based analytical devices have gained popularity, thanks to their low cost, ease of availability, fabrication, disposability, large-scale manufacturability, and biocompatibility (Pandey, Shahare et al. 2019; Rashiku and Bhattacharya 2019). There is a need for sensitive, specific, low-cost, and easy-to-use diagnostic devices, mostly in resource-limited environments and wearable integration for continuous monitoring, which can be satisfied using paper-based microfluidic analytical devices (Pandey, Tatiya, and Bhattacharya 2021).

3.3   APPLICATIONS OF IMPEDANCE-BASED BIOSENSORS

Impedance biosensors find applications in several biological and nonbiological fields because of their high utility value. This chapter focuses on the discussion of microfluidic devices primarily inclined toward applications like cancer biomarker detection, microorganism identification, water/food quality monitoring, and disease detection.

Cancer is among the prevalent and life-threatening illnesses that result from irreversible changes to the genetic material. Exposure to carcinogenic chemicals, ionizing radiation, and oxidative stress are the leading causes of cancer. There are various conventional diagnostic techniques used for the diagnosis of cancer. One of the most promising areas requiring sensitive and accurate cancer detection on the basis of different biomarkers uses biosensors with microfluidic techniques integrated with various sensing strategies. Over the years, people have reported the detection of various cancer biomarkers like α-fetoprotein (AFP) (Yang et al. 2014), carcinoembryonic antigen (CEA) (Zhou et al. 2014), and MCF-7 breast cancer cells (Zhou et al. 2018).

The other significant application of biosensors is to detect microorganisms. There are multiple biosensors reported in the literature for sensing various pathogenic microorganisms and nonpathogenic microorganisms. Gómez et al. reported a microfluidic biosensor for the impedance-based sensing of *Listeria innocua*, *Listeria monocytogenes*, and *Escherichia coli* cells (Gómez, Bashir, and Bhunia 2002). An impedimetric sensor for the sensitive detection of the H5N1 virus has been described (Wang et al. 2009). Biosensors for detecting pathogenic bacteria like *E. coli* O157:H7

(Yang, Li, and Erf 2004) and *Salmonella* B and D (Liu et al. 2019) have been reported. Iliescu et al. reported an EIS-based biosensor to detect dead and live yeast cells (Iliescu et al. 2007).

Water contamination is a serious cause of concern due to the increase in wastewater discharge from various industries, agricultural waste, and other human and animal activities. Industrial wastewater can be toxic, carcinogenic, or flammable. The quality of surface water has severely deteriorated in most parts of the world due to contamination. Water contamination may be due to biological contamination or chemical contamination. Biological contaminants include harmful pathogenic variants of *E. coli* (Yang, Li, and Erf 2004), *Salmonella* (Liu et al. 2019), and *Shigella*. Dangerous strains of *E. coli* may cause diarrhea, stomach cramps, nausea, and vomiting. Pesticides such as atrazine (Ramón-Azcón et al. 2008), pharmaceuticals, nitrates, and heavy metal ions are some of the common chemical contaminants in water.

Detecting pathogenic microorganisms in food is vital to ensure food quality and safety (Bhattacharya et al. 2008). Various sensors based on EIS have been reported. *E. coli* O157:H7 (Yang, Li, and Erf 2004), *Vibrio parahaemolyticus* (Sharif et al. 2019), human norovirus (Baek et al. 2020), *Salmonella* B and D (Liu et al. 2019) are some of the pathogens detected in food items. *V. parahaemolyticus* is the principal cause of seafood-associated gastroenteritis. *E. coli* O157:H7 is of particular concern since *E. coli* O157:H7 infection may involve loss of life. Carcinogenic mycotoxins such as aflatoxin B_1 have been detected in food samples using a portable sensor based on impedance spectroscopy (Li et al. 2016).

The detection of proteins is essential from the standpoint of disease diagnosis. Various proteins can be used as biomarkers to detect various pathogenic diseases through serological assays and an early detection of multiple forms of cancers (Chuang et al. 2016). Cytotoxic effects of various drugs (Caviglia et al. 2015) and toxic chemicals (F. Liu et al. 2013) have also been studied using impedance-based biosensors.

These applications are discussed in detail in the following sections along with an elaborated discussion on various microfluidic fabrication aspects of the device to get the detection done.

3.3.1 IMPEDANCE-BASED SENSORS FOR CANCER DETECTION

This section will cover the various impedance biosensors developed over time for the timely detection of various types of cancers through the sensitive detection of specific cancer biomarkers. Biosensors capable of detecting cancer biomarkers, such as α-fetoprotein, carcinoembryonic antigen (CEA), and MCF-7 cancer cells, have been reported along with their fabrication details and their working principles.

Yang et al. fabricated a novel impedimetric sensor for detecting cancer biomarker α-fetoprotein (AFP) (Yang et al. 2014). Two enzymes, viz. horseradish peroxidase (HRP) and glucose oxidase (GOx), catalyzed the oxidation of 4-chloro-1-naphthol (4-CN) by H_2O_2 to produce 4-CN precipitates. The 4-CN deposits on the electrode surface led to a reduction in the electrode surface area. EIS was used to observe a reduction of the redox couple corresponding to a decrease in the electrode surface

(a)

(b)

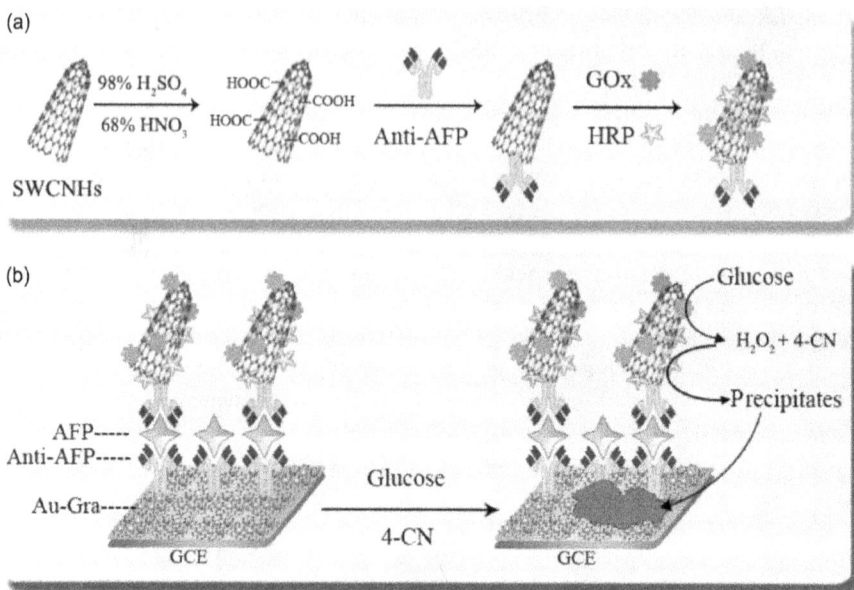

FIGURE 3.1 (A) Simplified illustration of the steps in fabricating SWCNHs/Ab2/GOx/HRP bionanocomposites and (B) steps in fabricating electrodes. [Reprinted with permission from Yang et al. (2014). Copyright (2014) Elsevier].

area. A linear response was observed between 0.001 and 60 ng/mL, and a detection limit of 0.33 pg/mL was reported.

The sensor was fabricated, as explained. Single-walled carbon nanohorns (SWCNHs) were first carboxylated by ultrasonicating in a solution containing 98% H_2SO_4, 68% HNO_3, and double-distilled water (1:3:6 ratio by volume) for 6 h. The solution was centrifuged at 8000 rpm, and the sediment was collected. The residue was rinsed with water and dried overnight at 50°C. After dissolving 2 mg of the carboxylated SWCNHs in 1X PBS (pH 7.4), the pH of the resulting solution was revived to its initial value. About 2.5 mL of a mixture of 1-ethyl-3-(3-dimethylaminopropyl) carbodsiimide (EDC) and N-hydroxysuccinimide in a ratio of 4:1 was mixed with the previous solution and vigorously stirred for 4 h at room temperature. The sediment was collected by centrifugation and mixed thoroughly in 1 mL of 1X PBS (pH 7.4). About 100 µL each of 1 mg/mL HRP, 1 mg/mL GOx, and anti-AFP antibody (Ab2) were introduced dropwise into the solution and kept at 4°C and agitated at 150 rpm for 12 h. A modified glassy carbon electrode (GCE) acted as the working electrode. The working electrode was subjected to sequential sonication with deionized water and ethanol, followed by drying. About 10 µL of gold nanoparticles/graphene nanosheet nano-composites (Au–Gra) was poured on the working electrode surface and air-dried. The modified GCE was submerged in 20 ng/mL solution of anti-AFP antibody (Ab_1) for antibody conjugation. Further, the electrode surface was cleaned with water to eliminate unconjugated antibodies. About 15 µL of bovine serum albumin (BSA) was added for 30 min to prevent nonspecific adsorption. A schematic diagram of the whole process is given in Figure 3.1.

Zhou et al. fabricated an EIS-based sensor for detecting carcinoembryonic anti-gen (CEA) utilizing a staphylococcal protein A (SPA) conjugated gold nanoparti-cle-enhanced electrode (Zhou et al. 2014). 1,6-hexanedithiol (HDT) was the linker molecule between the gold electrode and gold nanoparticles. Further, SPA was con-jugated to the gold nanoparticle surfaces. Next, the anti-carcinoembryonic antigen–antibody was attached to the electrode via the affinity of SPA to the Fc region of the antibodies. BSA was added to minimize nonspecific adsorption. EIS was used to quantify CEA. The attachment of the CEA antigen to the anti-carcinoembryonic antigen–antibody impeded the electron transfer of the redox probe, which affected the EIS curves. The fabrication process and the sensing principle are shown in Figure 3.2.

Zhou et al. reported a microfluidic device capable of measuring the mechanical and electrical characteristics of cancer cells at the same time (Zhou et al. 2018). Both the mechanical and electrical properties of cells are excellent for studying and characterizing cells. The time it takes for a single cell to travel via a striction with a size less than the cell size was correlated with the deformability of MCF-7 cancer cells. EIS was used to characterize undeformed cells and deformed cells. Traditional two-electrode design is less suited to detecting the impedance of undeformed cells and can only measure the impedance of deformed cells. In the conventional method, a cell passing through the narrow passage causes a change in impedance due to the replacement of the medium between the electrodes. The duration of the imped-ance change corresponds to the passage time. Total passage time comprises the time needed for the cell to deform to enter the striction and the time required for move-ment once within. Impedance measurement using the conventional two-electrode method cannot distinguish between the two time components.

An illustration depicting the design and working principle of the device is given in Figure 3.3. Four pairs of electrodes that were used for impedance measurements allow the quantification of movement times of a cell through different regions. The total passage time was measured using the first and last pair of electrodes. The elec-trodes received input voltage as shown in Figure 3.3; current was obtained from the corresponding electrodes, and impedance change was calculated. As the cell went through the electrode pair, a positive peak was observed in differential impedance

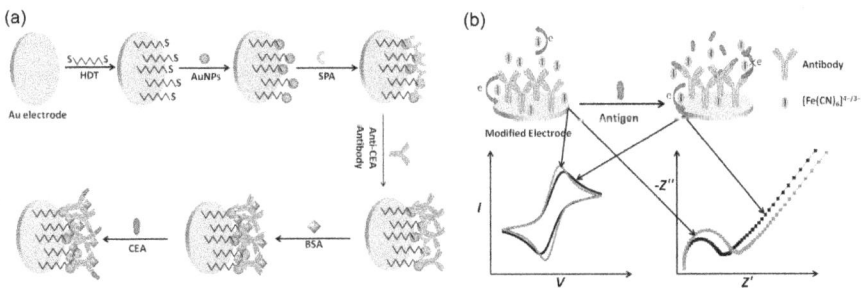

FIGURE 3.2 (A) Steps in constructing EIS-based sensor for detecting CEA; (B) the sensing principle of the electrochemical immunosensor. [Reprinted with permission from Zhou et al. (2014). Copyright (2014) Elsevier].

($\Delta|Z|$). As the cell went through the electrode pair, there was a negative peak in differential impedance ($\Delta|Z|$). Differential impedance ($\Delta|Z|$) remained zero when the cell traveled across the constriction. Thus the time delay between the two peaks give an idea about the total time passed. This setup could measure the entry and transit time by varying where the input signal was applied and where the current was measured, as shown in Figure 3.3.

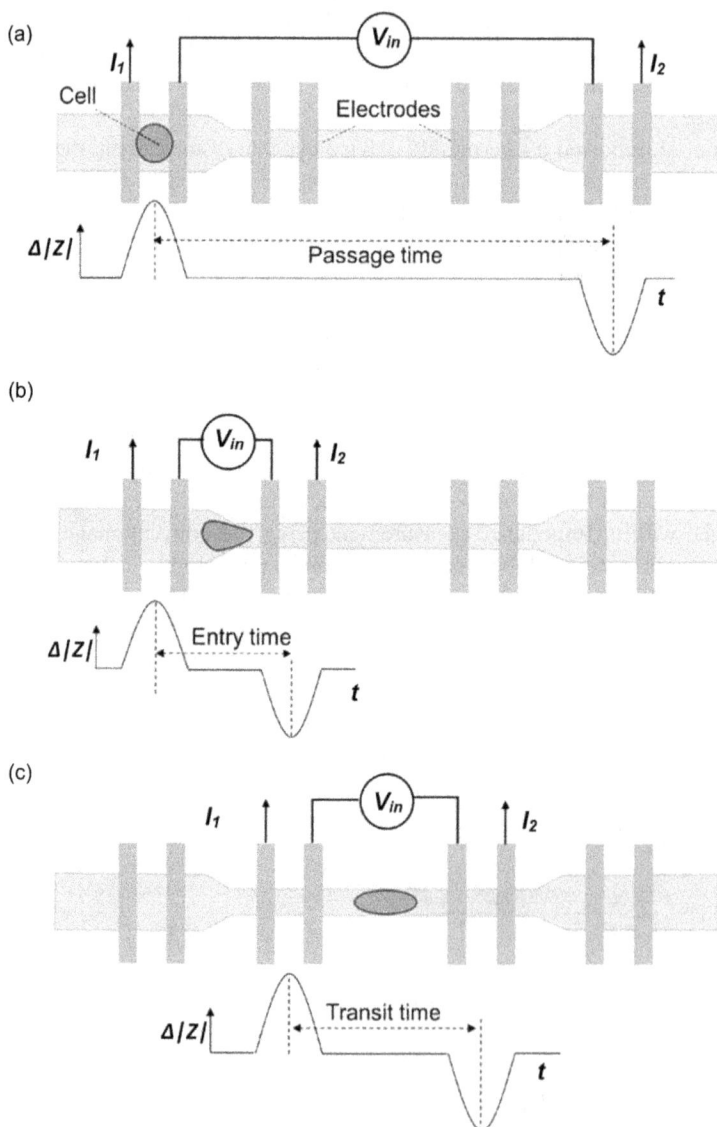

FIGURE 3.3 Diagram illustrating the working principle of a device. (A) Quantification of total passage time; (B) quantification of entry time; (C) quantification of transit time. [Reprinted with permission from Zhou et al. (2018). Copyright (2018) American Chemical Society].

Nguyen et al. demonstrated how to detect single-cell migration through electrical cell-substrate impedance sensing (ECIS) incorporated with a microfluidic system (Nguyen et al. 2013). Microfluidic principles were used for single-cell capture for culture and subsequent impedance quantification. The sensor consisted of micro-electrode arrays (MEAs), cell capture arrays (CCAs), and a microfluidic channel. The CCA was a structure with a V-shaped gap that trapped the cells while allowing hydrodynamic flow to pass through. MEAs consisted of small electrodes on both sides of a central counter electrode. The working electrodes had cells trapped on their surface, while the electrode on the opposite side with no trapped cell acted as the reference electrode. The migration of single MDA-MB-231 cells (highly meta-static) caused a significant change in impedance compared with the migration of MCF-7 cells. The device employed a technique to pin single cells on the surface of microelectrodes. Real-time tracking of individual cancer cell migration could be accomplished with the proposed device.

The fabrication process of the sensor involved the deposition of 10-nm thick chromium, 100-nm thick gold, and 100-nm thick titanium on a Pyrex wafer substrate through physical vapor deposition. The MEAs, the bonding pads, and the connecting lines were then created by a lift-off fabrication process using AZ 5214 photoresist. Next, a 500-nm thick layer of SiOx was coated using chemical vapor deposition. The SiOx coating acted as a passivation layer. The MEAs and the bonding pad areas were exposed using photolithography and reactive ion etching. The wafer was immersed in 1% HF solution to remove the titanium coating. Spin-coating and patterning of SU-8 negative photoresist were used to create the CCAs and a portion of the micro-fluidic channel walls. A PDMS layer was fabricated as shown in Figure 3.4D. The PDMS layer was pierced to provide the inlet and outlet ports. The PDMS layer was oriented correctly and bonded with the SU-8 channel. The fabrication steps are illustrated in Figure 3.4.

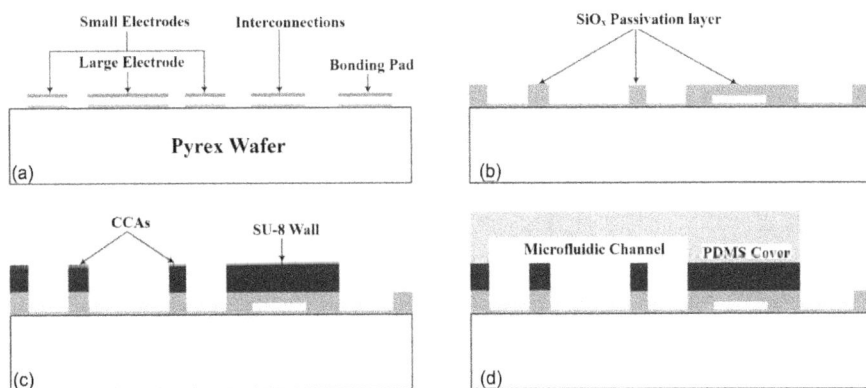

FIGURE 3.4 Illustration of the fabrication process. (A) Coating and patterning of Cr, Au, and Ti layers on the substrate; (B) deposition and patterning of SiOx; (C) patterning of CCAs and a portion of the microfluidic channel wall; (D) bonding of the PDMS cover to the SU-8 structure. [Reprinted with permission from Nguyen et al. (2013). Copyright (2013) American Chemical Society].

3.3.2 IMPEDANCE-BASED SENSORS FOR THE DETECTION OF MICROORGANISMS

A major chunk of the biosensors developed over time is dedicated to the detection of various pathogenic microorganisms and nonpathogenic microorganisms. This section provides details of fabrication techniques and performance of microfluidic biosensors for impedance-based sensing of *L. innocua*, *L. monocytogenes*, *E. coli* cells, H5N1 virus, *Salmonella* B and D, and yeast cells.

Iliescu et al. described a novel process for fabricating a device for the EIS-based characterization of cells (Iliescu et al. 2007). The device consisted of two glass wafers. The bottom glass wafer incorporated microfluidic channels for sample flow and two EIS electrodes. Inlet and outlet ports were provided on the top glass wafer. Glass offers significant advantages like a wide range of frequency spectrum, transparency, and hydrophilicity, which helps capillary-based movement, making pumping unnecessary and reducing the sample volume requirement. The fabrication process involved three steps. The first step was to cut inlet/outlet holes and via holes in the top wafer through wet etching. In the second step, microfluidic channels were etched into the bottom wafer, and electrodes were formed on top using lithography. Finally, the two wafers were bonded by selectively applying a skinny layer of SU8 adhesive on the bottom layer and joining the top and bottom wafers together. Figure 3.5 illustrates the design of the fabricated microfluidic device. The device successfully detected

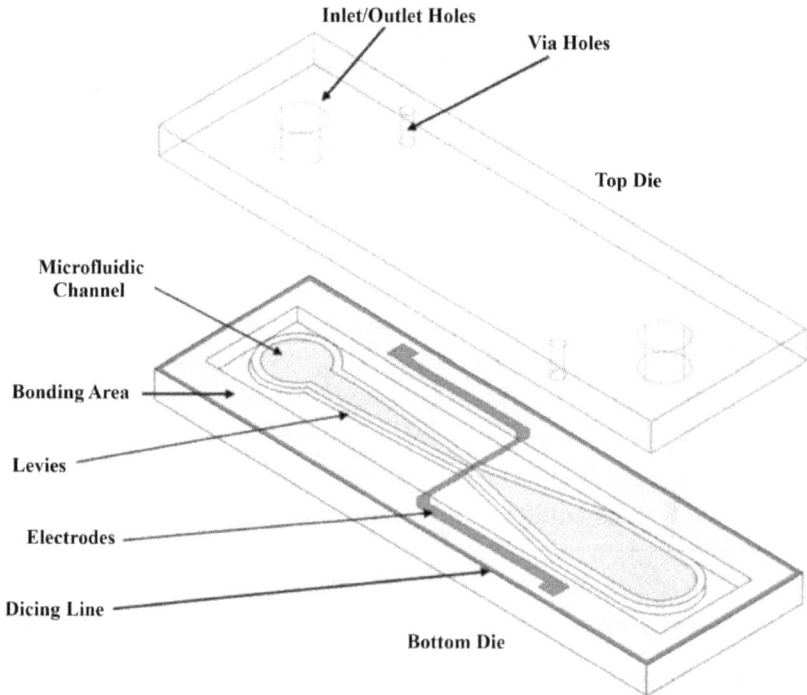

FIGURE 3.5   Schematic diagram of the EIS incorporated microfluidic biochip. [Reprinted with permission from Iliescu et al. (2007). Copyright (2007) Elsevier].

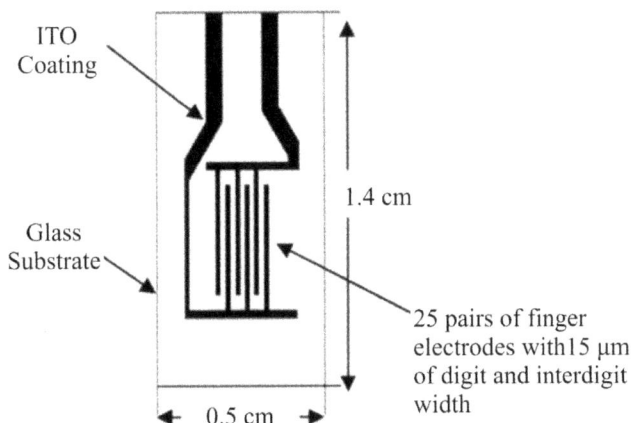

FIGURE 3.6 Design of the IDAM. [Reprinted with permission from Yang, Li, and Erf (2004). Copyright (2004) American Chemical Society].

live and dead yeast cells suspended in PBS. The device was capable of making a clear distinction between dead and live yeast cells.

Yang et al. demonstrated an electrochemical impedance immunosensor that rapidly detects *E. coli* O157:H7 (L. Yang, Li, and Erf 2004). Interdigitated array microelectrodes (IDAM) conjugated with anti-*E. coli* antibodies were utilized to fabricate the immunosensor. Indium tin oxide-coated glass substrates were employed for the realization of the sensor. The design of the IDAM is shown in Figure 3.6. An equivalent circuit was developed to interpret the impedance components of the IDAM system. The binding event of *E. coli* O157:H7 to the anti-*E. coli* antibodies caused an increase in electron-transfer resistance. The change in the electron-transfer resistance showed a linear fit with the log of the concentration of *E. coli* O157:H7 from 4.36×10^5 to 4.36×10^8 cfu/mL. The reported limit of detection was 10^6 cfu/mL.

Gómez et al. described a microfluidic biochip for EIS-based sensing of the metabolic activity of bacteria (Gómez, Bashir, and Bhunia 2002). Impedance change was observed due to the discharge of ionic species into the culture medium by the metabolizing bacteria. Using microfluidics, it is possible to rapidly detect a few cells by confining the cells to a small volume of the order of a few nanoliters. Impedance spectroscopy was performed between the frequency ranges of 100 Hz and 1 MHz. The biochip fabrication involved etching a matrix of microchannels and microchambers into a silicon substrate. Two serially connected groups of chambers were fabricated on a silicon substrate with an etch depth of 12 μm and electrically insulated with a layer of SiO_2 (0.45 μm thick). The channel sizes were varied to accommodate different sample volumes. Platinum electrodes in the interdigitated array format were formed at the bottom of the chambers for impedance measurement. The platinum electrodes were connected to pads on the periphery of the biochip. A glass slab was bonded to the top of the chip using a low-melting temperature spin-on-glass as adhesive. Microbore tubes were injected into trenches etched at the side for the fluid inlet and the fluid outlet. A chip design is schematically illustrated in Figure 3.7. The main challenge with the device was identifying the impedance signal produced by dead

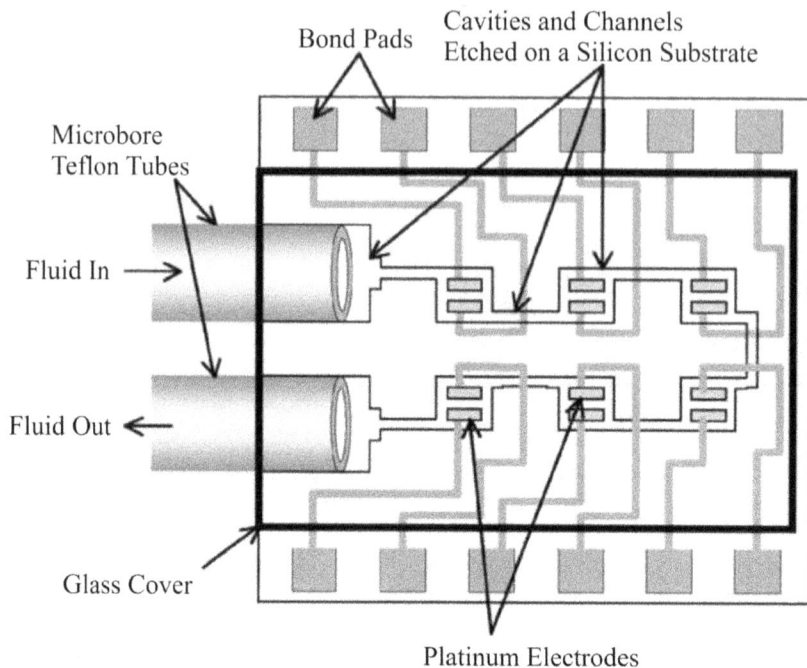

FIGURE 3.7   Schematic illustration of the biochip design. [Reprinted with permission from Gómez, Bashir, and Bhunia (2002). Copyright (2002) Elsevier].

cells. It was reported that the maximum ratio of dead cells to live cells for the correct detection of live cells was 10:1. The biochip proved effective for the sensing of *L. innocua*, *L. monocytogenes*, and *E. coli* cells.

Wang et al. developed an EIS-based immunosensor for sensing the H5N1 virus (Wang et al. 2009). Gold microelectrodes in the interdigitated array setup were used for creating the sensor. Polyclonal antibodies against surface antigen hemagglutinin were directionally bound to the gold microelectrode to capture H5N1 viruses. Directional immobilization of antibodies on the gold microelectrodes was achieved by binding protein A on the gold surface before antibody immobilization. The F_c region of antibody anti-H5N1 antibodies attach to protein A. The F_{ab} region of the H5N1 antibodies was thus directed away from the gold microelectrodes. BSA blocking solution was applied to suppress unbound sites. A change in impedance was detected upon avian influenza virus capture by the microelectrode. The virus capture was confirmed using atomic force microscopy as well. A linear graph was recorded between the response and the concentration of the avian influenza virus in the range of 10^3 and 10^7 EID$_{50}$/mL. The authors also did cross-reactivity studies against the Newcastle disease virus and infectious bronchitis virus and reported very little interference from both. It was also reported that the microelectrode used in the sensor could be reused after proper cleaning. For reuse, it had to be ensured that the impedance value of the bare electrode after cleaning and the impedance value before any antibody immobilization were similar.

3.3.3 IMPEDANCE-BASED SENSORS FOR THE DETECTION OF PROTEINS

This section details impedance-based sensors developed specifically for the detection of proteins.

Mok et al. (2014) described a novel biosensor integrating microfluidics and electrical impedance spectroscopy for protein biomarkers. The device employed a two-compartment design to isolate the capture/reaction area from the sensor area. Better sensitivity was reported since capture/reaction and detection steps could be individually refined. Target-specific probe molecules were immobilized on the capture/reaction chamber surface. The sensor chamber consisted of a micropore between two gold electrodes. The technology was based on the electrical impedance measurement of beads. Magnetic beads of size 2.8 μm were used because of their easy manipulation characteristic and no aggregation behavior. Antibodies specific to the protein were immobilized on the surface of the capture/reaction compartment and attached to the magnetic beads. The sample was first infused into the capture/reaction compartment, which was proceeded by an hour of incubation. The magnetic beads were infused into the capture/reaction compartment after emptying the sample. A sandwich structure was formed, and the beads were attached to the capture/reaction compartment surface depending upon the supply of the protein of interest present in the sample. The capture/reaction compartment was 300 μm wide, 4 mm long, and 30 μm tall. The device was capable of quantifying protein abundance and enzyme activity. Protein abundance of interleukin 6 was measured from concentrations up to 50 pM from a sample volume of 5 μL. The unbound beads were washed off, and the bound beads were extracted and transported to the sensor compartment. Each bead induced an impedance change, which in turn corresponded to the abundance of the target protein.

Chuang et al. reported an impedimetric immunosensor for sensitively detecting bladder cancer biomarker Galectin-1 (Gal-1) protein (Chuang et al. 2016). Nine individually addressable annular gold microelectrodes in the interdigitated array format were arranged in a three-by-three layout for fabricating the immunosensor. The fabrication involved cleaning the glass substrate, followed by the coating of 30-nm thick chromium and 70-nm thick gold using physical vapor deposition. Next, the photoresist (EPG512) was spin-coated on the electrode surface, and it was exposed to UV and developed. Wet etching was executed to selectively eliminate the electrode areas to form the annular microelectrode array layout. Next, the photoresist layer was stripped. Anti-Gal-1 antibodies conjugated to alumina nanoparticles (nanoprobes) were used to improve the sensitivity and the conjugation efficiency. The nanoprobes were attached to the microelectrode surface using dielectrophoresis. BSA was applied to reduce nonspecific adsorptions. A sample volume of 100 μL was added to the sensing area. The attachment of antigen (Gal-1) to the anti-Gal-1 antibodies caused an impedance change measured using EIS analysis. Impedance spectrum was obtained in the frequency range of 1–100 kHz.

3.3.4 IMPEDANCE-BASED SENSORS FOR FOOD AND WATER-QUALITY TESTING

Various biosensors based on EIS for food and water-quality testing have been reported. Biosensors for the sensitive and accurate detection of some of the common biological contaminants in food and water, such as *E. coli*, *Salmonella*, *Shigella*, *V. parahaemolyticus*, and human norovirus have been discussed.

Liu et al. demonstrated an impedance-based device for detecting *Salmonella* B and D in food samples (Liu et al. 2019). The authors reported a sensor capable of detecting up to 300 cells/mL within 1 h. The sensing regions for detecting *Salmonella* B and D were formed from IDAM incorporating 50-finger pairs. The optimum length, width, and electrode gap were determined by performing a simulation using COMSOL. Anti-*Salmonella* antibodies against *Salmonella* B or D were administered to the electrode via separate antibody entry points to functionalize the sensing areas and prevent cross-contamination. The microchannels were infused with antibody solutions for an hour to guarantee appropriate antibody adherence to the electrodes. To test the samples for *Salmonella* B or D, the sample solution was allowed to flow through, and flow was halted for 30 min after the microchannels were filled. During this time, the *Salmonella* antigens bound with the corresponding *Salmonella* antibodies. An illustration of the impedance-based microfluidic device is provided in Figure 3.8.

Ramón-Azcón et al. fabricated a novel biosensor comprising IDAM to detect atrazine pesticide (Ramón-Azcón et al. 2008). To realize the impedimetric immunosensor, the IDAMs were patterned with approximately 200-nm thick layers of gold and chromium. A thinner layer of chromium was deposited on the Pyrex 7740 glass substrate through sputtering, and then the gold layer was coated. The chromium layer was used to enhance the attachment of the gold layer to the substrate. The electrodes were patterned using photolithography followed by metal etching to provide an inter-electrode spacing of 6.8 μm. The IDAMs were activated with (3-glycidoxypropyl) trimethoxysilane. The IDAMs were functionalized with atrazine-haptenized protein. The impedance measurements were performed after exposing the electrodes to a

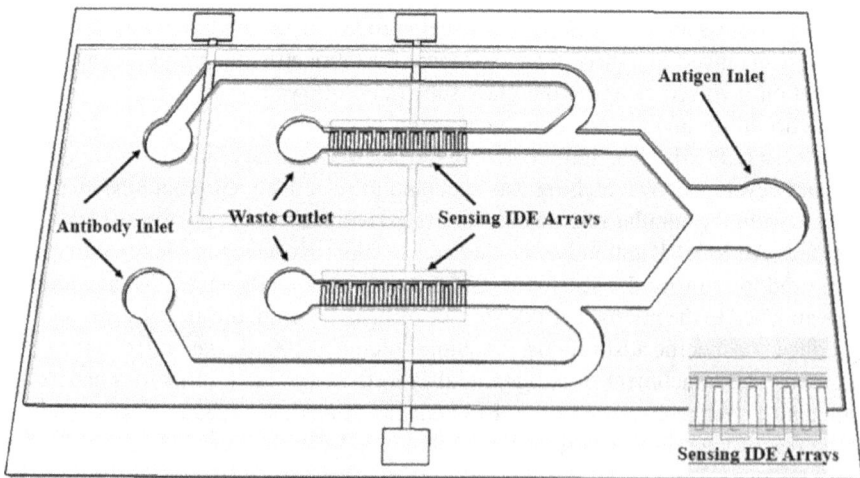

FIGURE 3.8 Illustration of the impedance-based microfluidic device. The magnified view of the IDAM is inserted at the bottom right. [Reprinted with permission from Liu et al. (2019). Copyright (2019) PLOS].

solution containing the atrazine-specific antibody and the sample to be tested for atrazine. If the atrazine concentration is low, more antibodies are available for binding to the immobilized antigen (atrazine-haptenized protein) and vice-versa. This way, the sensor employs the competitive reaction between atrazine and the immobilized atrazine-haptenized protein for the specific antibody. Impedance was measured without any redox probe. The limit of detection of atrazine reported was 0.04 µg/L. The authors demonstrated the use of the sensor by assessing its ability to detect pesticide residues in red wine. The sensor could detect atrazine in the red wine sample with a detection limit of 0.19 µg/L.

Baek et al. demonstrated a nanosensor incorporating EIS for the sensitive detection of the foodborne pathogen human norovirus (Baek et al. 2020). Ingestion of contaminated food containing human norovirus may lead to acute gastroenteritis, diarrhea, and food poisoning. For fabricating the biosensor, gold nanoparticles were immobilized on tungsten disulfide nanoflowers. The tungsten disulfide nanoflowers' surface was functionalized with 3-mercaptopropionic acid to enable the attachment of tungsten disulfide nanoflowers with gold nanoparticles. The addition of gold nanoparticles improved the electrochemical conductivity and provided sites for the immobilization of peptide. Norovirus affinity peptide was attached to the gold nanoparticles. The norovirus sample was added to the nanocomposite and then washed to remove any unbound viruses. The norovirus-modified nanocomposite was then added to a screen-printed carbon electrode and assessed using EIS. Norovirus has a protein that obstructs the charge transfer between the working electrode and the redox species, which leads to increased impedance. The device was tested with spiked norovirus samples and reported a lower detection limit of 2.37 copies/mL. The device was assessed for norovirus detection in oyster samples and reported a lower detection limit of 6.21 copies/mL. It was also tested with rotavirus samples and dengue virus samples to confirm the selectivity of the device and found that there was no signal developed.

Sharif et al. fabricated a novel microfluidic biosensor incorporating impedance measurement technique to detect loop-mediated isothermal amplification (LAMP) amplicons (Sharif et al. 2019). The microfluidic device contained an electrode chip composed of two disc gold electrodes of 2.18 mm diameter separated by 250 µm. A microfluidic cell was fabricated over the electrode chip using polymethyl methacrylate (PMMA). Carboxyl-modified magnetic beads were used for capturing amplicons of the LAMP reaction. About 25 µL of the amplicons were introduced to the rinsed magnetic beads and incubated for 20 min. Incubation with amplicons led to the electrostatic adsorption of amplicons on the surface of the magnetic beads. A magnetic stand was used to separate the produced amplicons–magnetic bead composites and re-suspended in 100 µL of deionized water. The suspension was pumped into the device to perform EIS measurements. The impedance of the amplicons–magnetic bead composites suspension in deionized water was measured by introducing an AC potential of 50 mV at a frequency range of 1 Hz to 900 kHz. The very low conductivity of the deionized water made it possible to detect even tiny changes in impedance. The negative charges of nucleic acids in the amplicons–magnetic bead composites caused a significant decrease in the impedance. The device successfully detected *E. coli* O157:H7, *V. parahaemolyticus*, *Staphylococcus aureus*, and *L. monocytogenes*.

Li et al. fabricated a portable sensor for rapidly detecting aflatoxin B_1 (AFB_1) in rice samples (Li et al. 2016). The portable device consisted of an impedance-based sensing unit and a USB-compatible sensor chip. A brief outline of the steps involved in the device fabrication is provided. First, the surface of screen-printed interdigitated microelectrodes (SPIMs) was immersed in 1 M NaOH solution for 5 min and then washed with water, and immersed in 1 M HCl solution for 2 min and washed with water. The cleaning efficiency was assessed using EIS. Next, the SPIM was dipped in a 2 mM solution of 3-dithiobis-(sulfosuccinimidyl-propionate) (DTSP) and incubated. The incubation time was optimized to maximize the sensor performance. Next, the SPIMs were rinsed with acetone to eliminate unattached DTSP. The SPIMs were soaked for 45 min in 1 mg/mL of protein G. The SPIMs were rinsed with ultrapure water to eliminate free protein G. The SPIMs were immersed in an optimized concentration of anti-AFB1 monoclonal antibody solution for 45 min. BSA was used to block the accessible locations on the SPIM surface and to reduce nonspecific adsorptions. The total detection time was reported to be approximately 1 h. The sensor detected AFB_1 in rice up to 5 ng/mL, which is lower than the permitted levels of AFB_1.

3.3.5 IMPEDANCE-BASED SENSORS FOR CELL VIABILITY OR CYTOTOXICITY ASSAYS

Cytotoxicity assays measure the ability of certain substances to cause cell damage or cell death. Cytotoxicity assays are used in research and drug discovery to screen for toxic chemicals. Herein, we discuss the working principle, and construction details of EIS-based sensors used for cytotoxicity analysis.

Caviglia et al. described a microfluidic cytotoxicity assay to assess the cytotoxic effects of the anticancer drugs doxorubicin and oxaliplatin (Caviglia et al. 2015). The microfluidic device facilitated electrochemical and optical detection. The data obtained through EIS-based detection was confirmed using fluorescent apoptosis assay. The microfluidic device employed peristaltic pumps to control fluid flow. The detection unit was composed of three electrode arrays, namely, the counter electrode, the reference electrode, and the interdigitated working electrode. The device proved effective in assessing the cytotoxic effects of anticancer medications.

Liu et al. combined quartz crystal microbalance (QCM) resonators with ECIS (Liu et al. 2013). In the dual-sensing system, the QCM upper electrode was the working electrode for ECIS. The QCM sensor was built on a thin quartz substrate with gold electrodes on either side. The semicircular counter electrode for ECIS was fabricated close to the upper electrode. The time-dependent electrical impedance variation between the working and the counter electrodes was assessed over 40 Hz to 100 kHz. The cell viability is unaffected owing to currents in the nano ampere (nA) range. Bovine aortic endothelial live cells were grown on the multiparametric biosensor system. The cells adhere to and expand across the electrode surface, increasing impedance. Impedance measurements reveal the shape and viability of the cell. The working principle of the device is demonstrated in Figure 3.9. The device was capable of simultaneously performing gravimetric and impedimetric measurements. The authors proposed to use the developed biosensor for water-quality testing in the future. Cell damage caused by toxic chemicals present in water would cause a reduction in impedance and an increase in the resonant frequency.

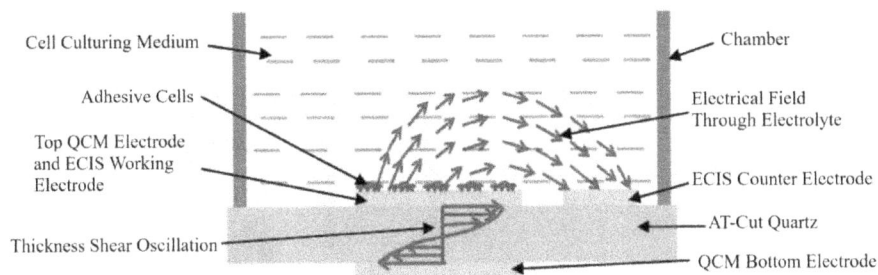

FIGURE 3.9 Illustration of the working principle of the hybrid biosensor. [Reprinted with permission from Liu et al. (2013). Copyright (2013) MDPI].

3.3.6 IMPEDANCE-BASED SENSORS FOR OTHER APPLICATIONS

Kustanovich et al. reported a dual sensor integrating surface acoustic wave resonance (SAR) sensor with EIS (Kustanovich et al. 2019). Both sensors shared interdigitated electrodes (IDE) but were operated at distinct frequency ranges using separate electrical interface ports. While EIS was performed at frequencies below 1 MHz, SAR was executed at about 185 MHz, leading to measured crosstalk below −60 dB. The SAR reflective gratings were reconstituted into IDE capacitive configuration, galvanically isolated from the SAR transducer. The two sensors provided complementary and non-interfering data. The SAR sensor detected changes in mass and viscoelasticity by monitoring changes in resonant frequency and the magnitude of the conductance peak of the SAR sensor. EIS sensor provided dielectric permittivity and dielectric loss (resistivity).

Lederer et al. described an impedance spectroscopy sensor integrated with digital microfluidics based on electrowetting on dielectrics (EWOD) for fluid flow actuation (Lederer et al. 2012). This integration was easy to fabricate due to the similar electric nature of the measurement (impedance spectroscopy) and the actuation (EWOD) principles. The dual setup was also reported to be suitable for real-time monitoring owing to its short measurement times.

Ben-Yoav et al. demonstrated an electrochemical LOC device with a 3 × 3 array of electrochemical sensors to analyze DNA hybridization (Ben-Yoav et al. 2015). A valve system was incorporated with the LOC device for automation and control of flow. The performance of the biosensor was assessed with an ssDNA probe assembly, followed by the introduction of ssDNA target. A theoretical detection limit of 1 nM was reported, which was attributed to the better control and reproducibility achieved with the incorporated valve system. Along with hybridization, direct nonlabeled/labeled amplified DNA has also been successfully detected (Bhatt, Mishra et al. 2019; Bhatt et al. 2021).

Zhang et al. integrated EIS with a digital microfluidic (DMF) device for the quantification of human peripheral blood mononuclear cell (PBMC) (Zhang and Liu 2022). The authors performed a dynamic cell capture assay using the droplet manipulation feature of the DMF device, which provided a 2.4 times signal enhancement over the conventional method of incubation in a static mode. The device exhibited a detection limit of 10^4 PBMCs/mL.

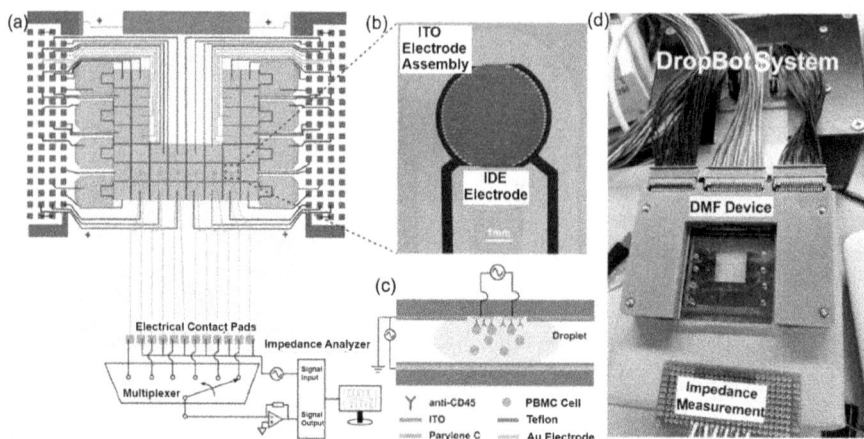

FIGURE 3.10 (A) DMF device with integrated IDEs; (B) microscopic image of IDEs deposited on the ITO glass substrate; (C) cross-sectional diagram of the biosensor; (D) a photo showing the integrated DMF device assembly. [Reprinted with permission from Zhang and Liu (2022). Copyright (2022) MDPI].

The design of the DMF device is shown in Figure 3.10. The DMF device is an electrowetting-on-dielectric (EWOD) based device with a top and a bottom plate. The top plate was an indium tin oxide (ITO) glass substrate containing an IDE-sensing array. Cr-coated glass substrate holding the actuation electrodes and the contact pads made up the bottom plate. The top and bottom plates were bonded by double-side electrically conductive tape. The droplets were actuated by a DropBot system and operated using MicroDrop software. Droplets were controlled by introducing sine wave potentials between the top plate and the bottom plate. Reservoirs on the right side were used for the storage of reagents. The reagents were dispensed as droplets by providing a voltage pulse between neighboring electrodes. Droplets were transported across the IDEs. Waste fluids were transported to the reservoir on the left side, which was subsequently removed using KimWipes.

3.4 CONCLUSION AND FUTURE PERSPECTIVES

Biosensors combine a biological component for detection and a physicochemical component for conversion into a measurable signal. This chapter provides the state of the art of the various impedimetric biosensors and the impact of the integration of microfluidic principles on the performance of impedimetric biosensors. Impedimetric biosensors have been used to detect cancer biomarkers, microorganisms, proteins, etc. Impedimetric biosensors for detecting different cancer biomarkers, such as α-fetoprotein (AFP) (Yang et al. 2014), carcinoembryonic antigen (CEA), and MCF-7 breast cancer cells, have been described with the methods and materials for their fabrication. Biosensors for detecting pathogens such as *L. innocua*, *L. monocytogenes*, *E. coli* O157:H7, avian influenza virus H5N1, and *Salmonella* have been reported. Biosensors for other applications, such as water-quality monitoring, food-quality testing, and cytotoxicity assays, have also been discussed.

REFERENCES

Baek, Seung Hoon, Chan Yeong Park, Thang Phan Nguyen, Min Woo Kim, Jong Pil Park, Changsun Choi, Soo Young Kim, Suresh Kumar Kailasa, and Tae Jung Park. 2020. "Novel Peptides Functionalized Gold Nanoparticles Decorated Tungsten Disulfide Nanoflowers as the Electrochemical Sensing Platforms for the Norovirus in an Oyster." *Food Control* 114 (March): 107225. https://doi.org/10.1016/j.foodcont.2020.107225.

Basu, Aviru Kumar, Shreyansh Tatiya, Geeta Bhatt, and Shantanu Bhattacharya. 2019. "Fabrication Processes for Sensors for Automotive Applications: A Review." In *Sensors for automotive and aerospace applications*, 123–42. https://doi.org/10.1007/978-981-13-3290-6_8.

Ben-Yoav, Hadar, Peter H. Dykstra, William E. Bentley, and Reza Ghodssi. 2015. "A Controlled Microfluidic Electrochemical Lab-on-a-Chip for Label-Free Diffusion-Restricted DNA Hybridization Analysis." *Biosensors and Bioelectronics* 64: 579–85. https://doi.org/10.1016/j.bios.2014.09.069.

Bhatt, Geeta, and Shantanu Bhattacharya. 2019. "Biosensors on Chip: A Critical Review from an Aspect of Micro/Nanoscales." *Journal of Micromanufacturing* 2 (2): 198–219. https://doi.org/10.1177/2516598419847913.

Bhatt, Geeta, Swati Gupta, Gurunath Ramanathan, and Shantanu Bhattacharya. 2021. "Integrated DEP Assisted Detection of PCR Products with Metallic Nanoparticle Labels through Impedance Spectroscopy." *IEEE Transactions on Nanobioscience* 21 (4): 502–10.

Bhatt, Geeta, Rishi Kant, Keerti Mishra, Kuldeep Yadav, Deepak Singh, Ramanathan Gurunath, and Shantanu Bhattacharya. 2017. "Impact of Surface Roughness on Dielectrophoretically Assisted Concentration of Microorganisms over PCB Based Platforms." *Biomedical Microdevices* 19 (2): 1–11. https://doi.org/10.1007/s10544-017-0172-5.

Bhatt, Geeta, Kapil Manoharan, Pankaj Singh Chauhan, and Shantanu Bhattacharya. 2019. "MEMS Sensors for Automotive Applications: A Review." In *Sensors for Automotive and Aerospace Applications*, 223–39. https://doi.org/10.1007/978-981-13-3290-6_12.

Bhatt, Geeta, Keerti Mishra, Gurunath Ramanathan, and Shantanu Bhattacharya. 2019. "Dielectrophoresis Assisted Impedance Spectroscopy for Detection of Gold-Conjugated Amplified DNA Samples." *Sensors and Actuators, B: Chemical* 288 (June): 442–53. https://doi.org/10.1016/j.snb.2019.02.081.

Bhattacharya, Shantanu, Avinash Kumar Agarwal, Om Prakash, Shailendra Singh, Mohit Pandey, and Rishi Kant. 2019. "Introduction to Sensors for Aerospace and Automotive Applications." In *Sensors for Automotive and Aerospace Applications*, edited by Shantanu Bhattacharya, Avinash Kumar Agarwal, Om Prakash, and Shailendra Singh, 1–6. Singapore: Springer Singapore. https://doi.org/10.1007/978-981-13-3290-6_1.

Bhattacharya, Shantanu, Shuaib Salamat, Dallas Morisette, Padmapriya Banada, Demir Akin, Yi-Shao Liu, Arun K. Bhunia, Michael Ladisch, and Rashid Bashir. 2008. "PCR-Based Detection in a Micro-Fabricated Platform." *Lab on a Chip* 8 (7): 1130. https://doi.org/10.1039/b802227e.

Caviglia, Claudia, Kinga Zór, Lucia Montini, Valeria Tilli, Silvia Canepa, Fredrik Melander, Haseena B. Muhammad et al. 2015. "Impedimetric Toxicity Assay in Microfluidics Using Free and Liposome-Encapsulated Anticancer Drugs." *Analytical Chemistry* 87 (4): 2204–12. https://doi.org/10.1021/ac503621d.

Chauhan, Pankaj Singh, Mohit Pandey, and Shantanu Bhattacharya. 2019. "Paper Based Sensors for Environmental Monitoring." In *Paper Microfluidics: Theory and Applications*, edited by Shantanu Bhattacharya, Sanjay Kumar, and Avinash K Agarwal, 165–81. Singapore: Springer Singapore. https://doi.org/10.1007/978-981-15-0489-1_10.

Choudhury, Sagnik Sarma, Mohit Pandey, and Shantanu Bhattacharya. 2021. "Recent Developments in Surface Modification of PEEK Polymer for Industrial Applications: A Critical Review." *Reviews of Adhesion and Adhesives* 9 (3): 401–33. https://doi.org/10.47750/RAA/9.3.03.

Chuang, Cheng Hsin, Yi Chun Du, Ting Feng Wu, Cheng Ho Chen, Da Huei Lee, Shih
    Min Chen, Ting Chi Huang, Hsun Pei Wu, and Muhammad Omar Shaikh. 2016.
    "Immunosensor for the Ultrasensitive and Quantitative Detection of Bladder Cancer
    in Point of Care Testing." *Biosensors and Bioelectronics* 84: 126–32. https://doi.
    org/10.1016/j.bios.2015.12.103.
Dixit, Chandra K., and Ajeet Kaushik. 2016. *Microfluidics for Biologists: Fundamentals and
    Applications*. Berlin: Springer, https://doi.org/10.1007/978-3-319-40036-5.
Dubey, Amarish, Himanshi Jangir, Mohit Pandey, Mayank Manjul Dubey, Shourya Verma,
    Manas Roy, Sushil Kumar Singh, Deepu Philip, Sabyasachi Sarkar, and Mainak Das.
    2018. "An Eco-Friendly, Low-Power Charge Storage Device from Bio-Tolerable Nano
    Cerium Oxide Electrodes for Bioelectrical and Biomedical Applications." *Biomedical
    Physics & Engineering Express* 4 (2): 25041. https://doi.org/10.1088/2057-1976/aaa282.
Gómez, Rafael, Rashid Bashir, and Arun K. Bhunia. 2002. "Microscale Electronic Detection
    of Bacterial Metabolism." *Sensors and Actuators, B: Chemical* 86 (2–3): 198–208.
    https://doi.org/10.1016/S0925-4005(02)00175-2.
Iliescu, Ciprian, Daniel P. Poenar, Mihaela Carp, and Felicia C. Loe. 2007. "A Microfluidic
    Device for Impedance Spectroscopy Analysis of Biological Samples." *Sensors and
    Actuators, B: Chemical* 123 (1): 168–76. https://doi.org/10.1016/j.snb.2006.08.009.
Jangir, Himanshi, Mohit Pandey, Rishabh Jha, Amarish Dubey, Shourya Verma, Deepu Philip,
    Sabyasachi Sarkar, and Mainak Das. 2018. "Sequential Entrapping of Li and S in a
    Conductivity Cage of N-Doped Reduced Graphene Oxide Supercapacitor Derived from
    Silk Cocoon: A Hybrid Li-S-Silk Supercapacitor." *Applied Nanoscience (Switzerland)* 8
    (3): 379–93. https://doi.org/10.1007/s13204-018-0641-z.
Kumar, Sanjay, Pulak Bhushan, Mohit Pandey, and Shantanu Bhattacharya. 2019. "Additive
    Manufacturing as an Emerging Technology for Fabrication of Microelectromechanical
    Systems (MEMS)." *Journal of Micromanufacturing* 2 (2): 175–97. https://doi.
    org/10.1177/2516598419843688.
Kustanovich, Kiryl, Ventsislav Yantchev, Baharan Ali Doosti, Irep Gözen, and Aldo Jesorka.
    2019. "A Microfluidics-Integrated Impedance/Surface Acoustic Resonance Tandem
    Sensor." *Sensing and Bio-Sensing Research* 25 (June): 100291. https://doi.org/10.1016/j.
    sbsr.2019.100291.
Lederer, Thomas, Stefan Clara, Bernhard Jakoby, and Wolfgang Hilber. 2012. "Integration
    of Impedance Spectroscopy Sensors in a Digital Microfluidic Platform." *Microsystem
    Technologies* 18 (7–8): 1163–80. https://doi.org/10.1007/s00542-012-1464-6.
Li, Zhanming, Zunzhong Ye, Yingchun Fu, Yonghua Xiong, and Yanbin Li. 2016. "A Portable
    Electrochemical Immunosensor for Rapid Detection of Trace Aflatoxin B1 in Rice."
    Analytical Methods 8 (3): 548–53. https://doi.org/10.1039/c5ay02643a.
Liu, Fei, Fang Li, Anis Nurashikin Nordin, and Ioana Voiculescu. 2013. "A Novel Cell-Based
    Hybrid Acoustic Wave Biosensor with Impedimetric Sensing Capabilities." *Sensors,
    Switzerland* 13 (3): 3039–55. https://doi.org/10.3390/s130303039.
Liu, Jiayu, Ibrahem Jasim, Zhenyu Shen, Lu Zhao, Majed Dweik, Shuping Zhang, and Mahmoud
    Almasri. 2019. "A Microfluidic Based Biosensor for Rapid Detection of Salmonella in
    Food Products." *PLoS ONE* 14 (5): 1–18. https://doi.org/10.1371/journal.pone.0216873.
Mok, Janine, Michael N. Mindrinos, Ronald W. Davis, and Mehdi Javanmard. 2014. "Digital
    Microfluidic Assay for Protein Detection." *Proceedings of the National Academy of
    Sciences of the United States of America* 111 (6): 2110–15. https://doi.org/10.1073/
    pnas.1323998111.
Nguyen, Tien Anh, Tsung I. Yin, Diego Reyes, and Gerald A. Urban. 2013. "Microfluidic Chip
    with Integrated Electrical Cell-Impedance Sensing for Monitoring Single Cancer Cell
    Migration in Three-Dimensional Matrixes." *Analytical Chemistry* 85 (22): 11068–76.
    https://doi.org/10.1021/ac402761s.

Pandey, Mohit, Mohammed Rashiku, and Shantanu Bhattacharya. 2021. "Recent Progress in the Development of Printed Electronic Devices." In *Chemical Solution Synthesis for Materials Design and Thin Film Device Applications*, edited by Soumen Das and Sandip B T - Chemical Solution Synthesis for Materials Design and Thin Film Device Applications Dhara, 349–68. Elsevier. https://doi.org/https://doi.org/10.1016/B978-0-12-819718-9.00008-X.

Pandey, Mohit, Krutika Shahare, Mahima Srivastava, and Shantanu Bhattacharya. 2019. "Paper-Based Devices for Wearable Diagnostic Applications." In *Paper Microfluidics: Theory and Applications*, edited by Shantanu Bhattacharya, Sanjay Kumar, and Avinash K. Agarwal, 193–208. Singapore: Springer Singapore. https://doi.org/10.1007/978-981-15-0489-1_12.

Pandey, Mohit, Mahima Srivastava, Krutika Shahare, and Shantanu Bhattacharya. 2019. "Paper Microfluidic-Based Devices for Infectious Disease Diagnostics." In *Paper Microfluidics: Theory and Applications*, edited by Shantanu Bhattacharya, Sanjay Kumar, and Avinash K. Agarwal, 209–25. Singapore: Springer Singapore. https://doi.org/10.1007/978-981-15-0489-1_13.

Pandey, Mohit, Poonam Sundriyal, Shreyansh Tatiya, and Shantanu Bhattacharya. 2022. "Polymer-Based Electrolytes for Solid-State Batteries: Current Status and Future Challenges in Emerging Applications." In *Trends in Applications of Polymers and Polymer Composites*: 5.1–5.22. Melville, New York: AIP Publishing LLC. https://doi.org/10.1063/9780735424555_005.

Pandey, Mohit, Shreyansh Tatiya, Shantanu Bhattacharya, and Shailendra Singh. 2019a. "Sensors in Assembly Shop in Automobile Manufacturing." In *Sensors for Automotive and Aerospace Applications*, edited by Shantanu Bhattacharya, Avinash Kumar Agarwal, Om Prakash, and Shailendra Singh, 193–207. Singapore: Springer Singapore. https://doi.org/10.1007/978-981-13-3290-6_10.

———. 2019b. "Sensors in the Joining and Welding Process in Automobile Manufacturing." In *Sensors for Automotive and Aerospace Applications*, edited by Shantanu Bhattacharya, Avinash Kumar Agarwal, Om Prakash, and Shailendra Singh, 241–56. Singapore: Springer Singapore. https://doi.org/10.1007/978-981-13-3290-6_13.

Pandey, Mohit, Shreyansh Tatiya, and Shantanu Bhattacharya. 2021. "Design and Development of MEMS-Based Sensors for Wearable Diagnostic Applications." In *MEMS Applications in Biology and Healthcare:* 10.1–10.34. Melville, New York: AIP Publishing LLC. https://doi.org/10.1063/9780735423954_010.

Ramón-Azcón, Javier, Enrique Valera, Ángel Rodríguez, Alejandro Barranco, Begoña Alfaro, Francisco Sanchez-Baeza, and M. Pilar Marco. 2008. "An Impedimetric Immunosensor Based on Interdigitated Microelectrodes (IDµE) for the Determination of Atrazine Residues in Food Samples." *Biosensors and Bioelectronics* 23 (9): 1367–73. https://doi.org/10.1016/j.bios.2007.12.010.

Rashiku, Mohammed, and Shantanu Bhattacharya. 2019. "Fabrication Techniques for Paper-Based Microfluidic Devices." In *Paper Microfluidics: Theory and Applications*, edited by Shantanu Bhattacharya, Sanjay Kumar, and Avinash K. Agarwal, 29–45. Singapore: Springer Singapore. https://doi.org/10.1007/978-981-15-0489-1_3.

Sharif, Sumaira, Yixian Wang, Zunzhong Ye, Zhen Wang, Qimin Qiu, Shengna Ying, and Yibin Ying. 2019. "A Novel Impedimetric Sensor for Detecting LAMP Amplicons of Pathogenic DNA Based on Magnetic Separation." *Sensors and Actuators, B: Chemical* 301 (August): 127051. https://doi.org/10.1016/j.snb.2019.127051.

Sundriyal, Poonam, Mohit Pandey, and Shantanu Bhattacharya. 2020. "Plasma-Assisted Surface Alteration of Industrial Polymers for Improved Adhesive Bonding." *International Journal of Adhesion and Adhesives* 101: 102626. https://doi.org/10.1016/j.ijadhadh.2020.102626.

Tatiya, Shreyansh, Mohit Pandey, Shantanu Bhattacharya, and Shailendra Singh. 2019. "Sensors Used in Automotive Paint Shops." In *Sensors for Automotive and Aerospace Applications*, edited by Shantanu Bhattacharya, Avinash Kumar Agarwal, Om Prakash, and Shailendra Singh, 257–64. Singapore: Springer Singapore. https://doi. org/10.1007/978-981-13-3290-6_14.

Tatiya, Shreyansh, Mohit Pandey, and Shantanu Bhattacharya. 2020. "Nanoparticles Containing Boron and Its Compounds—Synthesis and Applications: A Review." *Journal of Micromanufacturing* 3 (2): 159–73. https://doi.org/10.1177/2516598420965319.

Wang, Ronghui, Yun Wang, Kentu Lassiter, Yanbin Li, Billy Hargis, Steve Tung, Luc Berghman, and Walter Bottje. 2009. "Interdigitated Array Microelectrode Based Impedance Immunosensor for Detection of Avian Influenza Virus H5N1." *Talanta* 79 (2): 159–64. https://doi.org/10.1016/j.talanta.2009.03.017.

Yang, Fan, Jing Han, Ying Zhuo, Zhehan Yang, Yaqin Chai, and Ruo Yuan. 2014. "Highly Sensitive Impedimetric Immunosensor Based on Single-Walled Carbon Nanohorns as Labels and Bienzyme Biocatalyzed Precipitation as Enhancer for Cancer Biomarker Detection." *Biosensors and Bioelectronics* 55: 360–65. https://doi.org/10.1016/j. bios.2013.12.040.

Yang, Liju, Yanbin Li, and Gisela F. Erf. 2004. "Interdigitated Array Microelectrode-Based Electrochemical Impedance Immunosensor for Detection of Escherichia Coli O157:H7." *Analytical Chemistry* 76 (4): 1107–13. https://doi.org/10.1021/ac0352575.

Zhang, Yuqian, and Yuguang Liu. 2022. "A Digital Microfluidic Device Integrated with Electrochemical Impedance Spectroscopy for Cell-Based Immunoassay." *Biosensors* 12 (5): 1–14. https://doi.org/10.3390/bios12050330.

Zhou, Jie, Liping Du, Ling Zou, Yingchang Zou, Ning Hu, and Ping Wang. 2014. "An Ultrasensitive Electrochemical Immunosensor for Carcinoembryonic Antigen Detection Based on Staphylococcal Protein A - Au Nanoparticle Modified Gold Electrode." *Sensors and Actuators, B: Chemical* 197: 220–27. https://doi.org/10.1016/j.snb.2014.02.009.

Zhou, Ying, Dahou Yang, Yinning Zhou, Bee Luan Khoo, Jongyoon Han, and Ye Ai. 2018. "Characterizing Deformability and Electrical Impedance of Cancer Cells in a Microfluidic Device." *Analytical Chemistry* 90 (1): 912–19. https://doi.org/10.1021/acs. analchem.7b03859.

Part II

Applications of impedance-based biological sensors

4 Characterization of bioanalytical applications using impedance spectroscopy

Keerti Mishra

CONTENTS

4.1 INTRODUCTION

Biological interactions of certain micro/macromolecules, such as DNA and proteins, and microorganisms at the solid–support interface promote detection events for biological recognition (Patolsky et al. 1999; Kant et al. 2017). Micro/macromolecular interactions exhibit changes in the optical and electrical properties at the interface due to changes in the integrity of the molecule and its interaction with each other and with that of the interface. The alterations in the electrical properties are reflected as changes in voltage, current, refractive index, capacitance, or impedance (Geeta Bhatt and Bhattacharya 2018). Therefore, understanding of these interactions with reference to interfacial interactions plays an essential role in various biosensor technologies, cell adhesion, and surface wetting (Bhatt et al. 2016).

As depicted in Table 4.1, EIS can be efficiently utilized in various bioanalytical applications, including the detection of disease, pathogens, drugs, toxins, and cells, through various biomolecular chemistries, such as antibody–antigen interaction,

DOI: 10.1201/9781003358091-6

TABLE 4.1

Application of EIS in characterization of molecules (K'Owino and Sadik 2005)

Type	Species interacting	Application
Immunoassay	Ab–Ag binding	Disease detection like human mammary tumor
DNA	DNA–DNA	Pathogen detection
DNA	DNA–micromolecules	Drug and toxin tests
Cells	Cells–substrate	Real-time observation of cell spreading
Microorganism	Cell division	Real-time quantification of the microbe population
Enzyme	Bio complexes	Toxic molecules detection

DNA hybridization/amplification/molecular interactions, cells adhesion, and enzyme catalysis (Kant et al. 2017). These biomolecular interactions lead to a particular change in the present analyte and further govern the change in the electrical component associated with them.

4.2   BASIC THEORY AND INSTRUMENTATION

4.2.1   GENERAL PRINCIPLE

EIS measurement of an analyte comprises the process of stimulating the electrodes through a small sinusoidal voltage and measuring the overall current and the impedance associated with it. Impedance in general consists of three constituents, namely resistance, capacitance, and inductance, but in biological systems, resistance and capacitance are the only significant contributing elements. Faradaic EIS measurements utilize a redox probe and record its electrochemical property changes as a function of impedance, while non-faradaic EIS measurements do not use any redox probe for impedance change recording but record activities of an analyte directly (Liu et al. 2008). The overall impedance in a system is expressed in terms of Randles circuit having a circuit combination of R_s solution resistance, R_{ct} charge transfer resistance, C_d double layer capacitance, and Z_w Warburg impedance (detailed in Chapter 2).

In impedance corresponding to a particular biorecognition element, R_{ct} and C_d are the summation of the overall component resultant through the electrode and adsorbed biological species. R_{ct} is given as the sum of charge transfer resistance on a bare electrode (R_{el}) and through adsorbed species/biorecognition event (R_{ad}).

$$R_{ct} = R_{el} + R_{ad} \tag{4.1}$$

Further, C_d as a combination of electrode capacitance (C_{el}) and capacitance associated with the adsorbed biological species (C_{ad}) is given as:

$$\frac{1}{C_d} = \frac{1}{C_{el}} + \frac{1}{C_{ad}} \tag{4.2}$$

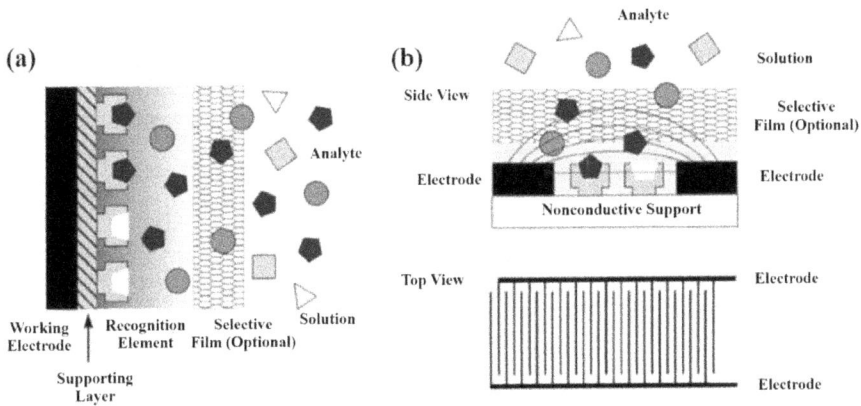

FIGURE 4.1 (a) Schematic of basic interactions in EIS measurements (conducting solid substrate); (b) basic electrochemical interactions in patterned substrate (interdigitated electrodes). [Reprinted with permission from Lisdat and Schäfer (2008). Copyright (2008) Springer].

It should be noted that EIS measurements can be carried on a solid substrate as well in the patterned electrode format. Figure 4.1 shows the various configurations and corresponding interactions associated with the individual format. In a solid substrate, a recognition event takes place on the substrate, where the signal change is directly associated with the substrate, while in the case of patterned electrodes, the recognition elements usually take place in spacing between the electrodes, and the corresponding electrical change is recorded through the nearby electrodes.

4.2.2 INSTRUMENTATION

Figure 4.2 shows a typical three-electrode (reference: rf, working: w, and counter: c) impedance measurement setup. It shows that the assembly comprises an AC waveform generator, a lock-in amplifier, an AC/DC voltmeter, and a variable resistor connected in a specific format, further connected to the measuring cell/electrode setup.

It is observed that the power generator sends a sinusoidal AC voltage to the potentiostat, which further applies an appropriate potential to the counter electrode maintaining the required potential difference between the reference and the working electrode. The corresponding current value of the system estimates the approximate impedance components in the electrical system.

4.3 IMPEDANCE SPECTROSCOPY IN BIOLOGY

EIS measurements can be efficiently utilized in characterizing biological and nonbiological systems for extracting system information. As compared with other optical/mechanical detection techniques, these systems can be quantized easily and can be efficiently utilized in diverse contexts (*in vitro/in vivo*) at low cost as well as in any solution format. The solution requirement is not at all limited to the water-based

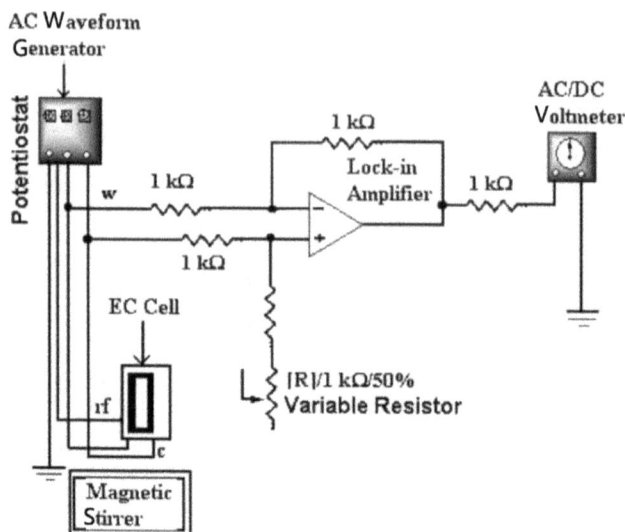

FIGURE 4.2   Basic instrumentation utilized in impedance sensing. [Reprinted with permission from K'Owino and Sadik (2005). Copyright (2005) Wiley].

solution with a particular ionic concentration, rather can be effectively exploited in any kind of solution, including water, oil, gel, and polymer. Hence, EIS can be applied to several application fields. This chapter primarily focuses on various bioanalytical applications, including biomedical engineering, monitoring cell culture, tracking food quality, detection/differentiation of bacteria, human health tracking/analysis, biosensors, and DNA/RNA detection.

4.3.1   EIS IN BIOMEDICAL ENGINEERING

Proteins and their interaction with solid surfaces showcase key components of bioengineering, such as protein purification, tissue engineering, cell immobilization, and drug discovery (Makohliso et al. 1999; Zhu and Snyder 2003). For monitoring the protein adsorption on conducting polymer solid substrates, EIS has been shown as an effective, noninvasive, and information-rich tool. EIS has been used to study the interaction of insulin on platinum-based electrodes for the development of artificial pancreas (Wright et al. 2004), α-lactoalbumin and β-casine on stainless steel (Cosman, Fatih, and Roscoe 2005), and HAS and IgG on a gold electrode (Moulton et al. 2004) through analyzing the increase in charge transfer and decrease in double layer capacitance. However, during analysis, surface defects are caused by the applied potential causing disturbed adsorption on the metal surface. Prevention of such disturbances is achieved by self-assembled monolayer facilitating the restoration of bodily motor function (Chang et al. 2004). Therefore, while designing the tissue implant devices, minimizing the inflammatory effect becomes essential. Resistance to protein adsorption can significantly reduce the inflammatory effect that can be achieved by replacing hydrophobic octadecanethiol on gold electrodes

(Zhang et al. 2000) with hydrophilic cysteine molecules, which further aids tissue healing by enhancing fibronectin (Faucheux et al. 2004).

4.3.2 EIS IN MONITORING THE PERFORMANCE OF 3D CULTURE

In vitro models are extensively utilized in understanding molecular metabolism and pathogenesis. Cell-based models are currently popular for replacing laboratory animals and are also cost-effective in the field of drug discovery (Poloznikov et al. 2018). Traditionally used 2D cell culture models are also diminished by 3D cell culture and organ-on-a-chip scaffold (Torras et al. 2018; Marx et al. 2020) because they mimic the *in vivo* conditions more closely, providing a comparatively better overview of the experimental progress. These models provide a higher possibility of personalized testing without using several rodents and other typical laboratory animals whose physiological and biological processes differ from the human system. Monitoring of these cell states using end-point assessment requires the introduction of the label molecules that are expensive, laborious, and often cause the destruction of the sample after every use (Single et al. 2015).

Impedance technology, on the other hand, is label-free and helps in the quantification of cellular properties in real-time. EIS is commonly used for accessing the quality of *in vitro* models of barrier tissues. The epithelial and endothelial cells form tight junctions, and this provides a vital barrier to the tissue and regulates its physiological functionality and permeability for the diffusion and transport of chemical substances (Benson, Cramer, and Galla 2013). The delivery of therapeutic agents needs penetration into the barrier without destroying its integrity. Therefore, in an organ-on-a-chip model, the permeability is controlled by measuring transepithelial resistance (TEER) (Samatov et al. 2015; Srinivasan et al. 2015). In a traditional TEER unit, the cells are grown in a semipermeable membrane and the electrodes are separated by the monolayer. An alternating low-frequency current is required in many cases for generating uniform current density across the membrane.

The most commonly used 3D cultures are known as spheroids (Zanoni et al. 2016) in suspension culture and several reports have shown to measure various properties of spheroids using impedance spectroscopy. Thielecke et al. investigated the effect of bioactive substances on spheroids using a circular planar electrode at several distance values from the spheroids and designed an optimal spheroid/electrode interphase for sensing the effect of a drug (Thielecke, Mack, and Robitzki 2001).

Microcavity array chips containing multiple square cavities on rectangular electrodes were developed to measure the increased impedance in the presence of a spheroid (Eichler et al. 2015). The formulated microchip platform can be utilized to measure the cytotoxicity of chemotherapeutic drugs. Another device made with the combination of the electrochemical transistor and a microfluidic trapping device was prepared to access the resistance of the spheroid (Curto et al. 2018). Later, several automated devices for measuring the impedance of spheroid in microfluidic settings were created, which alongside measuring the cytotoxic effect can also measure the spheroids made of cardiomyocytes (Bürgel et al. 2016). Another platform for measuring the size of the spheroid was developed by Schmid as a hanging drop platform (Schmid et al. 2016). The above-stated methods have proven efficient in measuring

(a)

| Microcavity | Microfluidic Channel | Hanging Drop |

| Size Measurement | Cytotoxicity | Internal Structure/Cell Type |

(b)

| Thick Hydrogel Construct | Thin Hydrogel Layer | Perfusion/Microfluidic Settings |

| Proliferation, Colony Size, and Spatial Distribution | Cytotoxicity | Cell Type |

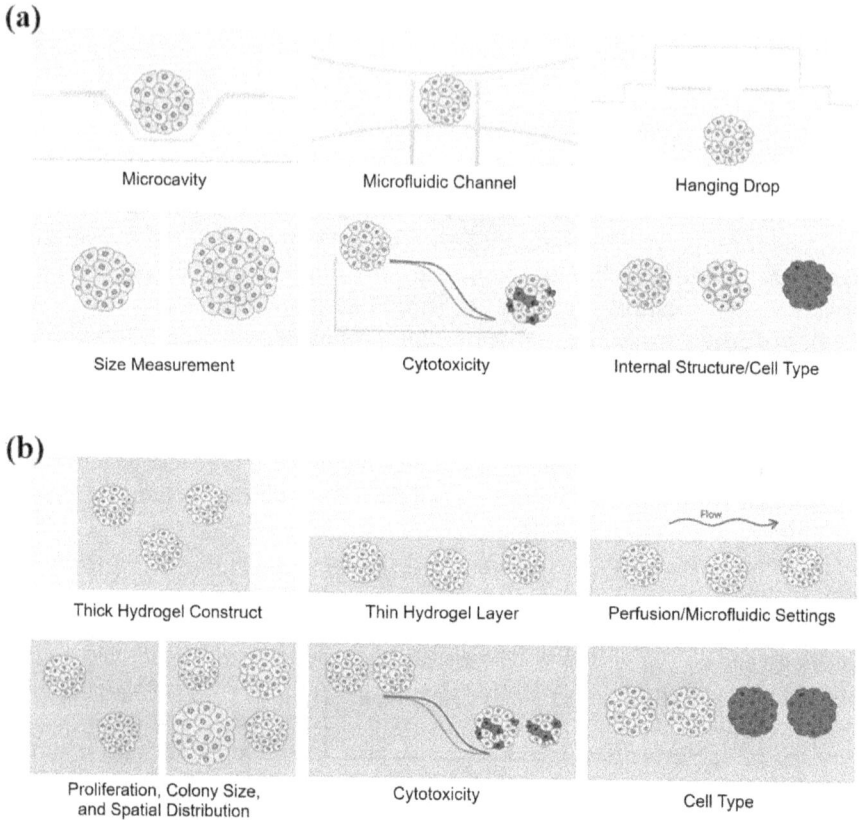

FIGURE 4.3 Application of impedance spectroscopy in three-dimensional culture model in (a) Scaffold-free spheroid; (b) Scaffold-based culture. [Reprinted with permission from Gerasimenko et al. (2020). Copyright (2020) Frontiers].

the size of the spheroid in different experimental conditions explaining the properties of spheroids in reference to the experimental requirements.

On the other hand, in another approach where growing the spheroids in the hydrogel or on the porous scaffold is a more suitable choice for primary organoids, the impedance of the cells was measured directly in PEG hydrogel, although the precision, in this case, was questionable (Lin, Kyriakides, and Chen 2009). Later, the cells were grown in the microporous scaffold, which not only measures the cell concentration but their morphologic variations were also easily accessible (Bagnaninchi et al. 2003). The drug profiling on the cancer cells was a more advanced approach using impedance microscopy in a three-dimensional (3D) scaffold. The drug-sensitive cells show a steep drop in the impedance showing the scope of the drug as an anticancer agent (Lei et al. 2014). Figure 4.3 shows various applications of impedance spectroscopy in a 3D culture model, viz. microcavity, microfluidic channel, and hanging drop, where the various cell features like size, cytotoxicity, and cell type/ internal structure are studied.

4.3.3 EIS IN TRACKING FOOD QUALITY

There is an urgent need for accessing food quality without destroying the many growth benefits in a time-efficient manner in order to prevent substantial food waste (Vanoli and Buccheri 2012). Compared with classical methods, the nondestructive quality evaluation and ripening test can be performed pre- and postharvest and is also environment-friendly, accurate, rapid, and provides real-time results (Li, Lecourt, and Bishop 2018). One such example of a nondestructive method is EIS, which is a reliable exchange for classical methods (Muñoz-Huerta et al. 2014). The interaction of the dielectric dipole moment of the sample and the electric field from an external source is quantitatively characterized by EIS. The chemical composition and structure of the sample plays a critical role in choosing the input frequency range as the specific frequency of the different types of materials correlates impedance with the quality parameter (El Khaled et al. 2017).

The structure of fruits and vegetables is such that their cytoplasm is laminated by a phospholipid bilayer. Within the plasmalemma, there is a nucleus in the center and vacuoles on the inner boundary. Plants' response to electric current flow has been understood through several models. The simplest model among various proposed models is the Cole model, which considers extracellular cell wall resistance, cytoplasm resistance, and membrane capacitance.

Harker and Maindonald studied the change in impedance with the ripening of nectarines in 1994 with the peeled nectarine tissue implanted on silver electrodes in a radial position. Frequency from 50 Hz to 1 MHz was used for characterization. Ripening of the fruit reduces the resistance at low frequencies and the higher frequency range response remains unchanged. This indicates that these are measuring the changes in the cell wall. They also showed that when fruits are kept in cold temperatures for several days and then are allowed to ripen, the changes at low frequency were observed to be limited (Harker and Maindonald 1994). A similar experimental setup has been used in later studies on persimmons (Harker and Forbes 1997), kiwis, and apples (Bauchot, Harker, and Arnold 2000) (Jackson and Harker 2000). In these experiments, the range of frequency was from the small to high range, and, mostly, the ripening and bruising can be observed in the smaller frequency range. No change in impedance frequency was seen in the case of kiwi in either frequency range. These results were significantly important as they showed a prominent effect of the physiological state of the fruit against the change in the electronic signaling (Figure 4.4). In recent years, several studies and nondestructive attempts have been made to analyze intact-skinned fruits like mangoes and bananas using different electrical models.

Along with fruits, EIS spectroscopy can also be used for the quality measurement of vegetable oils and similar products. The use of oil after several rounds of high-temperature exposure is hazardous to human health as they undergo degradation simultaneously. Vegetable oils are very good insulators, and, therefore, the low-frequency capacitance can be very low (Prevc et al. 2013). Therefore, the experiments

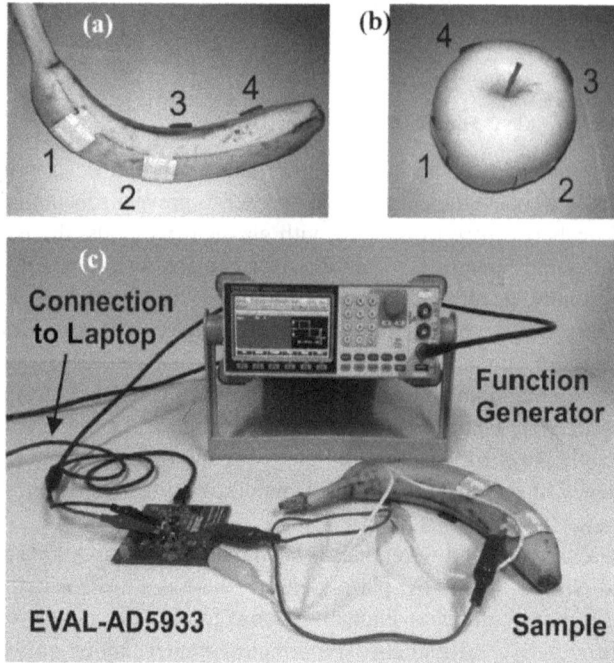

FIGURE 4.4 Electrode placement for fruits for impedance measurement. [Reprinted with permission from Ibba et al. (2020). Copyright (2020) Elsevier].

are mostly conducted in the range of 100 Hz to 1 MHz. The most useful parameter in the case of vegetable oils is the relative dielectric constant, which depends on the fatty acid content and moisture with changing temperatures (Lizhi, Toyoda, and Ihara 2008). Several plate capacitors are added parallelly to avoid the signal-to-noise ratio and to increase capacitance. Studies have also been performed on frying oil degradation with different conditions like constant heating in comparison with repeated heating of soybean oil (Stevan et al. 2015) and frying with different doughs of varying moisture levels (Yang, Zhao, and He 2016). Adulteration of high-quality extra virgin olive oil with other oils belonging to the linoleic, linolenic, and oleic acid family can also be measured using this technique (Lizhi, Toyoda, and Ihara 2010).

The EIS technique has also been widely used for dairy product detection. In the case of milk, lactose does not play a significant role in the conductivity of the milk, but it is mainly detected by its salt and fat contents. At a higher frequency range like 100 kHz, milk conductivity provides significant information on milk quality. Milk samples are measured with gold electrodes in the frequency range of 5 Hz to 1 MHz at 8°C (Mabrook and Petty 2003). The adulteration of milk samples with water, sodium hydroxide, hydrogen peroxide, and urea can be detected using this technique with varying electrodes having an accuracy of up to 94.9% (Durante et al. 2016). Similarly, cream, ice cream, and yogurt have been tested for their quality using the EIS technique.

4.3.4 EIS IN DETECTION AND DIFFERENTIATION OF BACTERIA

G.N. Stewart first demonstrated that the change in bacterial population altered the growth medium conductivity and hence concluded that the microorganism population could be measured through electrical measurements (Stewart 1899) where an increase in electrical conductivity was observed with an increased bacterial population. The use of impedance spectroscopy to measure concentration, however, did not achieve its full potential until the late twentieth century. Impedance microscopy exhausts information on the conversion of weakly charged molecules into charged compounds through bacterial metabolism. These charged molecules alter the electrical properties of the growth medium. These measurements are usually taken directly in the growth medium using an inert material electrode.

The detection of bacteria in solutions holds great importance in water management. The monitoring of the microbes continuously can provide better control for keeping up water quality. Measurement of the microbial bacteria present in the running water needs a highly sensitive sensor to differentiate bacteria from solid particles, measure the bacterial concentration in the water sample by the CFU counting method, and eventually differentiate the bacterial genus by detecting different signatures (Clausen et al. 2018). The conventional method of differentiating different bacteria into Gram-positive and Gram-negative categories is usually done by the Gram staining method, which is a time-consuming one as compared with the EIS method. Rima et al. developed a tool for real-time differentiation of bacteria at a concentration of 1 µg/µL at 5248 and 158,489 Hz for Gram-negative and Gram-positive, respectively (Gnaim et al. 2020). Therefore, EIS not only detects the presence but also plays a useful role in distinguishing the types of bacteria.

The infections caused by pathogenic bacteria are likely to be creating major challenges in the upcoming future and have been posing tremendous health threats that are expected to increase in the future. A big reason behind the increasing threat is the increasing resistance to antibiotic agents. Biofilms are one of the few reasons contributing to drug resistance in bacteria, which forms a physical barrier to prevent the penetration of the drug. Microbes are embedded in a self-generated extracellular matrix containing proteins, polysaccharides, and DNA in a biofilm that can develop chronic infections that are usually persistent for years (Van Duuren et al. 2017). Duuren et al. showed the use of EIS in monitoring the growth of *Pseudomonas aeruginosa* in a 96-well plate for a span of 72 hours. The EIS detects pellicle formation at the liquid–air interphase. The impairment and breaking of the biofilm treated with different antibiotics were also observed using the single-frequency impedance. These experiments represent EIS as an additional benchmark for the detection of bacteria and the development of corresponding small-molecule drug testing (Van Duuren et al. 2017).

4.3.5 EIS IN THE ANALYSIS OF HUMAN HEALTH ASPECTS

EIS is a cost-effective, fast, and noninvasive technique to measure human body composition. As shown in Figure 4.5, the human body has fat-free mass and fat mass (Mialich et al. 2014). Bone minerals, extracellular and intracellular water, and

(a)   (b)   (c)

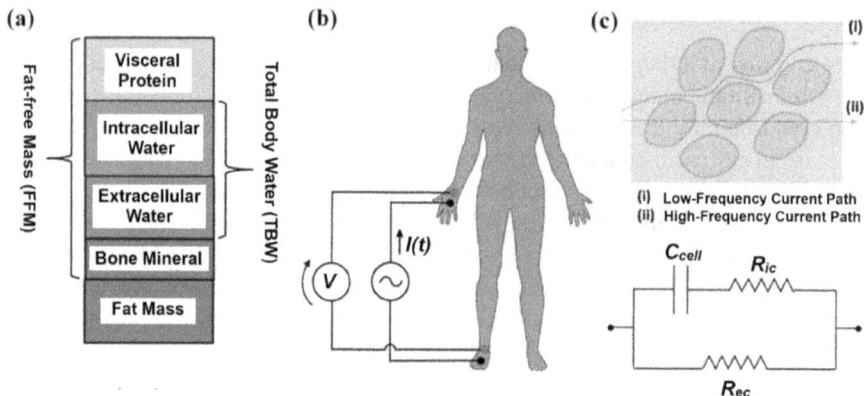

FIGURE 4.5   (a) Compartments of human body; (b) bioelectric impedance measurement configuration; (c) used equivalent electric circuit for data interpretation. [Reprinted with permission from Grossi and Ricco (2017). Copyright (2017) CP].

visceral protein are subdivisions of fat-free mass. Fat tissue and fatless lean tissue can be differentiated based on their electrical conductivity and thereby impedance. Electrical conductivity is higher in lean tissue due to the presence of the high electrolyte content (Bera 2014). Discussion about human health and wellness cannot go without cancer detection; the early detection of cancer is a challenging phenomenon. Early detection of cancer can help in precise therapeutic approaches. The techniques currently in use are fluorescence measurements and flow cytometry, which are widespread approaches, yet are very costly, time-consuming, and require ultrasophisticated instrumentation. The use of EIS has been considered as electrochemical cytosensors and different nanoprobes are used to improve the detection in a simplistic, rapid, and cost-effective way (Bera 2014). Several analytical methods like endoscopy, tomography, radiology, luminescence, and several other electrochemical approaches have been used for the diagnosis of cancer cells. Among the electrochemical approaches, EIS has received great attention (Han et al. 2016). It measures biological interactions at a very small range of nondestructive potential. The load transfer of the immune cells provides convenient monitoring of cancer cells (Wang et al. 2012) although the specific detection is a triggering drawback of this technique.

An unlabeled impedance-induced cancer cell-specific cytosensor where octapeptides in the melt phage were immobilized on the electrode surface was displayed by Han et al. It was observed that the Nyquist diagrams slowly changed with the lectin modification of the gold surface (Muñoz, Montes, and Baeza 2017). A lectin-based free electrochemical method was shown to detect liver cancer. A carbohydrate lectin is used for glycopeptide detection on the surface of cancer cells and provides a precise approximation of the body's cancer. The lectin-induced sensor is incubated with the cancer cell sample and the cell adhesion to the probe causes the change in the resistance of the charge transfer and the change in the impedances is measured as the concentration of the cancer cells. The resistance change due to electron transfer across the surface can be accessed by the Nyquist diagram.

Besides cancer, the leading cause of distress to human health is the increasing invasion of infectious diseases: the Covid 19 pandemic is one massive example. In the past decade, several infectious diseases like HIV, ZIKV, hepatitis, and chikungunya or CHIKV have been very prevalent. In the case of Covid 19, a silica electrode with functional graphene showed the detection of the viral RNA. RNA-specific complementary DNA primers can hybridize with the RNA and the difference in transition resistance can be examined by the EIS test. The modification in the complementary DNA primers can be used to access a broad range of proteins specific to viral entities and therefore plays an important role in diagnosis (Khan et al. 2020). Hepatitis B virus, another infectious disease-causing virus leading to liver failure, can be detected using a magnetic nanoparticle coated with streptavidin as a DNA sensor through the EIS method. Gold nanorods and colloidal gold have also been used as electrochemical DNA biosensors to detect the Hepatitis B virus.

4.3.6 EIS FOR DNA AND RNA DETECTION

Biosensors are precise, intuitive, and rapid tools for tissue engineering applications. Several factors such as the lack of accurate measuring of the low current, specificity, safe working environment while working with the living tissue, and poor scalability have limited the use of impedance in biomaterial applications. Yet, in comparison with other electrochemical techniques, it surpasses the capacity for probing interfacial properties of biomolecular films (Alfonta et al. 2001).

To detect DNA and RNA, conventional techniques such as sequencing, ELISA, northern blotting, polymerase chain reaction (PCR), and microarrays are currently used. However, each of these strategies has its own set of limitations, such as complicated processes, high instrumental and energy demands, error-prone amplification, and a lack of controls (Roychoudhury, Dear, and Bachmann 2022). For example, in the case of quantitative PCR (qPCR) or sequencing, the use of a fluorescent tag or fluoroprobe as an additional analyte tagging to detect the specific DNA and RNA (Bhattacharya et al. 2007, 2008; Nayak et al. 2013). EIS-based biosensors are capable of direct detection of the target analyte using a label-free method (Geeta Bhatt et al. 2019, 2021). In this section, some of the recent applications of EIS in the detection of DNA and RNA have been discussed.

Recently, a sensitive impedimetric biosensor was used to detect the miRNA 122 and its three different isoforms to a detection limit of 1 nM (Roychoudhury, Dear, and Bachmann 2022). The EIS signal depended on spatial orientation and length of overhangs toward the electrode surface. Similarly, a label-free DNA sensor was used to determine a synthetic 15-base oligonucleotide of *Mycobacterium tuberculosis* (MTB) with a detection limit of 1.24 nM. The quantitative measurement was achieved by monitoring the change in surface resistance in the presence of target DNA using EIS (Teengam et al. 2018). Furthermore, nucleic repeats are the cause of genetically transferable disease and they are difficult to detect using traditional electrochemical methods due to their complex structure and length of DNA repeat (Depienne and Mandel 2021). The peptide nucleic acid (PNA) probe distinguished between normal and abnormal CAG trinucleotide repeats in total RNA isolated from cells. In comparison with the DNA microprobe, the PNA probe showed an improved signal by 2.5-fold in a 100-fold lower sample concentration (Asefifeyzabadi et al. 2022).

4.4 CONCLUSION AND FUTURE PERSPECTIVES

The striking feature of impedance spectroscopy is its high throughput format with high temporal resolution. It is envisioned to be incorporated into the industrial format in the upcoming future due to its automated and low labor assays. EIS monitors the progress of the biological processes instead of the end-point evaluation. The enhanced precision of the measurement is deemed for the improvement of the expansion of the application that can be achieved by the increased sophistication of the electrode. On the other hand, EIS proves to be an important tool for the testing and validation of viral and bacterial strains, which provides us with the heads up to be better prepared for the upcoming challenges of emerging infectious diseases. The development requires the incorporation of nanotechnology and other multidisciplinary approaches. The immobilization of different biological receptors like nucleic acid, antibodies, cells, and bacteria on the electrode surface is carried out and the properties of the sensors are observed on the electrode interface due to insulation properties. However, the immobilization of the receptor on the electrode to create the biosensor is the most essential step. Thus, the EIS method is a highly sensitive method for monitoring the cognitive interaction of the biological molecule at the electrode surface.

Compared with other optic and electroanalysis techniques, EIS provides a significantly reliable characterization of modified electrode properties at the interfacial level. However, low sensitivity is the bottleneck for EIS. Therefore, improvement in the technique to enhance sensitivity with regard to interactions is required. Better sensitivity in the screening of pathogens and enhanced use of dielectric spectroscopy for pathogen screening can be very useful in the food industry. The issue of sensitivity can be addressed by the use of nanomaterials, which can later form cost-effective portable impedance devices. Furthermore, the noise ratio associated with reduced sensitivity can be improved by modifying the immobilization chemistry compatible with dielectric spectroscopy. To develop a system compatible with diagnostic assays, clinicians, engineers, physicists, and chemists need to come together.

Currently, 96 well plates with membrane inserts are available commercially but the currently available EIS system for monitoring the barrier function is only suitable to work with 24 membrane inserts; therefore, a compatible format for increased efficiency is required. Another interesting extension can be considered toward the measurement of another parameter like the change in pH and oxygen levels, throughout the measurement, which can also functionalize the electrode, and, therefore, measurements of these parameters can also be developed. The inclusion of flow cytometry with EIS in a microfluidic device can help understand the heterogeneity of cancer cells. An extension of the same for hybrid cell analysis is another future prospect. The monitoring of the cell culture using EIS has been discussed in the chapter and its extension in the cell culture incubators can be helpful in monitoring real-time culture expansion and its varying properties can be foreseen.

REFERENCES

Alfonta, Lital, Amos Bardea, Olga Khersonsky, Eugenii Katz, and Itamar Willner. 2001. "Chronopotentiometry and Faradaic Impedance Spectroscopy as Signal Transduction Methods for the Biocatalytic Precipitation of an Insoluble Product on Electrode Supports: Routes for Enzyme Sensors, Immunosensors and DNA Sensors." *Biosensors and Bioelectronics* 16 (9–12): 675–87. https://doi.org/10.1016/S0956-5663(01)00231-7.

Asefifeyzabadi, Narges, Grace Durocher, Kizito Tshitoko Tshilenge, Tanimul Alam, Lisa M. Ellerby, and Mohtashim H. Shamsi. 2022. "PNA Microprobe for Label-Free Detection of Expanded Trinucleotide Repeats." *RSC Advances* 12 (13): 7757–61. https://doi.org/10.1039/d2ra00230b.

Bagnaninchi, Pierre Olivier, Maria Dikeakos, Teodor Veres, and Maryam Tabrizian. 2003. "Towards On-Line Monitoring of Cell Growth in Microporous Scaffolds: Utilization and Interpretation of Complex Permittivity Measurements." *Biotechnology and Bioengineering* 84 (3): 343–50. https://doi.org/10.1002/bit.10770.

Bauchot, Anne D., F. Roger Harker, and W. Michael Arnold. 2000. "The Use of Electrical Impedance Spectroscopy to Assess the Physiological Condition of Kiwifruit." *Postharvest Biology and Technology* 18 (1): 9–18. https://doi.org/10.1016/S0925-5214(99)00056-3.

Benson, Kathrin, Sandra Cramer, and Hans Joachim Galla. 2013. "Impedance-Based Cell Monitoring: Barrier Properties and Beyond." *Fluids and Barriers of the CNS* 10 (1): 1–11. https://doi.org/10.1186/2045-8118-10-5.

Bera, Tushar Kanti. 2014. "Bioelectrical Impedance Methods for Noninvasive Health Monitoring: A Review." *Journal of Medical Engineering*: 1–28. https://doi.org/10.1155/2014/381251.

Bhatt, Geeta, and Shantanu Bhattacharya. 2018. "DNA-Based Sensors." In *Environmental, Chemical and Medical Sensors*, edited by Shantanu Bhattacharya, Avinash Kumar Agarwal, Nripen Chanda, Asok Pandey, and Ashis Kumar Sen, 343–70. Springer. https://doi.org/10.1007/978-981-10-7751-7_15.

Bhatt, Geeta, Swati Gupta, Gurunath Ramanathan, and Shantanu Bhattacharya. 2021. "Integrated DEP Assisted Detection of PCR Products with Metallic Nanoparticle Labels through Impedance Spectroscopy." *IEEE Transactions on Nanobioscience* 21 (4): 502–10.

Bhatt, Geeta, Keerti Mishra, Gurunath Ramanathan, and Shantanu Bhattacharya. 2019. "Dielectrophoresis Assisted Impedance Spectroscopy for Detection of Gold-Conjugated Amplified DNA Samples." *Sensors and Actuators, B: Chemical* 288 (June): 442–53. https://doi.org/10.1016/j.snb.2019.02.081.

Bhatt, Geeta, Sanjay Kumar, Poonam Sundriyal, Pulak Bhushan, Aviru Basu, Jitendra Singh, and Shantanu Bhattacharya. 2016. In *Microfluidics Overview. Microfluidics for Biologists: Fundamentals and Applications*. 33–83, https://doi.org/10.1007/978-3-319-40036-5_2.

Bhattacharya, Shantanu, Yuanfang Gao, Venumadhav Korampally, Maslina T. Othman, Sheila A. Grant, Steven B. Kleiboeker, Keshab Gangopadhyay, and Shubhra Gangopadhyay. 2007. "Optimization of Design and Fabrication Processes for Realization of a Optimization of Design and Fabrication Processes for Realization of a PDMS-SOG-Silicon." *Journal of Microelectromechanical Systems* 16 (2) (April): 404–10.

Bhattacharya, Shantanu, Shuaib Salamat, Dallas Morisette, Padmapriya Banada, Demir Akin, Yi-Shao Liu, Arun K. Bhunia, Michael Ladisch, and Rashid Bashir. 2008. "PCR-Based Detection in a Micro-Fabricated Platform." *Lab on a Chip* 8 (7): 1130. https://doi.org/10.1039/b802227e.

Bürgel, Sebastian C., Laurin Diener, Olivier Frey, Jin Young Kim, and Andreas Hierlemann. 2016. "Automated, Multiplexed Electrical Impedance Spectroscopy Platform for Continuous Monitoring of Microtissue Spheroids." *Analytical Chemistry* 88 (22): 10876–83. https://doi.org/10.1021/acs.analchem.6b01410.

Chang, Cheng Hung, Jiunn Der Liao, Jia Jin Jason Chen, Ming Shaung Ju, and Chou Ching K. Lin. 2004. "Alkanethiolate Self-Assembled Monolayers as Functional Spacers to Resist Protein Adsorption upon Au-Coated Nerve Microelectrode." *Langmuir* 20 (26): 11656–63. https://doi.org/10.1021/la040097t.

Clausen, Casper Hyttel, Maria Dimaki, Christian Vinther Bertelsen, Gustav Erik Skands, Romen Rodriguez-Trujillo, Joachim Dahl Thomsen, and Winnie E. Svendsen. 2018. "Bacteria Detection and Differentiation Using Impedance Flow Cytometry." *Sensors (Switzerland).* https://doi.org/10.3390/s18103496.

Cosman, Nicholas P., Khalid Fatih, and Sharon G. Roscoe. 2005. "Electrochemical Impedance Spectroscopy Study of the Adsorption Behaviour of α-Lactalbumin and β-Casein at Stainless Steel." *Journal of Electroanalytical Chemistry* 574 (2): 261–71. https://doi.org/10.1016/j.jelechem.2004.08.007.

Curto, V. F., M. P. Ferro, F. Mariani, E. Scavetta, and R. M. Owens. 2018. "A Planar Impedance Sensor for 3D Spheroids." *Lab on a Chip* 18 (6): 933–43. https://doi.org/10.1039/c8lc00067k.

Depienne, Christel, and Jean Louis Mandel. 2021. "30 Years of Repeat Expansion Disorders: What Have We Learned and What Are the Remaining Challenges?" *American Journal of Human Genetics* 108: 764–85. https://doi.org/10.1016/j.ajhg.2021.03.011.

Durante, Gabriel, Wesley Becari, Felipe A. S. Lima, and Henrique E. M. Peres. 2016. "Electrical Impedance Sensor for Real-Time Detection of Bovine Milk Adulteration." *IEEE Sensors Journal* 16 (4): 861–65. https://doi.org/10.1109/JSEN.2015.2494624.

Duuren, Jozef B. J. H. Van, Mathias Müsken, Bianka Karge, Jürgen Tomasch, Christoph Wittmann, Susanne Häussler, and Mark Brönstrup. 2017. "Use of Single-Frequency Impedance Spectroscopy to Characterize the Growth Dynamics of Biofilm Formation in Pseudomonas Aeruginosa." *Scientific Reports* 7 (1): 5223. https://doi.org/10.1038/s41598-017-05273-5.

Eichler, Marie, Heinz Georg Jahnke, Dana Krinke, Astrid Müller, Sabine Schmidt, Ronny Azendorf, and Andrea A. Robitzki. 2015. "A Novel 96-Well Multielectrode Array Based Impedimetric Monitoring Platform for Comparative Drug Efficacy Analysis on 2D and 3D Brain Tumor Cultures." *Biosensors and Bioelectronics* 67: 582–89. https://doi.org/10.1016/j.bios.2014.09.049.

Faucheux, N., R. Schweiss, K. Lützow, C. Werner, and T. Groth. 2004. "Self-Assembled Monolayers with Different Terminating Groups as Model Substrates for Cell Adhesion Studies." *Biomaterials* 25 (14): 2721–30. https://doi.org/10.1016/j.biomaterials.2003.09.069.

Gerasimenko, Tatiana, Sergey Nikulin, Galina Zakharova, Andrey Poloznikov, Vladimir Petrov, Ancha Baranova, and Alexander Tonevitsky. 2020. "Impedance Spectroscopy as a Tool for Monitoring Performance in 3D Models of Epithelial Tissues." *Frontiers in Bioengineering and Biotechnology* 7: 474. https://doi.org/10.3389/fbioe.2019.00474.

Gnaim, Rima, Alexander Golberg, Julia Sheviryov, Boris Rubinsky, and César A. González. 2020. "Detection and Differentiation of Bacteria by Electrical Bioimpedance Spectroscopy." *BioTechniques* 69 (1): 27–36. https://doi.org/10.2144/BTN-2019-0080.

Grossi, Marco, and Bruno Ricco. 2017. "Electrical Impedance Spectroscopy (EIS) for Biological Analysis and Food Characterization: A Review." *Journal of Sensors and Sensors Systems* 6: 303–25.

Han, Lei, Pei Liu, Valery A. Petrenko, and Ai Hua Liu. 2016. "A Label-Free Electrochemical Impedance Cytosensor Based on Specific Peptide-Fused Phage Selected from Landscape Phage Library." *Scientific Reports* 6: 22199. https://doi.org/10.1038/srep22199.

Harker, F. Roger, and Shelley K. Forbes. 1997. "Ripening and Development of Chilling Injury in Persimmon Fruit: An Electrical Impedance Study." *New Zealand Journal of Crop and Horticultural Science* 25 (2): 149–57. https://doi.org/10.1080/01140671.1997.9514001.

Harker, F. R., and J. H. Maindonald. 1994. "Ripening of Nectarine Fruit." *Plant Physiology* 106: 165–71.

Ibba, Pietro, Aniello Falco, Biresaw Demelash Abera, Giuseppe Cantarella, Luisa Petti, and Paolo Lugli. 2020. "Bio-Impedance and Circuit Parameters: An Analysis for Tracking Fruit Ripening." *Postharvest Biology and Technology* 159: 110978. https://doi. org/10.1016/j.postharvbio.2019.110978.

Jackson, Phillipa J., and F. Roger Harker. 2000. "Apple Bruise Detection by Electrical Impedance Measurement." *HortScience* 35 (1): 104–07. https://doi.org/10.21273/hortsci.35.1.104.

Kant, Rishi, Geeta Bhatt, Poonam Sundriyal, and Shantanu Bhattacharya. 2017. "Relevance of Adhesion in Fabrication of Microarrays in Clinical Diagnostics." In *Adhesion in Pharmaceutical, Biomedical and Dental Fields*, edited by K. L. Mittal and F. M. Etzler, 257–98. Scrivener Publishing LLC.

Khaled, D. El, N. N. Castellano, J. A. Gazquez, R. M. García Salvador, and F. Manzano-Agugliaro. 2017. "Cleaner Quality Control System Using Bioimpedance Methods: A Review for Fruits and Vegetables." *Journal of Cleaner Production*. https://doi. org/10.1016/j.jclepro.2015.10.096.

Khan, M. Z. H., M. R. Hasan, S. I. Hossain, M. S. Ahommed, and M. Daizy. 2020. "Ultrasensitive Detection of Pathogenic Viruses with Electrochemical Biosensor: State of the Art." *Biosensors and Bioelectronics* 166: 112431. https://doi.org/10.1016/j.bios.2020.112431.

K'Owino, Isaac O., and Omowunmi A. Sadik. 2005. "Impedance Spectroscopy: A Powerful Tool for Rapid Biomolecular Screening and Cell Culture Monitoring." *Electroanalysis* 17 (23): 2101–13. https://doi.org/10.1002/elan.200503371.

Lei, Kin Fong, Min Hsien Wu, Che Wei Hsu, and Yi Dao Chen. 2014. "Real-Time and Non-Invasive Impedimetric Monitoring of Cell Proliferation and Chemosensitivity in a Perfusion 3D Cell Culture Microfluidic Chip." *Biosensors and Bioelectronics* 51: 16–21. https://doi.org/10.1016/j.bios.2013.07.031.

Li, Bo, Julien Lecourt, and Gerard Bishop. 2018. "Advances in Non-Destructive Early Assessment of Fruit Ripeness towards Defining Optimal Time of Harvest and Yield Prediction—a Review." *Plants* 7 (1): 3. https://doi.org/10.3390/plants7010003.

Lin, Shu Ping, Themis R. Kyriakides, and Jia Jin J. Chen. 2009. "On-Line Observation of Cell Growth in a Three-Dimensional Matrix on Surface-Modified Microelectrode Arrays." *Biomaterials* 30 (17): 3110–17. https://doi.org/10.1016/j.biomaterials.2009.03.017.

Lisdat, F., and D. Schäfer. 2008. "The Use of Electrochemical Impedance Spectroscopy for Biosensing." *Analytical and Bioanalytical Chemistry* 391 (5): 1555–67. https://doi. org/10.1007/s00216-008-1970-7.

Liu, Yi Shao, Padmapriya P. Banada, Shantanu Bhattacharya, Arun K. Bhunia, and Rashid Bashir. 2008. "Electrical Characterization of DNA Molecules in Solution Using Impedance Measurements." *Applied Physics Letters* 92 (14): 143902. https://doi. org/10.1063/1.2908203.

Lizhi, Hu, K. Toyoda, and I. Ihara. 2008. "Dielectric Properties of Edible Oils and Fatty Acids as a Function of Frequency, Temperature, Moisture and Composition." *Journal of Food Engineering* 88 (2): 151–58. https://doi.org/10.1016/j.jfoodeng.2007.12.035.

———. 2010. "Discrimination of Olive Oil Adulterated with Vegetable Oils Using Dielectric Spectroscopy." *Journal of Food Engineering* 96 (2): 167–71. https://doi.org/10.1016/j. jfoodeng.2009.06.045.

Mabrook, M. F., and M. C. Petty. 2003. "Effect of Composition on the Electrical Conductance of Milk." *Journal of Food Engineering* 60 (3): 321–25. https://doi.org/10.1016/S0260-8774(03)00054-2.

Makohliso, S. A., D. Léonard, L. Giovangrandi, H. J. Mathieu, M. Ilegems, and P. Aebischer. 1999. "Surface Characterization of a Biochip Prototype for Cell-Based Biosensor Applications." *Langmuir* 15 (8): 2940–46. https://doi.org/10.1021/la980688h.

Marx, Uwe, Takafumi Akabane, Tommy B. Andersson, Elizabeth Baker, Mario Beilmann, Sonja Beken, Susanne Brendler-Schwaab et al. 2020. "Biology-Inspired Microphysiological Systems to Advance Patient Benefit and Animal Welfare in Drug Development." *ALTEX* 37 (3): 365–94. https://doi.org/10.14573/altex.2001241.

Mialich, Mirele Savegnago, Juliana Maria, Faccioli Sicchieri, Alceu Afonso, and Jordao Junior. 2014. "Analysis of Body Composition : A Critical Review of the Use of Bioelectrical Impedance Analysis." *International Journal of Clinical Nutrition* 2 (1): 1–10.

Moulton, S. E., J. N. Barisci, A. Bath, R. Stella, and G. G. Wallace. 2004. "Studies of Double Layer Capacitance and Electron Transfer at a Gold Electrode Exposed to Protein Solutions." *Electrochimica Acta* 49 (24): 4223–30. https://doi.org/10.1016/j.electacta.2004.03.034.

Muñoz-Huerta, Rafael F., Antonio de J. Ortiz-Melendez, Ramon G. Guevara-Gonzalez, Irineo Torres-Pacheco, Gilberto Herrera-Ruiz, Luis M. Contreras-Medina, Juan Prado-Olivarez, and Rosalia V. Ocampo-Velazquez. 2014. "An Analysis of Electrical Impedance Measurements Applied for Plant N Status Estimation in Lettuce (Lactuca Sativa)." *Sensors (Switzerland)* 14 (7): 11492–503. https://doi.org/10.3390/s140711492.

Muñoz, Jose, Raquel Montes, and Mireia Baeza. 2017. "Trends in Electrochemical Impedance Spectroscopy Involving Nanocomposite Transducers: Characterization, Architecture Surface and Bio-Sensing." *TrAC - Trends in Analytical Chemistry* 97: 201–15. https://doi.org/10.1016/j.trac.2017.08.012.

Nayak, Monalisha, Deepak Singh, Himanshu Singh, Rishi Kant, Ankur Gupta, Shashank Shekhar Pandey, Swarnasri Mandal, Gurunath Ramanathan, and Shantanu Bhattacharya. 2013. "Integrated Sorting, Concentration and Real Time PCR Based Detection System for Sensitive Detection of Microorganisms." *Scientific Reports* 3: 3266. https://doi.org/10.1038/srep03266.

Patolsky, Fernando, Eugenii Katz, Amos Bardea, and Itamar Willner. 1999. "Enzyme-Linked Amplified Electrochemical Sensing of Oligonucleotide-DNA Interactions by Means of the Precipitation of an Insoluble Product and Using Impedance Spectroscopy." *Langmuir* 15 (11): 3703–06. https://doi.org/10.1021/la981682v.

Poloznikov, Andrey, Irina Gazaryan, Maxim Shkurnikov1, Sergey Nikulin, Oxana Drapkina, Ancha Baranova, and Alexander Tonevitsky. 2018. "In Vitro and in Silico Liver Models: Current Trends, Challenges and Opportunities." *Altex* 35 (3): 397–412. https://doi.org/10.14573/altex.1803221.

Prevc, Tjaša, Blaž Cigić, Rajko Vidrih, Nataša Poklar Ulrih, and Nataša Šegatin. 2013. "Correlation of Basic Oil Quality Indices and Electrical Properties of Model Vegetable Oil Systems." *Journal of Agricultural and Food Chemistry* 61 (47): 11355–62. https://doi.org/10.1021/jf402943b.

Roychoudhury, Appan, James W. Dear, and Till T. Bachmann. 2022. "Proximity Sensitive Detection of MicroRNAs Using Electrochemical Impedance Spectroscopy Biosensors." *Biosensors and Bioelectronics* 212: 114404.

Samatov, Timur R., Maxim U. Shkurnikov, Svetlana A. Tonevitskaya, and Alexander G. Tonevitsky. 2015. "Modelling the Metastatic Cascade by in Vitro Microfluidic Platforms." *Progress in Histochemistry and Cytochemistry* 49: 21–29. https://doi.org/10.1016/j.proghi.2015.01.001.

Schmid, Yannick R. F., Sebastian C. Bürgel, Patrick M. Misun, Andreas Hierlemann, and Olivier Frey. 2016. "Electrical Impedance Spectroscopy for Microtissue Spheroid Analysis in Hanging-Drop Networks." *ACS Sensors* 1 (8): 1028–35. https://doi.org/10.1021/acssensors.6b00272.

Single, Andrew, Henry Beetham, Bryony J. Telford, Parry Guilford, and Augustine Chen. 2015. "A Comparison of Real-Time and Endpoint Cell Viability Assays for Improved Synthetic Lethal Drug Validation." *Journal of Biomolecular Screening* 20 (10): 1286–93. https://doi.org/10.1177/1087057115605765.

Srinivasan, Balaji, Aditya Reddy Kolli, Mandy Brigitte Esch, Hasan Erbil Abaci, Michael L. Shuler, and James J. Hickman. 2015. "TEER Measurement Techniques for In Vitro Barrier Model Systems." *Journal of Laboratory Automation* 20 (2): 107–26. https://doi.org/10.1177/2211068214561025.

Stevan, Sergio Luiz, Leandro Paiter, José Ricardo Galvão, Daniely Vieira Roque, and Eduardo Sidinei Chaves. 2015. "Sensor and Methodology for Dielectric Analysis of Vegetal Oils Submitted to Thermal Stress." *Sensors (Switzerland)* 15 (10): 26457–77. https://doi.org/10.3390/s151026457.

Stewart, G. N. 1899. "The Changes Produced by the Growth of Bacteria in the Molecular Concentration and Electrical Conductivity of Culture Media." *Journal of Experimental Medicine* 4 (2): 235–43.

Teengam, Prinjaporn, Weena Siangproh, Adisorn Tuantranont, Tirayut Vilaivan, Orawon Chailapakul, and Charles S. Henry. 2018. "Electrochemical Impedance-Based DNA Sensor Using Pyrrolidinyl Peptide Nucleic Acids for Tuberculosis Detection." *Analytica Chimica Acta* 1044: 102–9. https://doi.org/10.1016/j.aca.2018.07.045.

Thielecke, Hagen, Alexandra Mack, and Andrea Robitzki. 2001. "Biohybrid Microarrays - Impedimetric Biosensors with 3D in Vitro Tissues for Toxicological and Biomedical Screening." *Analytical and Bioanalytical Chemistry* 369 (1): 23–29. https://doi.org/10.1007/s002160000606.

Torras, Núria, María García-Díaz, Vanesa Fernández-Majada, and Elena Martínez. 2018. "Mimicking Epithelial Tissues in Three-Dimensional Cell Culture Models." *Frontiers in Bioengineering and Biotechnology* 6: 197. https://doi.org/10.3389/fbioe.2018.00197.

Vanoli, Maristella, and Marina Buccheri. 2012. "Overview of the Methods for Assessing Harvest Maturity." *Stewart Postharvest Review* 8 (1): 1–11. https://doi.org/10.2212/spr.2012.1.4.

Wang, Ruimin, Jing Di, Jie Ma, and Zhanfang Ma. 2012. "Highly Sensitive Detection of Cancer Cells by Electrochemical Impedance Spectroscopy." *Electrochimica Acta* 61: 179–84. https://doi.org/10.1016/j.electacta.2011.11.112.

Wright, Jennifer E.I., Nicholas P. Cosman, Khalid Fatih, Sasha Omanovic, and Sharon G. Roscoe. 2004. "Electrochemical Impedance Spectroscopy and Quartz Crystal Nanobalance (EQCN) Studies of Insulin Adsorption on Pt." *Journal of Electroanalytical Chemistry* 564 (1–2): 185–97. https://doi.org/10.1016/j.jelechem.2003.10.031.

Yang, J., K. S. Zhao, and Y. J. He. 2016. "Quality Evaluation of Frying Oil Deterioration by Dielectric Spectroscopy." *Journal of Food Engineering* 180: 69–76. https://doi.org/10.1016/j.jfoodeng.2016.02.012.

Zanoni, Michele, Filippo Piccinini, Chiara Arienti, Alice Zamagni, Spartaco Santi, Rolando Polico, Alessandro Bevilacqua, and Anna Tesei. 2016. "3D Tumor Spheroid Models for in Vitro Therapeutic Screening: A Systematic Approach to Enhance the Biological Relevance of Data Obtained." *Scientific Reports* 6. https://doi.org/10.1038/srep19103.

Zhang, Y., Q. J. Xie, A. H. Zhou, and S. Z. Yao. 2000. "In Situ Monitoring of Lysozyme Adsorption onto Bare and Cysteine- or 1-Octadecanethiol-Modified Au Electrodes Using an Electrochemical Quartz-Crystal Impedance System." *Analytical Sciences* 16 (8): 799–805. https://doi.org/10.2116/analsci.16.799.

Zhu, Heng, and Michael Snyder. 2003. "Protein Chip Technology." *Current Opinion in Chemical Biology* 7 (1): 55–63. https://doi.org/10.1016/S1367-5931(02)00005-4.

5 Impedance spectroscopy and environmental monitoring

Ranamay Saha, Nitish Katiyar, Sagnik Sarma Choudhury, and Kapil Manoharan

CONTENTS

5.1 INTRODUCTION

Electrochemical methods have been extensively used to recognize and/or measure compounds with clinical, environmental, and industrial applications. In Germany, washing clothes needed more than 1 billion liters of water and 5,80,000 tons of detergent in 2010 (Gruden and Kanoun 2013). Especially, water quality is a crucial concern. Using impedance spectroscopy, the detergent concentration of suds can be determined. Understanding the soil's water content is essential for enhancing the efficiency of irrigation methods, in terms of costs and environmental factors (Tetyuev and Kanoun 2006). For environmental control, a precise in situ moisture quantification of soil is crucial additionally for several uses in civil engineering (Chauhan, Pandey, and Bhattacharya

DOI: 10.1201/9781003358091-7

2019). Because they are inexpensive and simple to use, capacitive sensors are sometimes preferred. Semiconducting metal-oxide nanoparticles have recently piqued the curiosity of scholars as a novel way to detect harmful gases as an electrochemical biosensor. Production of electrical signals in the system is caused by the interaction between immobilized molecules and target analytes, which, in this case, are airborne pathogens. The resulting electrical signal may be read using a variety of methods, such as voltammetry [square wave voltammetry (SWV), cyclic voltammetry (CW), differential pulse voltammetry (DPV)], amperometry, and electrochemical impedance spectroscopy (EIS), depending on the detection mechanism and application.

EIS has attained a significant amount of interest as an assessment technique for a wide variety of analytes, primarily since it is a nondestructive method that allows in situ observations. EIS comes in two types: Faradaic and not Faradaic. In a faradaic EIS, electro-active entities undergo redox reactions at the electrode, which produce an electrical current. The electrical characteristics of non-faradaic EIS are generated by double layer capacitance, making it a DC-based impedance (Dorledo de Faria et al. 2019). The non-Faradaic sensor, although being less sensitive, is a viable contender for real-time applications because redox couples are not required for its operation. Its working principle is such that applying an alternating potential in an electrochemical system as a tiny excitation signal, changes in the cell can be tracked by measuring the current response. Figure 5.1a depicts a basic interaction of the components required to build an EIS system: the signal processing unit, the sensing unit, and the data analysis unit comprise the three major subsystems of an impedance spectroscopy system. Plot (i) in Figure 5.1a shows the Nyquist plot, which is usually used to study resistive processes. In Figure 5.1a, plot (ii) shows the Bode plot acquired during the EIS method, between phase angle and frequency to determine the capacitance of an electrochemical system. Figure 5.1b illustrates a few of the factors to be considered while designing the impedance sensor.

Figure 5.2 comprises photolithographically constructed microelectrodes with straight parallel fingers that are mesh-meshed to produce a rectangular sensing space. For some applications, adjusting the interdigit width and spacing might improve sensitivity. The huge gap area between the electrodes is one of the advantages offered by interdigitated electrodes (IDE), which are used in sensor applications. This space

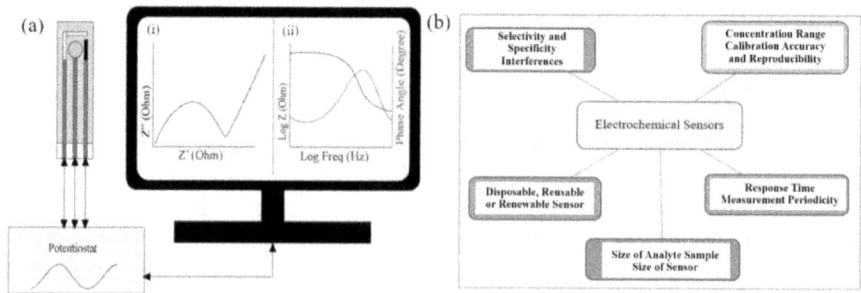

FIGURE 5.1 (a) Overview of electrochemical impedance spectroscopy setup. [Reprinted with permission from Strong et al. (2021). Copyright (2021) Elsevier]; (b) essential criteria to consider while selecting electrochemical sensors for environmental monitoring.

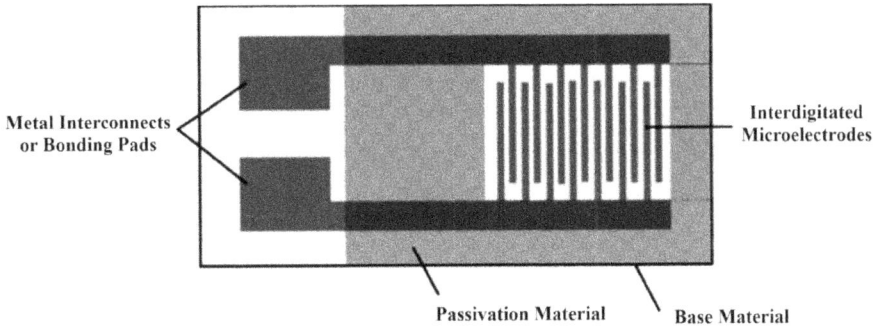

FIGURE 5.2 Schematic diagram of a planar interdigitated microelectrode. [Reprinted with permission from Varshney and Li (2009). Copyright (2009) Elsevier].

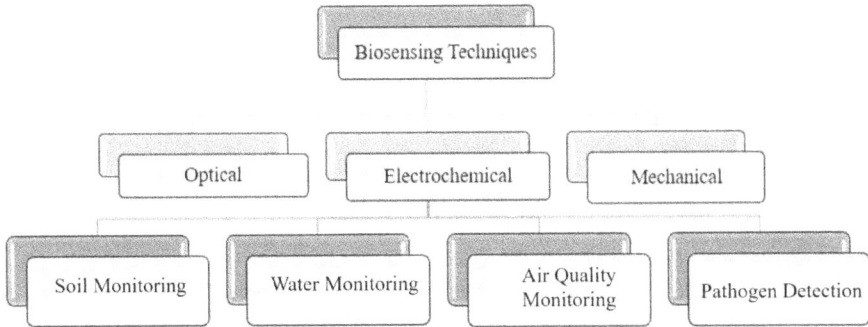

FIGURE 5.3 Electrochemical impedance spectroscopy applications.

will be occupied by the sensing material, which can result in a significant number of additional binding locations between the sensing material and the analyte.

EIS can be used to improve agricultural production by motoring soil, water, and air quality along with detecting harmful pathogens in air to improve human health. Humidity control can be a crucial factor to enhance the life of various instruments that are used in medical, automated production systems, and this can be done using EIS. Figure 5.3 depicts EIS applications throughout various fields.

5.2   AIR MONITORING

Each day, a fit individual breathes about 1,000 liters of air (Kant and Bhattacharya 2018) and the existence of humans on our planet has a fundamental dependence on the quality of the air that they breathe. The current pattern of industrialization and modernization of human civilization has led to a crisis scenario in people's lives. This has been the case as the number of air pollutants has been rising each day. These contaminants in the air have the potential to enter human systems directly, where they might have an adverse effect on organs, such as the lungs, kidneys, heart, and nervous system. The exact monitoring and management of air pollution is becoming increasingly important in today's environment.

To accurately identify and measure harmful emissions, sensors should have good sensitivity, selectivity, fast response, reversibility, lower manufacturing expense, reduced energy consumption, and ease of handling. With the help of environmental monitoring, lawmakers, international organizations, and the public can be informed about environmental trends and conditions as well as assistance policy development.

5.2.1 AIRBORNE PATHOGENS

The pandemic of the airborne disease SARS-CoV-2 has created an unfavorable condition that the entire world is currently dealing with. It is true that prevention is preferable to treatment; hence, the quick identification of airborne infections is essential because it may help control epidemics and also save many people. There is a serious risk to human health and economic development from microorganisms comprising bacteria (e.g., *Mycobacterium*), viruses (e.g., Influenza), and fungus (e.g., *Aspergillus niger*) (M. Pandey, Srivastava et al. 2019; Sivakumar and Lee 2022). Anything that can be breathed in or landed on by an organism is a potential threat.

5.2.2 TOXIC GASES DETECTION USING EIS

Gas detection and monitoring refers to the process of continuously monitoring changes in concentrations of different atmospheric constituents. Air quality is measured by means of an indicator known as the Air Quality Index (AQI). The AQI displays changes in air pollution. In 1984, the city of Bhopal in the Indian state of Madhya Pradesh had a chemical leak known as the Bhopal tragedy. It was the worst industrial disaster ever, where a pesticide facility operated by Union Carbide India, a part of the American company Union Carbide Corporation, leaked about 45 tons of the toxic chemical methyl isocyanate on December 3, 1984 (Singh and Rehalia 2016). Hazardous gas leakage accidents are the main cause of fatalities of workers in industries that primarily use chemicals (Figure 5.4).

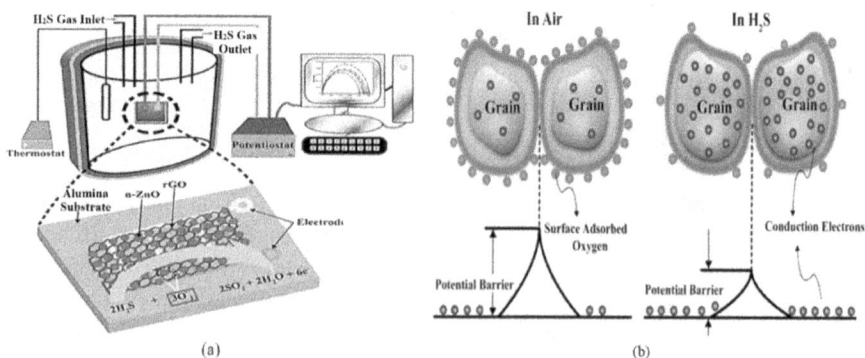

FIGURE 5.4   (a) A diagrammatic representation of the experimental apparatus; (b) H_2S gas sensor schematic. [Reprinted with permission from Balasubramani, Sureshkumar, Rao et al. (2019). Copyright (2019) American Chemical Society].

TABLE 5.1

Some important gas-sensing applications through electrochemical impedance spectroscopy technique

Sensitive substances	Analyte gases	Concentration/LODs (ppm)	Ref
$MnNb_2O_6$	SO_2	5/50 ppb	Liu et al. (2017)
$Li_2CO_3/BaCO_3$	CO_2	1.5 vol%	Wierzbicka et al. (2007)
PANI/PVS	NH_3	0–20	Santos et al. (2017)
rGO/ZnO	H_2S	2–100	Balasubramani et al. (2019)
SnO_2/rGO	H_2S	1–100	White et al. (2008)
Mn-doped ZnO	H_2S	2–100	Sureshkumar et al. (2019)

Typical toxic gases in the oil and gas industries are hydrogen sulfide (H_2S), carbon monoxide (CO), oxygen depletion (O_2), and carbon dioxide (CO_2). There are many other gas-sensing techniques, but in this chapter, we focus on one that uses impedance spectroscopy technique for detection. Gas sensing based on EIS has been regarded as an ultrasensitive method for identifying and quantifying various harmful gases. The sensors' electrodes were created by a variety of methods including spin coating, sputtering, and spray coatings and electrophoretic deposition. Initially, impedance spectroscopy methods were used to solid electrolytes of yttria-stabilized zirconia, laying the groundwork for the impedance-detecting method. Since then, scientists have developed sensors that can detect NO_x, water vapor, hydrocarbons, and carbon monoxide using the impedance change approach.

Balasubramani et al. (2019) have created a reduced graphene oxide (rGO) by integrating it with ZnO (rGO/ZnO) and used impedance spectroscopy to create a H_2S gas sensor. Screen printing method was employed to produce the electrode on an alumina substrate. Figure 5.4a,b demonstrates the experimental setup, Nyquist impedance, and an equivalent circuit for ZnO/rGO composites. The created sensor was shown to be most sensitive to H_2S gas at 90°C, where concentrations between 2 and 100 ppm were tested. n-ZnO/rGO composites as compared with pure n-ZnO showed a significantly enhanced response. Two reasons were proposed as the causes of the elevated H_2S reaction. First, rGO produces H_2S adsorption sites and second, rGO's superior electrical conductivity relative to n-ZnO allows the active transport of electrons through H_2S gas to the sensor layer, as a consequence of which the gas responsiveness at 90°C is enhanced. A survey of numerous impedance-based gas sensors is summarized in Table 5.1. Sensor's accuracy and quality can be measured by limit of detection (LOD) as well as limit of quantification.

5.2.3 HUMIDITY SENSORS

Another characteristic that poses notorious challenges to obtaining accurate measurements is humidity. Many applications rely on humidity, including instrumentation, automated manufacturing systems, agriculture, weather, biological, and medicinal systems (Farahani et al. 2014). In order to accurately measure humidity, a sensor needs to have certain qualities. These include high sensitivity, a quick reaction and recovery, a long lifetime, and outstanding stability. At this time, three distinct humidity sensors are available: impedance, capacitance, and piezoelectric. Impedance-based humidity sensors can be used to measure humidity by looking at the variation in impedance values of moisture-sensing materials on the surface of the IDEs. $BaTiO_3$, SnO_2, ZnO, and TiO_2 are major humidity-sensing nanoparticles. They have accomplished extraordinary success in terms of the humidity-sensing properties they possess. Humidity sensors that are based on TiO_2 provide relatively superior performance. Nevertheless, there are still some issues that cannot be avoided, such as a lengthy period of recovery time (Biju and Jain 2007).

Based on their research, Wu et al. (2022) reported that the performance of TiO_2 film-based humidity sensors could be improved, by $N3(C_{26}H_{16}N_6O_8RuS_2)$ molecular modification and that the problem of slow recovery time could be solved. An Al_2O_3 substrate was used with an IDE made of gold. Fabrication of TiO_2 humidity sensor is

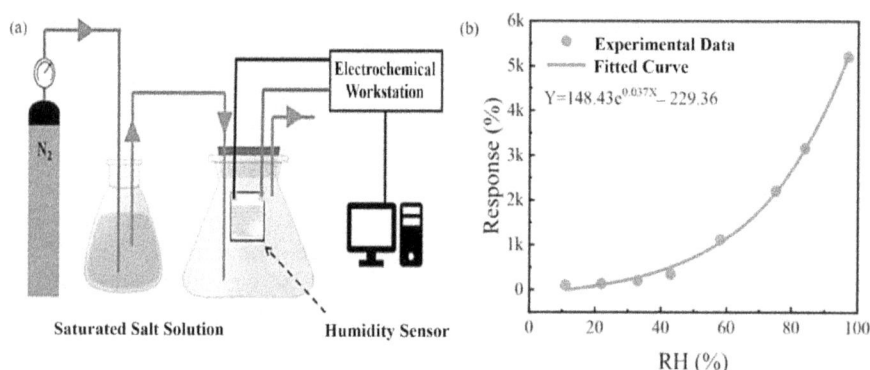

FIGURE 5.5 (a) Schematic representation of the humidity measuring device; (b) nano-TiO_2 humidity-sensing system's calibration curve. [Reprinted with permission from Wu et al. (2022). Copyright (2022) Elsevier].

TABLE 5.2

Comparison of response and recovery times (R_s, R_r) for electrochemical impedance spectroscopy based humidity sensors made of different materials

Sensing material	Response time (R_s)	Recovery time (R_r)	Ref
SnO_2/SBA-15	15	21	Tomer and Duhan (2015)
$K_{0.5}Na_{0.5}NbO_3$	8	18	Yuan et al. (2015)
$ZnSnO_3$	7	16	Bauskar, Kale, and Patil (2012)
MoS_2	10	60	Zhao et al. (2017)
$CH_3NH_3PbI_{0.2}Cl_{28}$	24	24	Ren et al. (2017)
$Cs_2BiAgBr_6$	1.78	0.45	Weng et al. (2019)
$CsPb_2Br_5$-$BaTiO_3$	5	5	Cho et al. (2020)
$CsPbBr_3$ nano particles	2.8	9.7	Wu et al. (2022)

done by preparing a paste of TiO_2, and this paste is deposited uniformly by using the screen printing method. Sintering was performed inside a muffle furnace at 400°C, and a nanoporous coating of TiO_2 is formed on the IDE's surface. Figure 5.5 demonstrates the humidity-sensing setup developed by Wu et al. (2022). The results of this experiment showed that the response time was 5 s, and the recovery time was 8 s with a sensitivity of 1.3% relative humidity. Table 5.2 summarizes some important results observed through the analysis of EIS-based humidity sensors made of different materials. Minimum response time of 1.78 s is observed with $Cs_2BiAgBr_6$ sensing material (Weng et al. 2019).

5.3 WATER ANALYSIS

Water ecology regulates food chain, nutrient cycling, habitat preservation, flood control, and soil formation. Ecosystems are affected when water is contaminated by various chemicals. Transmissible diseases begin to spread due to the presence of biological and chemical impurities in tap and drinking water. Furthermore, the presence of antibiotics from pharmaceuticals in surface and ground water systems may lead to the formation of unnecessary antibiotic resistance within the body due to overconsumption. Again, cyanotoxins that amount to around 50–70% of substances produced by cyanobacterial blooms are abundantly found in water systems. Exposure to such toxins may inhibit protein biosynthesis, disrupt keratin filaments and change features of haptic ultrastructure (Zhang et al. 2017; Magro et al. 2021). Often traces of some contaminants are still found even after they have been outlawed. As a result, monitoring of natural water has become an absolute necessity. The in-use fixed monitoring stations often cost quite high and need skilled personnel to examine the received data. There has been an increase in interest in the creation of portable and user-friendly devices that might provide quick, accurate, and trustworthy information in order to save expenses and improve monitoring (Kivirand, Min, and Rinken 2019). In the same context, EIS is considered one of the most popular methods of surface characterization in biosensors. Table 5.3 summarizes some important applications of impedance-based sensors in water monitoring.

5.3.1 PESTICIDES AND TOXINS

Contamination of ground as well as surface water has been an alarming issue due to the increasing usage of pesticides and insecticides for crop control. Industrial or agricultural chemical discharges have the potential to create toxic effects on the environment and biological systems. Biosensors based on impedimetric analysis can be utilized for the detection of diazinon, an organophosphate pesticide in aqueous solution (Zehani et al. 2014). Two different types of lipase, candida rugosa lipase (CRL) and porcine pancreas lipase (PPL) were immobilized on thiol-functionalized gold electrode to catalyze the hydrolysis of diazinon. These biosensors exhibited a linearity range of up to 50 μM.

Insecticides can also be detected using EIS. However, sensitivity is a challenge. When the thickness of the sensor layer is low, there is a chance of surface exposure that may lead to low signal-to-noise (SN) ratio whereas a highly thick layer may

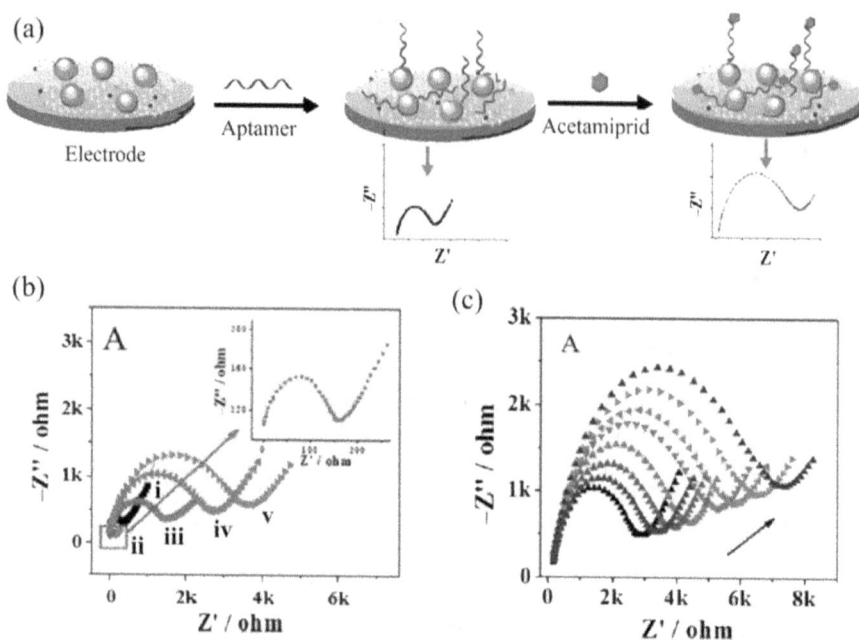

FIGURE 5.6 Electrochemical impedance spectroscopy (EIS)-based aptasensor for the detection of acetamiprid: (a) fabrication principle of the biosensor; (b) EIS responses of (i) bare electrode; (ii) Ag–NG electrode; (iii) aptamer/Ag–NG electrode; (iv) MCH/aptamer/Ag–NG electrode; (v) after 1×10^{-9} M acetamiprid capture; (c) EIS responses with increasing acetamiprid concentrations. [Reprinted with permission from Jiang et al. (2015). Copyright (2015) MDPI].

limit the electron transfer. Sensitivity can be improved by tagging nanoparticles. In a highly sensitive EIS aptasensor fabricated by Jiang et al. (2015) for the detection of acetamiprid insecticide, enhancement in sensitivity of the biosensor was ensured through silver nanoparticles deposited nitrogen-doped graphene (NG) nanocomposites (Ag/NG nanocomposites). One step thermal cycling was used to prepare silver nanoparticles deposited Ag/NG nanocomposites. Figure 5.6a demonstrates the fabrication mechanism of the biosensor. The aptamer/Ag–NG/GCE was treated with MCH to block nonspecific bindings. Figure 5.6b,c shows that impedance values decreased with Ag treatment and then increased with increasing concentrations of acetamiprid. For a frequency range of 10 kHz to 0.01 Hz with a sinusoidal potential of 5 mV amplitude, the biosensor showed a linear response in the range of 0.0001 nM to 5.0 nM and a low LOD of 0.03 pM for a SN ratio of 3. Again, Atrazine, a herbicide, was detected in drinking water, phosphate buffer solution, and river water samples through the immobilization of small molecules into nanoporous alumina membrane integrated on the printed circuit board (PCB) platform (Pichetsurnthorn, Vattipalli, and Prasad 2012). The signal output was 1.5 times stronger for the specific bindings as compared with the nonspecific bindings. Microcystin-LR is a toxin released by cyanobacteria. Electrochemical characterization of the biochemical activity on the electrode-specific anti-MC-LR monoclonal antibodies was carried out for the selective detection of MC-LR in drinking water.

Figure 5.7a–i schematically shows the synthesis of 3D graphene foam and the consequent fabrication of the 3D graphene-based sensing system (Zhang et al. 2017). Bisphenol A (BPA) is an industrial material widely used in the plastic manufacturing industry. Exposure to BPA may cause severe health issues through endocrine disruption. Functionalized carbon nanotubes have been efficiently used for the sensitive detection of BPA through EIS characterization (Azadbakht et al. 2016; Figure 5.7).

Change in electron transfer resistance corresponding to the formation of BPA – aptamer complex at the functionalized electrode sensor surface was used to sense the BPA concentration in three different environmental water samples. Using Prussian blue (PB) as the redox tag deposition of neatly arranged gold nanoparticles (AuNPs) and aptamer immobilization at the surface of AuNPs/PB/CNTs-COOH, this ultrasensitive aptasensor for BPA exhibited a low LOD of 0.045 and two linear ranges: one from 0.1 to 1 pM and the other from 10 pM to 10 nM. The proposed impedance-based biosensor demonstrated good selectivity, reproducibility, and stability. Another group of endocrine disruptors known as phthalate esters are abundant food toxins. Phthalates are mainly used as a plasticizer to PVC products and also in cosmetics, pesticides, paints, and pharmaceutical compounds. A real-time, noninvasive, label-free rapid detection scheme based on EIS was developed to carry out phthalates' detection in deionized water and fruit juices. Different concentrations of di (2-ethylhexyl) phthalate (DEHP) ranging from 0.002 to 2 ppm of DEHP in deionized water were introduced to specially designed interdigitated sensor platform by the dip-testing technique. EIS measurement data showed potential applicability of the sensor toward DEHP detection in deionized water and fruit juices. However, selectivity was a major challenge in this system (Zia et al. 2013).

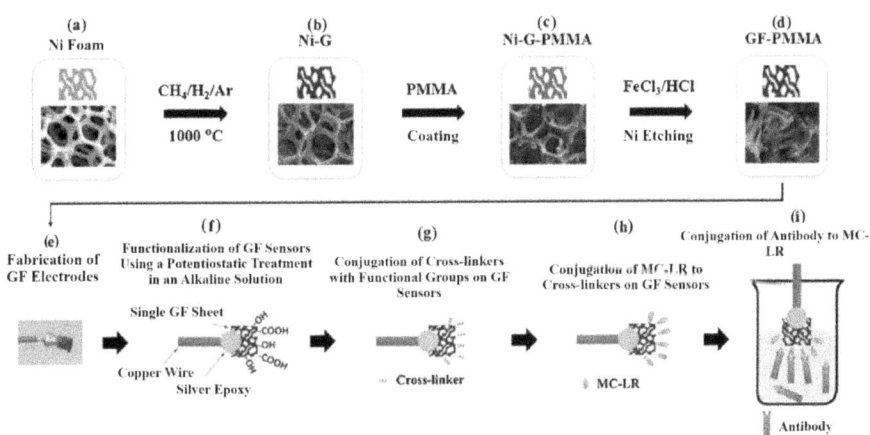

FIGURE 5.7 Schematic illustration of the sensor development for MC-LR detection in drinking water through electrochemical impedance spectroscopy (a–i). [Reprinted with permission from Zhang et al. (2017). Copyright (2017) Elsevier].

5.3.2 Ionic concentrations

EIS has been used for quite a time in water quality monitoring. Estimation of ions from metal residues, nutrients, and carbon traces in water may give important environmental information as well as water quality measures. Detection limit of anions such as acetate, chloride, cyanide, carbonate, sulfate, and bicarbonate with low ionic strength in water can be recorded through the EIS technique to investigate the effect of ion charge and size on the impedance measurement at 25°C. Detection limits at the parts per trillion (ppt) level have been obtained. It was shown that the charge of anions was more influential on impedance data than size (Scott and Alseiha 2017). Further, salt present in the solution under consideration can be estimated through the determination of total dissolved solids in the aqueous solution of sodium chloride and magnesium sulfate. The ratio of impedance data at two different frequencies was used to develop graphs through which the two solutions could be differentiated (De Beer and Joubert 2019). A lot of water gets wasted and detergents get accumulated in the environment due to excess dosage of detergents during washing processes. A novel detergent sensor was developed based on impedance spectroscopy and cyclic voltammetry, which could assess the amount of detergent to be added to washing water. Impedance spectroscopy could yield conductivity as a reciprocal of resistance, which in turn was correlated to the total hardness of the water sample and changes in the composition of the medium. These results along with individual ionic concentrations and carbonate hardness data were used to determine the water quality and concentration of detergent in suds so as to assess the amount of detergents to be added for the washing process (Gruden and Kanoun 2013).

5.3.3 Pharmaceutical residues

Substantial rise in the production and usage of pharmaceuticals has led to their presence and accumulation in various aquatic environments, such as freshwater and ground water sources, soil, sediments, and food products. This could be due to the continuous release of pharmaceuticals into the environment, low degradation rate of pharmaceuticals, and inability of waste water treatment plants to completely detect and remove pharmaceutical residues. Long exposure and intake of such residues by human and wildlife may lead to unintended negative impacts on ecosystems, morbidity, and mortality as well as modifications to physiology, cognition, and fertility.

Pharmaceuticals used for human health, such as hormones, antibiotics, pain killers, and antidepressants, as well as pharmaceuticals used in veterinary medicine are of particular concern. Therefore, monitoring and quantification of these pollutants is of utmost importance. Ceramic and glass support sensors consisting of gold IDEs coated with a thin film of five bilayers of positively charged polyethyleneimine (PEI) and anionic poly (sodium 4-styrenesulfonate) (PSS) thin films were fabricated by Magro et al. (2021) in order to monitor an antibiotic called clarithromycin concentrations in different aquatic samples. An array of such sensors, termed as 'electronic tongue' could successfully discriminate clarithromycin concentrations in the range of 10^{-15} to 10^{-5} M through impedance spectroscopy, when immersed in surface and mineral water samples.

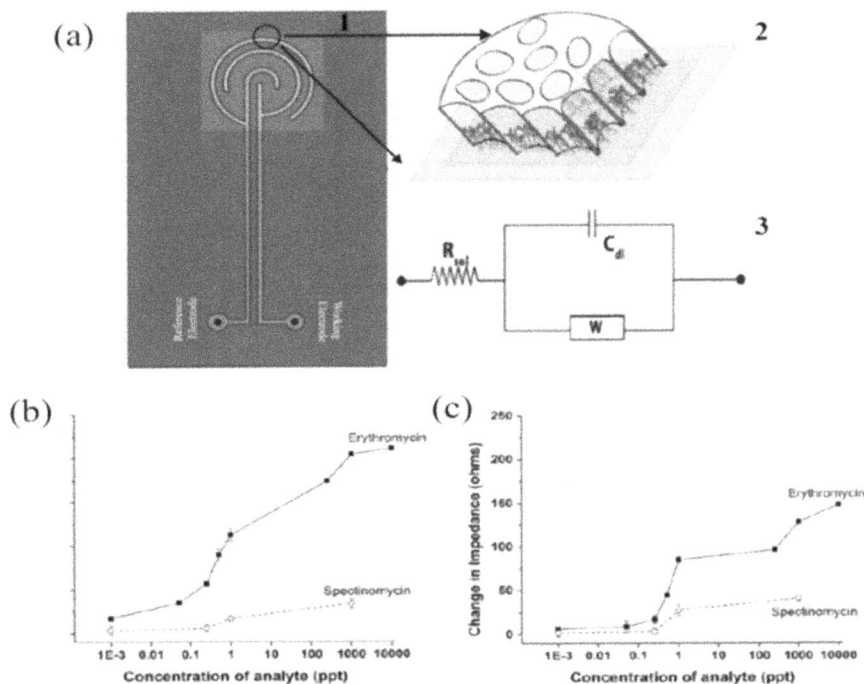

FIGURE 5.8 Erythromycin detection through EIS: (a) Sensor developed: (1) gold electro-plated PCB platform; (2) biochemical reactions in nanoporous membrane; (3) Randle's circuit for non-faradaic EIS assay; (b) impedance variation with analyte concentration for drinking water; (c) impedance variation with analyte concentration for river water. [Reprinted with permission from Jacobs et al. (2013). Copyright (2013) RSC].

However, as the antibiotic binding sites got saturated, further increase in concentrations did not significantly change the impedance (Figure 5.8b,c). The LOD in drinking and river water was close to 0.1 and 1 ppt, respectively. Because of the organic materials that can prevent erythromycin from adhering, the sensitivity was typically lower in river water. The overall impedance variation was nevertheless significant enough to indicate whether the level of erythromycin concentrations was appropriate or inappropriate to make the sample safe for drinking. Progesterone is a natural hormone playing a vital role in pregnancy, growth, and development. However, excessive consumption of progesterone through food or drinking water may lead to disruptive endocrine behavior and diverse health problems. In order to have a prior estimation of the progesterone levels, a simple, fast, selective, and highly sensitive detection scheme was formulated using magnetic graphene oxide as electrode casting nanomaterials and impedance spectroscopy (Disha et al. 2021). In an aptamer-based sensor for sensing progesterone P4, two types of aptamers, namely P4G13 and a cDNA oligonucleotide sequence, were immobilized on the gold electrode by self-assembly.

TABLE 5.3

Some important biosensing applications in water analysis through electrochemical impedance spectroscopy technique

Target analyte	Sample	Electrode	Linearity range	LOD	Reusability	Ref
Diazinon	River water	Functionalized gold electrode	0.01–50 µM (CRL) 0.1–50 µM (PPL)	0.01 µM (CRL) 0.1 µM (PPL)	2–5% (RSD)	Zehani et al. (2014)
Acetamiprid	Waste water	Ag/NG nanocomposites	0.0001–5 nM	0.03 pM	6.9% (RSD)	Jiang et al. (2015)
Atrazine	River and drinking water	Nanoporous alumina membrane integrated with PCB platform	10 fg mL^{-1} to ng mL^{-1}	10 fg mL^{-1}	—	Pichetsurnthorn, Vattipalli, and Prasad (2012)
Microcystin-LR (MC-LR)	Drinking water	Functionalized 3D Graphene electrode	0.05 and 20 mg L^{-1}	0.05 mg L^{-1}	6.9% inter- and 3.6% intra-assay coefficients of variability	Zhang et al. (2017)
NaCl and MgSO4	Water	Header pins	—	5–5000 ppm	—	De Beer and Joubert (2019)
Anions such as acetate, chloride, cyanide, carbonate, sulfate, and bicarbonate	Water	Inner and outer stainless steel cylinder in a two-electrode configuration	—	13 ppt	—	Scott and Alseiha (2017)
Clarithromycin	Surface (SW) and mineral water (MW)	Gold IDEs deposited on ceramic and glass BK7	10^{-15} M–10^{-5} M	10^{-15} M	Ceramic: reproducible for SW & MW Glass BK7: reproducible for SW	Magro et al. (2021)
Progesterone P4	Tap water	Gold electrode	10 and 60 ng mL^{-1}	0.90 ng mL^{-1}	1–7% (RSD)	Contreras Jiménez et al. (2015)

Aptamer-P4 binding showed a consequent change in the electron transfer resistance that could be assessed to determine the P4 concentration in tap water sample. The aptasensor demonstrated a linear range for P4 concentrations between 10 and 60 ng mL^{-1} with a LOD of 0.90 ng mL^{-1} (Contreras Jiménez et al. 2015).

Boumya et al. (2017) proposed a rapid EIS-based detection technique of aldehydes upon derivatization with 2,4-dinitrophenylhydrazine (DNPH) prior to impedance measurement. Formation of a stale complex upon aldehyde-DNPH reaction on the glassy carbon electrode led to an increase in charge transfer and hence the charge transfer resistance could be used as a quantification of the concentration of aldehyde in various samples. Under optimal conditions, a linear range was established between concentrations of 1000–0.05 µmol L^{-1}. The detection limits for various samples were between 0.097 and 0.0109 µmol L^{-1}. This study could be applied in the estimation of aldehydes in drinking water, orange juice, and apple cider vinegar samples with acceptable recovery rates.

5.4 SOIL MONITORING

Environmental monitoring of soil is generally concerned with the biological functions of soil, its characteristics, and the anthropogenic and natural variables, such as climate, influencing its quality. The microstructure, porosity, and conductivity of soil are very critical since they control the mechanical properties of structures and have a big impact on real-life engineering. Often a conservative approach is adopted in planning constructions, which may lead to inefficient and uneconomic designs. Soil monitoring through impedance spectroscopy allows the use of capacitive and resistive parts of soil impedance that eliminates the dependence on soil properties, and hence calibration error can be avoided. With a few limitations, such as the requirement for temperature adjustments to achieve high accuracy and soil-type-specific calibration, impedance-based approaches provide prospects for creating low-cost, field-applicable soil moisture sensors. Commercially available impedance or capacitance-based devices can measure the soil conductivity and moisture content with high accuracy within 5% (Tetyuev and Kanoun 2006; González-Teruel et al. 2019; Kashyap and Kumar 2021).

5.4.1 SOIL IONIC CONCENTRATIONS

Measurement of constituent ionic concentrations can give information about important soil properties such as salinity and conductivity. Estimation of nitrate content of soil may help in the regulation of soil nitrate concentrations and hence lead to the improvement of agricultural productions. Further, nitrogen leeching into underground water may also be controlled. Ionic concentration can be measured by using the EIS technique to gauge dielectric permittivity of soil through directly measured capacitance or impedance and then using dielectric mixing and relaxation models to estimate the individual ionic concentrations from a mixture. Increase in sodium nitrate content was found to simultaneously increase permittivity of the soil, which led to the corresponding increase in capacitance and conductance of the soil dielectric mixture (Pandey, Kumar, and Weber 2013). Such results were then used in dielectric mixing models and debye-type dielectric relaxation models to estimate ionic concentration.

EIS can also be used to understand the effect of different types of salts on the corrosiveness of sandy soil. Nyquist plots obtained from the EIS test are then used to analyze ionic concentrations in different sandy soil samples and hence predict the influence of those salts on the electrochemical behavior of the samples. A copper sheet of required size was used (Xie et al. 2021) to form the working and counter electrode while a calomel electrode was used as the reference electrode (Figure 5.9a). Different soil samples were characterized by different Nyquist plots. Potassium ions can also be detected using the impedance method (Figure 5.9b; Day et al. 2018).

A larger arc radius of the representing capacitive impedance arc signifies a very passive electrochemical reaction process, that is, lower salt concentrations and hence, low corrosiveness (Figure 5.10). Measurement of ionic concentrations from soil was also utilized to study the influence of salinity and water content to salt expansion of saline soil. At a maintained water content and varying salinity, the magnitude of impedance of saline soil decreases with increasing frequency. This is because the capacitance has the characteristics of passing high frequency and blocking low frequency. Furthermore, when salinity increases, the impedance of the sulfate-saline soil drops. Similarly, the region enclosed by the real and imaginary parts of impedance in characteristic Nyquist plots gradually shrinks as salinity increases. This is because the increase of salinity concentrates the ion percentage in the saline soil and

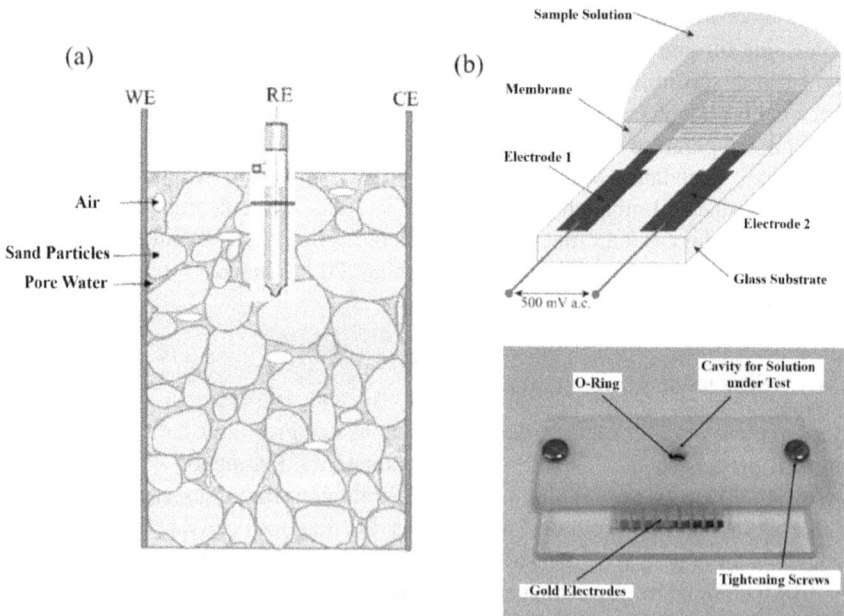

FIGURE 5.9 Electrode setup used for EIS of soil samples: (a) three-electrode test device [Reprinted with permission from Xie et al. (2021). Copyright (2021) ESG]; (b) gold IDEs system covered by membrane. [Reprinted with permission from Day et al. (2018). Copyright (2018) Elsevier].

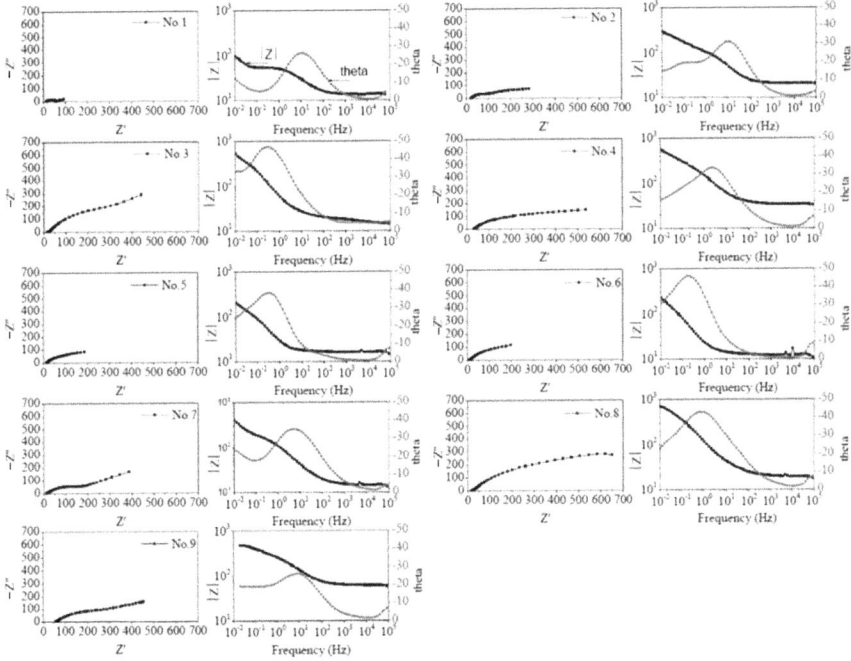

FIGURE 5.10 Impedance responses of different salt samples characterized by varied concentrations of ions (1–9): Nyquist plots (left) and Bode plots (right). [Reprinted with permission from Xie et al. (2021). Copyright (2021) ESG].

hence the resistance decreases. Also, the power supply time of capacitance shortens between the test copper sheet and soil sample with the increase of salinity leading to a reduction in the phase angle (Shuquan, Fan, and Ling 2019). The accuracy and universality of the equivalent model greatly depends on the choice of circuit elements. A proposed equivalent circuit model consisted of a pore fluid resistance (R_e), salt layer resistance (R_s), and a faradaic resistance (R_{ct}, W), where R_{ct} was related to the rate of the electrochemical reaction process while R_e and R_s indicated the double layer capacitance, that is, charge storage capacity of the soil. In another work, six different resistance and reactance elements were used in the integrated circuit model to deduce important soil properties (Han et al. 2015).

5.4.2 SOIL MOISTURE AND DENSITY

Changes in volume and density of soil due to transport of fine soils to coarse soil weaken the structural health of civil constructions. Therefore, an estimation of contact erosion is essential toward maintenance and monitoring of flood protection structures. Again, the development of nondestructive techniques to detect moisture and density changes has been currently the need of the hour. In this context, impedance-based soil sensors work on the impedance/capacitance responses out of a buried probe or electrode, which directly depends on the permittivity of the soil. These sensors have the capability to work in comparatively lower frequency ranges (10 s of MHz). EIS-measuring devices are used to estimate the moisture regain and electrical conductivity of soil. With the change in impedance magnitude ratio and phase difference, the coefficients relating to real and imaginary parts of the impedance change with increasing water percentage in the soil sample according to some linear/power function model.

At low moisture, soil sample contains more air than water, hence capacitance is more as compared with resistance. Semicircle diameter reduces with increasing soil moisture content due to a reduction in resistance since water has better conductivity than air. Fairly good results have been obtained with a mean error of 0.21% in impedance (Figure 5.11b; Umar and Setiadi 2015). The soil sensor in this case was made up of double brass electrodes that penetrated into deep soil for in situ moisture measurement (Figure 5.11a).

Comparison of such a change in measured impedance with the moisture regain values and conductivity values of the prepared ammonium nitrate aqueous extracts can yield acceptable correlation between the depth of soil sample extraction, moisture content, temperature, and salt concentrations to conductivity and impedance. Such correlations between the physical characteristics of soil rock mixture (SRM) and the impedance parameters have also been utilized to study the physical and mechanical properties of SRM. Change in density of SRM was monitored using the EIS technique (Dong et al. 2021). Density first increased and then dropped with the increase in water volume. Water volume and saturation threshold were 16% and 80 ± 2%, respectively. This work opened a new window to SRM research.

Often, a proctoring system is developed to simulate the actual density changes due to contact erosion. A proctoring cylinder was developed to simulate the actual erosion process, however, keeping the water content constant so as to estimate the volume change from the impedance data. Figure 5.12a shows the measured impedance data as a function of frequency for different hydraulic gradients. With change in volume of soil from 1.5% to 5.3%, a corresponding impedance change of 17.2% to 29.8% was observed. The linear graph fitted over the impedance–volume change plot revealed that there was a proportional relationship. However, significant material

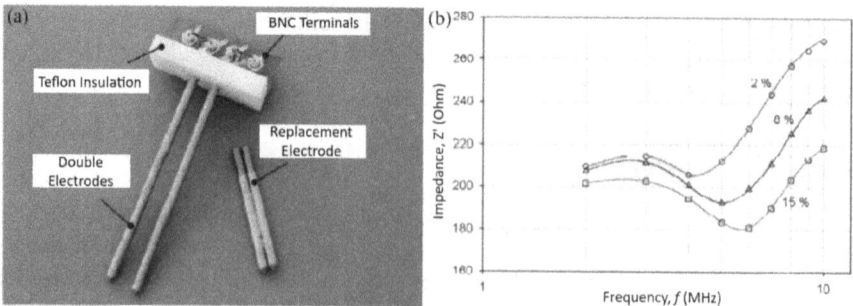

FIGURE 5.11   Soil moisture sensor: (a) double brass electrode sensor design; (b) soil impedance as a function of frequency at different moisture levels of soil. [Reprinted with permission from Umar and Setiadi (2015). Copyright (2015) AIP].

transport occurred after 1% change in volume (Figure 5.12c; Clemens et al. 2021). Further, flat sensor arrangements were found to be more effective against deformations and insufficient compactions. Out of the two types of flat sensors used (Figure 5.11b), flat coil arrangement was used for the analysis due to more defined impedance changes and effect of shifting resonance frequency.

Real-time wireless detection of water content and conductivity of soil has also been developed. Square excitation signal of known amplitude is applied, which is determined by a peak detector and transformed to a DC signal response proportional to the amplitude of the AC response of the soil electrodes. The response is then translated to soil moisture content from the measured permittivity. A low RF band and robust EIS-based underground soil sensor, capable of wireless sensing through a diplexer, was developed. The setup was developed on a PCB platform and consisted of a sensing electrode that also acted as the antenna, working at a transmission frequency of 433 MHz. The impedance value could be determined from the ratio of reflected (V_r) and incident signals (V_i) using Eq. (5.1) for a frequency range of 3–30 MHz:

$$\frac{V_r}{V_i} = \frac{|Z| - Z_o}{|Z| + Z_o} \mu \tag{5.1}$$

where $|Z|$ is the soil impedance to be measured and Z_o is the constant impedance of the transmission line linked to the electrode. The complex impedance when dissolved into the real and imaginary parts reflected the conductivity and the moisture content of the soil, respectively.

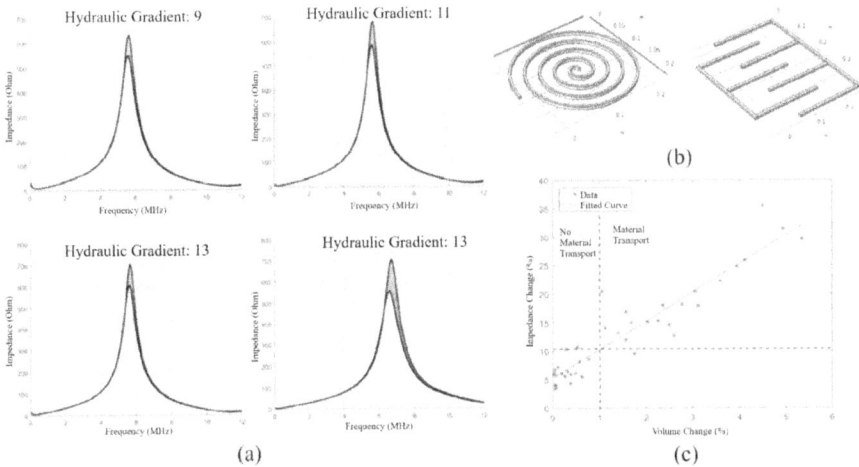

FIGURE 5.12 Soil density measurement through impedance spectroscopy: (a) impedance changes at different hydraulic gradients; (b) two different sensors used in the experiment; (c) impedance change as a function of volume change along with a fitted curve. [Reprinted with permission from Clemens et al. (2021). Copyright (2021) MDPI].

The impedance-based sensor functioned with an accuracy of more than 90%. This research was then further extended to dielectric mixing models to study soil salinity as well (Tetyuev and Kanoun 2006; Pandey et al. 2014; Rusu et al. 2019).

5.5 MONITORING OF PATHOGENIC BACTERIA THROUGH EIS

Global public health is being burdened by infectious diseases and their rising antimicrobial drug resistance. Issues such as food-borne pathogenic outbreaks due to the intake of contaminated food are still a major concern. Initiation of antimicrobial therapy without a firm diagnosis on the basis of empirical evidence leads to the overuse of broad-spectrum antibiotics and the consequent growth in antimicrobial resistance.

Enzyme-based biosensors (ELISA) have attracted public interest due to their reliability, fast response, high sensitivity, and selectivity. However, the signal detected in an ELISA measurement probes the reaction between the enzyme-linked small molecule and its substrate; therefore, it is not a direct measurement of antibody–analyte binding. As a result, nonspecific signals may be generated.

Conventional detection techniques yield accurate and reliable results. However, traditional bacterial detection techniques such as culture method are time-consuming and require higher sample volumes. Fluorescence-based methods are expensive and require trained personnel (Brosel-Oliu et al. 2015; Bhatt et al. 2017; Bhatt, Kant, and Bhattacharya 2019). However, advanced fabrication techniques, such as surface modifications by laser, plasma processing, paper microfluidics, and additive manufacturing (Kumar et al. 2019; Sundriyal, Pandey, and Bhattacharya 2020; Pandey, Rashiku, and Bhattacharya 2021; Pandey et al. 2022; Sarma Choudhury, Pandey, and Bhattacharya 2022) and sample collection techniques integrated with wearables are explored to minimize the time consumption and ease of detection (Pandey, Shahare et al. 2019; Pandey, Tatiya, and Bhattacharya 2021). These technologies require newer material and composites exploration to be used as base substrate as well as sensing part and are in the developmental phases and thus require rigorous testing and iteration before commercialization (Tatiya, Pandey, and Bhattacharya 2020).

As most of the bacterial cells are electrically charged, a change in electrical impedance is observed on the capture of bacteria on the electrode surface. Also, the immobilization of microbial particles on the electrode surface may also reduce the open surface area, which in turn may affect the net charge transfer between electrodes and the solution. This concept has been utilized for the development of EIS-based biosensors for label-free and rapid detection of pathogenic bacteria. Some important biosensing applications of bacterial detection through the EIS technique are summarized in Table 5.4.

5.5.1 DETECTION OF METABOLITES PRODUCED BY BACTERIAL CELLS

Metabolites are tiny ionic intermediates or end products of the metabolic process. As the large molecules, such as proteins, nucleic acids, and polysaccharides, are broken down into corresponding smaller units, such as amino acids, nucleotides, fatty acids, and sugars, by bacteria using catabolic pathways, there is a correlated change in the growth medium's ionic composition due to the formation of tiny, charged ionic metabolites, such as glutamic acid, ethanol, lactic acid, acetic acid, CO_2, ammonia, glycerol, and urea. This makes it possible to quantify these modifications and determine the correlations between the bacterial concentration and growth.

Impedance-based growth correlation schemes have been on a rise over the past few decades due to advantageous aspects such as simplicity of technique, sensitivity, and cost-effectiveness. A minimum concentration of microbes in the media is required to produce measurable modifications in the impedance that may be monitored. A threshold concentration, also known as LOD, is the quantity of bacterial cells that must be present. A low LOD is always desirable. Researchers have realized LOD in the range of 10^4 to 10^8 cell mL^{-1}. Both direct and indirect methods can be used to measure metabolic reactions. Electrodes submerged in the growth medium can evaluate changes in microbial activity occurring across the majority of the growth medium directly.

However, an indirect method identifies CO_2 generated by bacteria (Felice et al. 1999). Since bacteria metabolic products are majorly of acidic nature, pH changes (Ur and Brown 1975) of the growth media also take place along with the conductivity changes associated with the formation of ionic metabolites. However, to improve sensitivity and effectiveness of device, miniaturization of the detection systems is necessary. Rather than using the pH sensor and the reference electrode, interdigitated microelectrodes (Butler et al. 2019) are utilized for the detection of microbial growth. Figure 5.13 shows the experimental steps followed by Butler et al. to study the non-faradaic impedance response related to *Escherichia coli* metabolism in minute sample volume (1 µL). Agar layer was used to protect electrodes from potential degrading non-faradaic reactions. Using a frequency range of 1–10 MHz to control bacterial growth, *E. coli* was quantified, showing that the technique is capable of detecting quantities in the limit of 10^4 to 10^6 cells mL^{-1} in only 60 min. Detection in concentrations as low as 7 cells mL^{-1} has been achieved, however, with a longer detection time (Liu et al. 2015).

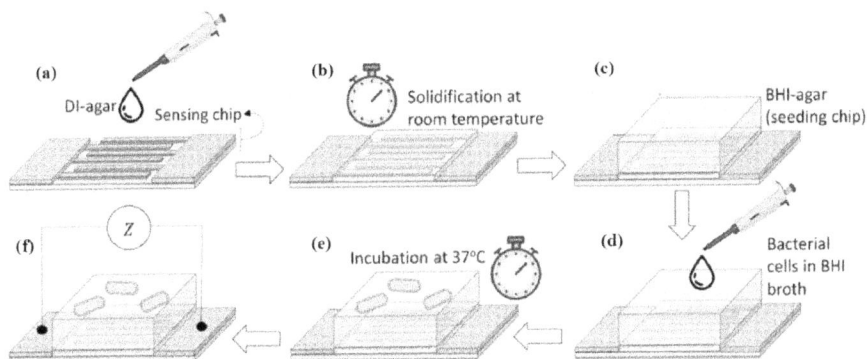

FIGURE 5.13 Schematic representation of the experimental method of bacterial metabolism monitoring through impedance response. [Reprinted with permission from Butler et al. (2019). Copyright (2019) Elsevier].

5.5.2   DETECTION THROUGH THE IMMOBILIZATION OF
BIORECOGNITION ELEMENTS ON THE ELECTRODE SURFACE

Bacterial capture on the electrode surface is mostly executed through the immobilization of biorecognition elements, such as antibodies, aptamers, peptides, and
lectins on the electrode surface. Interaction between biorecognition element and analyte leads to specific impedance responses that can be then utilized to estimate the
bacterial concentration. Sensitive, low-cost, and portable impedance-based biosensors have been a boon in the detection of food-borne pathogens and their toxins. The
electrodes can be functionalized with antibodies specific to target analytes.

Functionalized gold IDEs connected with a network of microfluidics was used to
detect *Listeria monocytogenes* in milk samples (Figure 5.14a; Chiriacò et al. 2018).
The microfluidic arrangement consisted of four peripheral inlets and one central inlet
intended to allow separate functionalization of the IDEs. Varying the concentrations
of *L. monocytogenes* from 2.2×10^1 to 2.2×10^3 cfu mL^{-1}, the impedance values
were observed to be increased as shown in Figure 5.14b. However, the stability of
antibodies on the electrode surface remains a matter of concern. Applications of
impedance spectroscopy has also been explored widely in the fabrication of immunosensors. The upper edge of such sensors has been in the specificity and sensitivity
of the detection of antigen–antibody interactions. Faradaic impedance is measured
when the functionalized electrodes are treated with target analyte. While monitoring
impedance of *Salmonella* for an ideal incubation period of 35 min, lowest detectable
concentration was observed to be 3 cfu mL^{-1}.

The major disadvantage of using aptamers is that DNA/RNA oligonucleotides are
susceptible to nuclease degradation. The immobilization of antibodies and insulating
bacterial cells increase the electron transfer resistance, allowing for the establishment
of a relationship between the electron transfer resistance and bacterial concentration.
Such a principle has been used to design functional EIS-based immunosensors (Yang
et al. 2004; Mantzila et al. 2008; Barreiros dos Santos et al. 2013) to identify bacterial species such as *Salmonella enterica serovar typhimurium* and *E. coli* in milk and
food samples and LOD as low as 2 cfu mL^{-1} has been obtained.

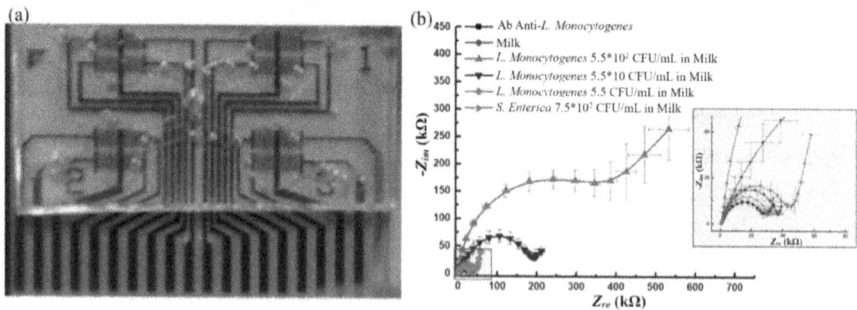

FIGURE 5.14   Fabrication of EIS biosensor for food-borne pathogenic bacteria detection: (a)
system assembly with IDEs aligned with microfluidic network at top; (b) impedance response
at different concentrations of bacteria in milk. [Reprinted with permission from Chiriacò
et al. (2018). Copyright (2018) Elsevier].

However, immunosensors bear the drawback of reusability. Single-stranded oligonucleotides called aptamers are also used for bacterial immobilization. Aptamer-based biosensors are popular for their specificity. Functionalization of electrodes with graphene oxide and gold nanoparticles enhance sensitivity and biocompatibility (Ma et al. 2014).

There are wide applications of impedance-based biosensors in the detection of pathogenic bacteria in food, water, and air and detailed reviewing has been carried out by different researchers on the same (Brosel-Oliu et al. 2015; Furst and Francis 2018; Magar et al. 2021).

5.5.3 Detection through nonspecific binding on electrode surface

Biorecognition element-based immobilization of bacterial cells has the limitation of repeatability and reproducibility as it is challenging to maintain the element density on the sensor surface. Further, the available functional sensor surface area reduces with the immobilization of biomolecules. The use of microbeads and nanoparticles has been proposed as a solution to this problem. Polystyrene bead was used as nonbiological reference on the surface to detect and enumerate *E. coli*/*S. aureus* from solid constituents in tap water and saline solution samples (Clausen et al. 2018). While bacteria concentration was continuously varied, the beads were kept constant at 2×10^6 mL^{-1}.

Peak differential current signal during a transition was used as the characterization technique of the microorganisms. Impedance spectra recorded from EIS measurements in the frequency range of 200 kHz to 7 MHz showed proportional relation to the enumeration of concentration done using particle transition time between electrodes. Gram-positive and Gram-negative bacteria can also be differentiated using

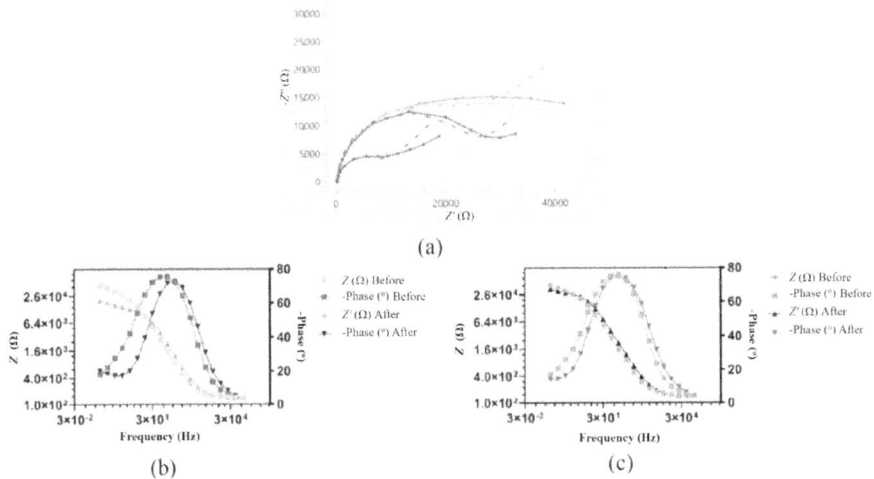

FIGURE 5.15 A rapid label-free pathogenic bacterial detection technique: (a) Nyquist plots for *S. carnosus* and *E. coli* and Bode plots for (b) *S. Carnosus*; (c) *E. coli*. [Reprinted with permission from Schulze et al. (2021). Copyright (2021) Elsevier].

TABLE 5.4

Some important biosensing applications in bacterial detection through electrochemical impedance spectroscopy technique

Target organism	Sample	Electrode	LOD	Ref
E. coli/S. aureus	Tap and saline water	Gold electrode	522 mL^{-1}	Clausen et al. (2018)
Gram-positive (S. epidermidis and Bacillus sp.) and Gram-negative (P. aeruginosa and Cobetia sp.)	Saline suspension	—	1 µg µL^{-1}	Gnaim et al. (2020)
S. Carnosus/E. coli	River and drinking water	Functionalized gold electrodes	—	Schulze et al. (2021)
Bacteria	Saline water	Coplanar electrodes on PCB	—	Bouzid et al. (2022)
Listeria monocytogenes	Milk	Functionalized gold IDEs	5.5 cfu mL^{-1}	Chiriacò et al. (2018)
S. typhimurium	Milk	Functionalized gold electrodes	<10^5 cfu mL^{-1}	Mantzila, Maipa, and Prodromidis (2008)
E. coli	Deionized water	IDA microelectrodes	10^6 cfu mL^{-1}	Yang, Li, and Erf (2004)
E. coli	—	Gold disc electrode	2 cfu mL^{-1}	Santos et al. (2017)

this technique. In order to include the simplicity of assay and rapid diagnosis, label-free EIS-based bacterial detection methods are considered.

Vancomycin-modified highly branched polymers are sometimes used as a new set of nonbiological binder molecules to functionalize screen-printed gold electrodes. Bode and Nyquist plots (Figure 5.15) are then used to characterize bacterial concentration. Incubation of *Staphylococcus carnosus* as well as *E. coli* on functionalized electrodes led to a reduction in charge transfer resistance (R_{ct}) (Schulze et al. 2021). These bacterial detection and phase differentiating EIS-based devices have been proved to work with a maximum error of 7% and 3% in impedance magnitude and phase, respectively. Quadrature excitation signal can be used for the phase-sensitive detectors to reduce power consumption and device complexity (Bouzid et al. 2022). Biofunctionalized microbeads and magnetic nanoparticle–antibody conjugates are other alternatives to the direct immobilization of biomolecules on the sensor surface (Varshney and Li 2007).

5.6　CONCLUSION AND FUTURE PERSPECTIVES

EIS-based sensors have found large applications in the sensing, detection, and monitoring of various toxic gases, pollutants, humidity, and pathogenic substances in the environment although selectivity is one of the major challenges faced in EIS-based sensors as real samples may contain nontarget molecules in higher levels than the target analyte. The potential of bacterial detection without the direct immobilization of biomolecules needs to be explored further since these biosensors show a lack of reusability as it is difficult and challenging to maintain element density on the sensor's surface. In real specimens where the proportion of the target analyte may be very low, small molecules alone typically produce very little recordable responses, which can be exceedingly challenging to measure. One way to enhance the identification of small molecules is to tag them to a bigger carrier molecule by using a functional group, such as a protein, or to fluorescent nanoparticles. Biorecognition element-mediated analyte capture is one of the most utilized biosensing techniques. However, the stability of biomolecules is a matter of concern as the majority of the biomolecules tend to lose stability rapidly.

A deeper inspection into miniaturization of EIS-based sensors is required to find solutions to challenges such as the requirement of higher SN ratio due to the scaling down of the signal with the sensor size and the difficulty in scaling down associated electronics. Wearable sensing techniques, which are the current essence elements need to be explored more widely to ease sample collection.

REFERENCES

Azadbakht, Azadeh, Mahmoud Roushani, Amir Reza Abbasi, and Zohreh Derikvand. 2016. "A Novel Impedimetric Aptasensor, Based on Functionalized Carbon Nanotubes and Prussian Blue as Labels." *Analytical Biochemistry* 512: 58–69. https://doi.org/10.1016/j.ab.2016.08.006.

Balasubramani, V., S. Sureshkumar, T. Subba Rao, and T. M. Sridhar. 2019. "Impedance Spectroscopy-Based Reduced Graphene Oxide-Incorporated ZnO Composite Sensor for H_2S Investigations." *ACS Omega* 4 (6): 9976–82. https://doi.org/10.1021/acsomega.9b00754.

Balasubramani, V., S. Sureshkumar, T. Subbarao, T. M. Sridhar, and R. Sasikumar. 2019. "Development of 2D SnO_2 /RGO Nano-Composites for H 2 S Gas Sensor Using Electrochemical Impedance Spectroscopy at Room Temperature." *Sensor Letters* 17 (3): 237–44. https://doi.org/10.1166/sl.2019.4074.

Barreiros dos Santos, M., J. P. Agusil, B. Prieto-Simón, C. Sporer, V. Teixeira, and J. Samitier. 2013. "Highly Sensitive Detection of Pathogen Escherichia Coli O157: H7 by Electrochemical Impedance Spectroscopy." *Biosensors and Bioelectronics* 45 (1): 174–80. https://doi.org/10.1016/j.bios.2013.01.009.

Bauskar, Dipak, B. B. Kale, and Pradip Patil. 2012. "Synthesis and Humidity Sensing Properties of ZnSnO3 Cubic Crystallites." *Sensors and Actuators B: Chemical* 161 (1): 396–400. https://doi.org/10.1016/j.snb.2011.10.050.

Beer, D. J. De, and T. H. Joubert. 2019. "Impedance Spectroscopy for Determination of Total Dissolved Solids in Aqueous Solutions of Sodium Chloride and Magnesium Sulphate." *Proceedings of IEEE Sensors* 2019-October. https://doi.org/10.1109/SENSORS43011.2019.8956627.

Bhatt, Geeta, Rishi Kant, Keerti Mishra, Kuldeep Yadav, Deepak Singh, Ramanathan Gurunath, and Shantanu Bhattacharya. 2017. "Impact of Surface Roughness on Dielectrophoretically Assisted Concentration of Microorganisms over PCB Based Platforms." *Biomedical Microdevices* 19 (2): 28. https://doi.org/10.1007/s10544-017-0172-5.

Bhatt, Geeta, Rishi Kant, and Shantanu Bhattacharya. 2019. "Enhanced Fluorescence-Based Detection of Vibrio Cells over Nanoporous Silica Substrate." In *Lecture Notes in Mechanical Engineering*, 1–9. Pleiades Publishing. https://doi.org/10.1007/978-981-13-6412-9_1.

Biju, Kuyyadi P., and Mahaveer K. Jain. 2007. "Effect of Polyethylene Glycol Additive in Sol on the Humidity Sensing Properties of a TiO 2 Thin Film." *Measurement Science and Technology* 18 (9): 2991–96. https://doi.org/10.1088/0957-0233/18/9/033.

Boumya, W., F. Laghrib, S. Lahrich, A. Farahi, M. Achak, M. Bakasse, and M. A. El Mhammedi. 2017. "Electrochemical Impedance Spectroscopy Measurements for Determination of Derivatized Aldehydes in Several Matrices." *Heliyon* 3 (10): e00392. https://doi.org/10.1016/j.heliyon.2017.e00392.

Bouzid, Karim, Partha Sarati Das, Denis Boudreau, Sandro Carrara, and Benoit Gosselin. 2022. "Portable Multi-Frequency Impedance-Sensing Device for Bacteria Classification in a Flowing Liquid." In *2022 20th IEEE Interregional Newcas Conference (Newcas)*, 514–18. https://doi.org/10.1109/newcas52662.2022.9841999.

Brosel-Oliu, Sergi, Naroa Uria, Natalia Abramova, and Andrey Bratov. 2015. "Impedimetric Sensors for Bacteria Detection." *Biosensors - Micro and Nanoscale Applications* (September 24): 257–88. https://doi.org/10.5772/60741.

Butler, Derrick, Nishit Goel, Lindsey Goodnight, Srinivas Tadigadapa, and Aida Ebrahimi. 2019. "Detection of Bacterial Metabolism in Lag-Phase Using Impedance Spectroscopy of Agar-Integrated 3D Microelectrodes." *Biosensors and Bioelectronics* 129 (September 2018): 269–76. https://doi.org/10.1016/j.bios.2018.09.057.

Chauhan, Pankaj Singh, Mohit Pandey, and Shantanu Bhattacharya. 2019. "Paper Based Sensors for Environmental Monitoring." In *Paper Microfluidics: Theory and Applications*, edited by Shantanu Bhattacharya, Sanjay Kumar, and Avinash K. Agarwal, 165–81. Singapore: Springer Singapore. https://doi.org/10.1007/978-981-15-0489-1_10.

Chiriacò, Maria Serena, Ilaria Parlangeli, Fausto Sirsi, Palmiro Poltronieri, and Elisabetta Primiceri. 2018. "Impedance Sensing Platform for Detection of the Food Pathogen Listeria Monocytogenes." *Electronics (Switzerland)* 7 (12): 1–11. https://doi.org/10.3390/electronics7120347.

Cho, Myung-Yeon, Sunghoon Kim, Ik-Soo Kim, Eun-Seong Kim, Zhi-Ji Wang, Nam-Young Kim, Sang-Wook Kim, and Jong-Min Oh. 2020. "Perovskite-Induced Ultrasensitive and Highly Stable Humidity Sensor Systems Prepared by Aerosol Deposition at Room Temperature." *Advanced Functional Materials* 30 (3): 1907449. https://doi.org/10.1002/adfm.201907449.

Clausen, Casper Hyttel, Maria Dimaki, Christian Vinther Bertelsen, Gustav Erik Skands, Romen Rodriguez-Trujillo, Joachim Dahl Thomsen, and Winnie E. Svendsen. 2018. "Bacteria Detection and Differentiation Using Impedance Flow Cytometry." *Sensors (Switzerland)* 18(10): 3496. https://doi.org/10.3390/s18103496.

Clemens, Christoph, Mario Radschun, Annette Jobst, Jörg Himmel, and Olfa Kanoun. 2021. "Detection of Density Changes in Soils with Impedance Spectroscopy." *Applied Sciences (Switzerland)* 11 (4): 1–14. https://doi.org/10.3390/app11041568.

Contreras Jiménez, Gastón, Shimaa Eissa, Andy Ng, Hani Alhadrami, Mohammed Zourob, and Mohamed Siaj. 2015. "Aptamer-Based Label-Free Impedimetric Biosensor for Detection of Progesterone." *Analytical Chemistry* 87 (2): 1075–82. https://doi.org/10.1021/ac503639s.

Day, C., S. Søpstad, H. Ma, C. Jiang, A. Nathan, S. R. Elliott, F. E. Karet Frankl, and T. Hutter. 2018. "Impedance-Based Sensor for Potassium Ions." *Analytica Chimica Acta* 1034: 39–45. https://doi.org/10.1016/j.aca.2018.06.044.

Disha, Disha, Poonam Kumari, Manoj Kumar Nayak, and Parveen Kumar. 2021. "Development of an Improved Immunosensor for Fast Electrochemical Determination of Progesterone." *ECS Meeting Abstracts* MA2021-01 (55): 1414–14. https://doi.org/10.1149/MA2021-01551414mtgabs.

Dong, Hui, Xianming Zhu, Xiuzi Jiang, Li Chen, and Qian Feng Gao. 2021. "Structural Characteristics of Soil-Rock Mixtures Based on Electrochemical Impedance Spectroscopy." *Catena* 207 (August): 105579. https://doi.org/10.1016/j.catena.2021.105579.

Dorledo de Faria, Ricardo Adriano, Luiz Guilherme Dias Heneine, Tulio Matencio, and Younès Messaddeq. 2019. "Faradaic and Non-Faradaic Electrochemical Impedance Spectroscopy as Transduction Techniques for Sensing Applications." *International Journal of Biosensors & Bioelectronics* 5 (1): 29–31. https://doi.org/10.15406/ijbsbe.2019.05.00148.

Farahani, Hamid, Rahman Wagiran, and Mohd Hamidon. 2014. "Humidity Sensors Principle, Mechanism, and Fabrication Technologies: A Comprehensive Review." *Sensors* 14 (5): 7881–939. https://doi.org/10.3390/s140507881.

Felice, Carmelo J., Rossana E. Madrid, Juan M. Olivera, Viviana I. Rotger, and Max E. Valentinuzzi. 1999. "Impedance Microbiology: Quantification of Bacterial Content in Milk by Means of Capacitance Growth Curves." *Journal of Microbiological Methods* 35 (1): 37–42. https://doi.org/10.1016/S0167-7012(98)00098-0.

Furst, Ariel L., and Matthew B. Francis. 2018. "Impedance-Based Detection of Bacteria." https://doi.org/10.1021/acs.chemrev.8b00381.

Gnaim, Rima, Alexander Golberg, Julia Sheviryov, Boris Rubinsky, and César A. González. 2020. "Detection and Differentiation of Bacteria by Electrical Bioimpedance Spectroscopy." *BioTechniques* 69 (1): 27–36. https://doi.org/10.2144/BTN-2019-0080.

González-Teruel, Juan D., Roque Torres-Sánchez, Pedro J. Blaya-Ros, Ana B. Toledo-Moreo, Manuel Jiménez-Buendía, and Fulgencio Soto-Valles. 2019. "Design and Calibration of a Low-Cost SDI-12 Soil Moisture Sensor." *Sensors (Switzerland)* 19 (3): 491. https://doi.org/10.3390/s19030491.

Gruden, R., and O. Kanoun. 2013. "A8.1- Water Quality Assessment by Combining Impedance Spectroscopy Measurement with Cyclic Voltammetry." *Proceedings SENSOR 2013* May 14: 164–69. https://doi.org/10.5162/sensor2013/a8.1.

Han, Peng ju, Ya feng Zhang, Frank Y. Chen, and Xiao hong Bai. 2015. "Interpretation of Electrochemical Impedance Spectroscopy (EIS) Circuit Model for Soils." *Journal of Central South University* 22 (11): 4318–28. https://doi.org/10.1007/s11771-015-2980-1.

Jacobs, Michael, Vinay J. Nagaraj, Tim Mertz, Anjan Panneer Selvam, Thi Ngo, and Shalini Prasad. 2013. "An Electrochemical Sensor for the Detection of Antibiotic Contaminants in Water." *Analytical Methods* 5 (17): 4325–29. https://doi.org/10.1039/c3ay40994e.

Jiang, Ding, Xiaojiao Du, Qian Liu, Lei Zhou, Liming Dai, Jing Qian, and Kun Wang. 2015. "Silver Nanoparticles Anchored on Nitrogen-Doped Graphene as a Novel Electrochemical Biosensing Platform with Enhanced Sensitivity for Aptamer-Based Pesticide Assay." *Analyst* 140 (18): 6404–11. https://doi.org/10.1039/c5an01084e.

Kant, Rishi, and Shantanu Bhattacharya. 2018. "Sensors for air monitoring." In *Environmental, Chemical and Medical Sensor*, 9–30, Springer, https://doi.org/10.1007/978-981-10-7751-7_2.

Kashyap, Bhuwan, and Ratnesh Kumar. 2021. "Sensing Methodologies in Agriculture for Soil Moisture and Nutrient Monitoring." *IEEE Access* 9: 14095–121. https://doi.org/10.1109/ACCESS.2021.3052478.

Kivirand, Kairi, Mart Min, and Toonika Rinken. 2019. "Challenges and Applications of Impedance-Based Biosensors in Water Analysis." *Biosensors for Environmental Monitoring* (October 2): 2–4. https://doi.org/10.5772/intechopen.89334.

Kumar, Sanjay, Pulak Bhushan, Mohit Pandey, and Shantanu Bhattacharya. 2019. "Additive Manufacturing as an Emerging Technology for Fabrication of Microelectromechanical Systems (MEMS)." *Journal of Micromanufacturing* 2 (2): 175–97. https://doi.org/10.1177/2516598419843688.

Liu, Fangmeng, Yinglin Wang, Bin Wang, Xue Yang, Qingji Wang, Xishuang Liang, Peng Sun, Xiaohong Chuai, Yue Wang, and Geyu Lu. 2017. "Stabilized Zirconia-Based Mixed Potential Type Sensors Utilizing MnNb2O6 Sensing Electrode for Detection of Low-Concentration SO2." *Sensors and Actuators B: Chemical* 238 (January): 1024–31. https://doi.org/10.1016/j.snb.2016.07.145.

Liu, Jen Tsai, Kalpana Settu, Jang Zern Tsai, and Ching Jung Chen. 2015. "Impedance Sensor for Rapid Enumeration of E. Coli in Milk Samples." *Electrochimica Acta* 182: 89–95. https://doi.org/10.1016/j.electacta.2015.09.029.

Ma, Xiaoyuan, Yihui Jiang, Fei Jia, Ye Yu, Jie Chen, and Zhouping Wang. 2014. "An Aptamer-Based Electrochemical Biosensor for the Detection of Salmonella." *Journal of Microbiological Methods* 98 (1): 94–98. https://doi.org/10.1016/j.mimet.2014.01.003.

Magar, Hend S., Rabeay Y. A. Hassan, and Ashok Mulchandani. 2021. "Electrochemical Impedance Spectroscopy (Eis): Principles, Construction, and Biosensing Applications." *Sensors* 21 (19): 6578. https://doi.org/10.3390/s21196578.

Magro, Cátia, Tiago Moura, Paulo A. Ribeiro, Maria Raposo, and Susana Sério. 2021. "Smart Sensing for Antibiotic Monitoring in Mineral and Surface Water: Development of an Electronic Tongue Device." *Chemistry Proceedings* 5 (1): 58. https://doi.org/10.3390/csac2021-10606.

Mantzila, Aikaterini G., Vassiliki Maipa, and Mamas I. Prodromidis. 2008. "Development of a Faradic Impedimetric Immunosensor for the Detection of Salmonella Typhimurium in Milk." *Analytical Chemistry* 80 (4): 1169–75. https://doi.org/10.1021/ac071570l.

Pandey, Gunjan, Ratnesh Kumar, and Robert J. Weber. 2013. "Determination of Soil Ionic Concentration Using Impedance Spectroscopy." *Sensing Technologies for Global Health, Military Medicine, and Environmental Monitoring III* 8723: 872317. https://doi.org/10.1117/12.2021969.

———. 2014. "A Low RF-Band Impedance Spectroscopy Based Sensor for in Situ, Wireless Soil Sensing." *IEEE Sensors Journal* 14 (6): 1997–2005. https://doi.org/10.1109/JSEN.2014.2307001.

Pandey, Mohit, Mohammed Rashiku, and Shantanu Bhattacharya. 2021. "Chapter 10- Recent Progress in the Development of Printed Electronic Devices." In *Chemical Solution Synthesis for Materials Design and Thin Film Device Applications*, edited by Soumen Das and Sandip Dhara, 349–68. Elsevier. https://doi.org/https://doi.org/10.1016/B978-0-12-819718-9.00008-X.

Pandey, Mohit, Krutika Shahare, Mahima Srivastava, and Shantanu Bhattacharya. 2019. "Paper-Based Devices for Wearable Diagnostic Applications." In *Paper Microfluidics: Theory and Applications*, edited by Shantanu Bhattacharya, Sanjay Kumar, and Avinash K Agarwal, 193–208. Singapore: Springer Singapore. https://doi.org/10.1007/978-981-15-0489-1_12.

Pandey, Mohit, Mahima Srivastava, Krutika Shahare, and Shantanu Bhattacharya. 2019. "Paper Microfluidic-Based Devices for Infectious Disease Diagnostics." In *Paper Microfluidics: Theory and Applications*, edited by Shantanu Bhattacharya, Sanjay Kumar, and Avinash K. Agarwal, 209–25. Singapore: Springer Singapore. https://doi.org/10.1007/978-981-15-0489-1_13.

Pandey, Mohit, Poonam Sundriyal, Shreyansh Tatiya, and Shantanu Bhattacharya. 2022. "Polymer-Based Electrolytes for Solid-State Batteries: Current Status and Future Challenges in Emerging Applications." *Trends in Applications of Polymers and Polymer Composites*: 5–22. https://doi.org/10.1063/9780735424555_005.

Pandey, Mohit, Shreyansh Tatiya, and Shantanu Bhattacharya. 2021. "Design and Development of MEMS-Based Sensors for Wearable Diagnostic Applications." In *MEMS Applications in Biology and Healthcare, AIP*: 10–34. https://doi.org/10.1063/9780735423954_010.

Pichetsurnthorn, Pie, Krishna Vattipalli, and Shalini Prasad. 2012. "Nanoporous Impedemetric Biosensor for Detection of Trace Atrazine from Water Samples." *Biosensors and Bioelectronics* 32 (1): 155–62. https://doi.org/10.1016/j.bios.2011.11.055.

Ren, Kuankuan, Le Huang, Shizhong Yue, Shudi Lu, Kong Liu, Muhammad Azam, Zhijie Wang, Zhongming Wei, Shengchun Qu, and Zhanguo Wang. 2017. "Turning a Disadvantage into an Advantage: Synthesizing High-Quality Organometallic Halide Perovskite Nanosheet Arrays for Humidity Sensors." *Journal of Materials Chemistry C* 5 (10): 2504–08. https://doi.org/10.1039/C6TC05165K.

Rusu, Cristina, Anatol Krozer, Christer Johansson, Fredrik Ahrentorp, Torbjörn Pettersson, Christian Jonasson, and John Rösevall et al. 2019. "Miniaturized Wireless Water Content and Conductivity Soil Sensor System." *Computers and Electronics in Agriculture* 167 (October): 105076. https://doi.org/10.1016/j.compag.2019.105076.

Santos, M. C., A. G. C. Bianchi, D. M. Ushizima, F. J. Pavinatto, and R. F. Bianchi. 2017. "Ammonia Gas Sensor Based on the Frequency-Dependent Impedance Characteristics of Ultrathin Polyaniline Films." *Sensors and Actuators A: Physical* 253 (January): 156–64. https://doi.org/10.1016/j.sna.2016.08.005.

Sarma Choudhury, Sagnik, Mohit Pandey, and Shantanu Bhattacharya. 2022. "Recent Developments in Surface Modification of PEEK Polymer for Industrial Applications: A Critical Review." *Reviews of Adhesion and Adhesives* 9: 401–33. https://doi.org/10.47750/RAA/9.3.03.

Schulze, Holger, Harry Wilson, Ines Cara, Steven Carter, Edward N. Dyson, Ravikrishnan Elangovan, Stephen Rimmer, and Till T. Bachmann. 2021. "Label-Free Electrochemical Sensor for Rapid Bacterial Pathogen Detection Using Vancomycin-Modified Highly Branched Polymers." *Sensors* 21 (5): 1–14. https://doi.org/10.3390/s21051872.

Scott, Dane W., and Yahya Alseiha. 2017. "Determining Detection Limits of Aqueous Anions Using Electrochemical Impedance Spectroscopy." *Journal of Analytical Science and Technology* 8 (1): 1–5. https://doi.org/10.1186/s40543-017-0126-9.

Shuquan, Peng, Wang Fan, and Fan Ling. 2019. "Study on Electrochemical Impedance Response of Sulfate Saline Soil." *International Journal of Electrochemical Science* 14 (9): 8611–23. https://doi.org/10.20964/2019.09.30.

Singh, Harmandeep, and Arvind Rehalia. 2016. "Case Study : Bhopal Gas Tragedy." *International Journal of Advanced Engineering Research and Applications* 2 (6): 367–70.

Sivakumar, Rajamanickam, and Nae Yoon Lee. 2022. "Recent Advances in Airborne Pathogen Detection Using Optical and Electrochemical Biosensors." *Analytica Chimica Acta* 1234: 340297. https://doi.org/10.1016/j.aca.2022.340297.

Strong, Madison E., Jeffrey R. Richards, Manuel Torres, Connor M. Beck, and Jeffrey T. La Belle. 2021. "Faradaic Electrochemical Impedance Spectroscopy for Enhanced Analyte Detection in Diagnostics." *Biosensors and Bioelectronics* 177 (May 2020): 112949. https://doi.org/10.1016/j.bios.2020.112949.

Sundriyal, Poonam, Mohit Pandey, and Shantanu Bhattacharya. 2020. "Plasma-Assisted Surface Alteration of Industrial Polymers for Improved Adhesive Bonding." *International Journal of Adhesion and Adhesives* 101: 102626. https://doi.org/https://doi.org/10.1016/j.ijadhadh.2020.102626.

Sureshkumar, S, B. Venkatachalapathy, and T. M. Sridhar. 2019. "Enhanced H 2 S Gas Sensing Properties of Mn Doped ZnO Nanoparticles—an Impedance Spectroscopic Investigation." *Materials Research Express* 6 (7): 075009. https://doi.org/10.1088/2053-1591/ab0eef.

Tatiya, Shreyansh, Mohit Pandey, and Shantanu Bhattacharya. 2020. "Nanoparticles Containing Boron and Its Compounds—Synthesis and Applications: A Review." *Journal of Micromanufacturing* 3 (2): 159–73. https://doi.org/10.1177/2516598420965319.

Tetyuev, Andrey, and Olfa Kanoun. 2006. "Method of Soil Moisture Measurement by Impedance Spectroscopy with Soil Type Recognition for In-Situ Applications." *Technisches Messen* 73 (7–8): 404–12. https://doi.org/10.1524/teme.2006.73.7-8.404.

Tomer, Vijay K., and Surender Duhan. 2015. "In-Situ Synthesis of SnO2/SBA-15 Hybrid Nanocomposite as Highly Efficient Humidity Sensor." *Sensors and Actuators B: Chemical* 212 (June): 517–25. https://doi.org/10.1016/j.snb.2015.02.054.

Umar, Lazuardi, and Rahmondia N. Setiadi. 2015. "Low Cost Soil Sensor Based on Impedance Spectroscopy for In-Situ Measurement." *AIP Conference Proceedings* 1656 (June 2016). https://doi.org/10.1063/1.4917112.

Ur, A., and D. F.J. Brown. 1975. "Impedance Monitoring of Bacterial Activity." *Journal of Medical Microbiology* 8 (1): 19–28. https://doi.org/10.1099/00222615-8-1-19.

Varshney, Madhukar, and Yanbin Li. 2007. "Interdigitated Array Microelectrode Based Impedance Biosensor Coupled with Magnetic Nanoparticle-Antibody Conjugates for Detection of Escherichia Coli O157:H7 in Food Samples." *Biosensors and Bioelectronics* 22 (11): 2408–14. https://doi.org/10.1016/j.bios.2006.08.030.

———. 2009. "Interdigitated Array Microelectrodes Based Impedance Biosensors for Detection of Bacterial Cells." *Biosensors and Bioelectronics* 24 (10): 2951–60. https://doi.org/10.1016/j.bios.2008.10.001.

Weng, Zhenhua, Jiajun Qin, Akrajas Ali Umar, Jiao Wang, Xin Zhang, Haoliang Wang, Xiaolei Cui, Xiaoguo Li, Lirong Zheng, and Yiqiang Zhan. 2019. "Lead-Free Cs 2 BiAgBr 6 Double Perovskite-Based Humidity Sensor with Superfast Recovery Time." *Advanced Functional Materials* 29 (24): 1902234. https://doi.org/10.1002/adfm.201902234.

White, Briggs, Enrico Traversa, and Eric Wachsman. 2008. "Investigation of La[Sub 2] CuO[Sub 4]/YSZ/Pt Potentiometric NO[Sub x] Sensors with Electrochemical Impedance Spectroscopy." *Journal of The Electrochemical Society* 155 (1): J11. https://doi.org/10.1149/1.2799768.

Wierzbicka, M., P. Pasierb, and M. Rekas. 2007. "CO_2 Sensor Studied by Impedance Spectroscopy." *Physica B: Condensed Matter* 387 (1–2): 302–12. https://doi.org/10.1016/j.physb.2006.04.020.

Wu, Zongjian, Weiqing Liu, Jing Shi, Baoshuo Han, Datian Li, Xiaobo Xu, and Wenhao Chen. 2022. "Renewable and Fast Response Humidity Sensors Based on Multiple Construction of Water Graftable Molecules Highly Sensitive Surface." *Surfaces and Interfaces* 31 (February): 102035. https://doi.org/10.1016/j.surfin.2022.102035.

Xie, Ruizhen, Yating Xie, Boqiong Li, Pengju Han, Bin He, Baojie Dou, and Xiaohong Bai. 2021. "Electrochemical Impedance Spectroscopy of Sandy Soil Containing Cl-, SO42- and HCO3-." *International Journal of Electrochemical Science* 16: 1–11. https://doi.org/10.20964/2021.12.42.

Yang, Liju, Yanbin Li, and Gisela F. Erf. 2004. "Interdigitated Array Microelectrode-Based Electrochemical Impedance Immunosensor for Detection of Escherichia Coli O157:H7." *Analytical Chemistry* 76 (4): 1107–13. https://doi.org/10.1021/ac0352575.

Yuan, Mengjiao, Yong Zhang, Xuejun Zheng, Bin Jiang, Peiwen Li, and Shuifeng Deng. 2015. "Humidity Sensing Properties of K0.5Na0.5NbO3 Powder Synthesized by Metal Organic Decomposition." *Sensors and Actuators B: Chemical* 209 (March): 252–57. https://doi.org/10.1016/j.snb.2014.11.118.

Zehani, Nedjla, Sergei V. Dzyadevych, Rochdi Kherrat, and Nicole J. Jaffrezic-Renault. 2014. "Sensitive Impedimetric Biosensor for Direct Detection of Diazinon Based on Lipases." *Frontiers in Chemistry* 2 (July): 1–7. https://doi.org/10.3389/fchem.2014.00044.

Zhang, Wei, Changseok Han, Baoping Jia, Christopher Saint, Mallikarjuna Nadagouda, Polycarpos Falaras, Labrini Sygellou, Vasileia Vogiazi, and Dionysios D. Dionysiou. 2017. "A 3D Graphene-Based Biosensor as an Early Microcystin-LR Screening Tool in Sources of Drinking Water Supply." *Electrochimica Acta* 236: 319–27. https://doi.org/10.1016/j.electacta.2017.03.161.

Zhao, Jing, Na Li, Hua Yu, Zheng Wei, Mengzhou Liao, Peng Chen, Shuopei Wang, Dongxia Shi, Qijun Sun, and Guangyu Zhang. 2017. "Highly Sensitive MoS 2 Humidity Sensors Array for Noncontact Sensation." *Advanced Materials* 29 (34): 1702076. https://doi.org/10.1002/adma.201702076.

Zia, Asif I., A. R. Mohd Syaifudin, S. C. Mukhopadhyay, P. L. Yu, I. H. Al-Bahadly, Chinthaka P. Gooneratne, Jrgen Kosel, and Tai Shan Liao. 2013. "Electrochemical Impedance Spectroscopy Based MEMS Sensors for Phthalates Detection in Water and Juices." *Journal of Physics: Conference Series* 439 (1): 012026. https://doi.org/10.1088/1742-6596/439/1/012026.

6 Application of noninvasive impedance spectroscopy techniques

Vinay Kishnani and Ankur Gupta

CONTENTS

6.1 INTRODUCTION

Noninvasive procedures generally refer to medical treatments that do not involve cutting or tearing the skin. Beyond a certain penetration, no indication of an interior bodily cavity is seen; instead, nontraditional techniques like sonography and lithotripsy are used, which employ waves capable of piercing into the human body without injuring the patient or rupturing the skin.

The detection process can be integrated with spectroscopy, a relatively recent subject that studies the simultaneous emission and absorption of electromagnetic waves, which are naturally used in the form of light. As a result, the electromagnetic wave

DOI: 10.1201/9781003358091-8

spectrum is divided into different bands of light. Diffraction grating is used to help with light dispersion, which then aids in the creation of 2D spectra, which are then used to acquire 1D spectra to provide a meaningful collection of data. One method would be to project a range of wavelengths onto the object and then investigate patterns and aberrations using the wavelength that corresponds to the known spectrum.

Impedance and resistance are identical, with the exception that the former is frequently employed in complex circuits that take into consideration resistance, inductance, and capacitance. Electrical impedance is the complete opposite of current flow in a circuit because of the interaction between resistance and reactance. Resistance in the circuit is caused by the collision of the atoms in the charged particles carrying the current, whereas reactance is another barrier to the flow of current caused by the production of magnetic and electric fields because of the movement of electric charge. Impedance is a new word that is created by combining the two effects mentioned above. The impact of reactance disappears if we assume a constant direct current; at this point, impedance equals resistance. It is mathematically equivalent to the maximum current flowing through the circuit divided by the maximum potential difference applied across the circuit. The application of impedance spectrometry is diverse, but, in this chapter, the entire focus will be on noninvasive biosensing applications.

In general, a sensor, a measuring circuit, and an appropriate algorithm are needed to get results by utilizing this approach for detection. Adopting noninvasive technologies for human health detection plays a feasible role because impedance signature is a function of tissue structure, composition, and health status. These days, it finds applications in fields like diabetes estimation, assessment of radiation injury, determination of body composition, cancer, and more.

An electrochemical technique for observing how a system responds to a disturbance at a steady state is called impedance spectrometry. Impedance is assessed using the current's reaction to the imposed alternating voltage's fluctuating frequency across a large range, and it is divided into real and imaginary components (Iwakura, Inoue, and Nohara 2001). In EIS, the sample under test (SUT) is exposed to low-amplitude electrical signals at various frequencies, and the electrical impedance at each frequency point is determined by means of Ohm's law (Figure 6.1).

SUT attached with a linear array of electrodes measures the surface potentials [$V(f)$] generated for a constant current injection [$I(f)$] at the boundary in order to estimate the complex electrical impedance [$Z(f)$] and its phase angle [$\theta(f)$] of the SUT at various frequency points f_i (f_i: $f_1, f_2, f_3, \ldots f_n$). In EIS, a two-electrode approach or a four-electrode method is used to inject a frequency-dependent constant amplitude sinusoidal current [$I(f)$]. The pair of electrodes (surface) that are used for injecting current signal are termed as driving or current electrodes. In 2- and 4-electrode-based EIS procedures, voltage electrodes or sensing electrodes are the terms used to describe the electrodes, which are used to monitor the frequency-dependent ac potential [$V(f_i)$]. As a result, the frequency-dependent electrical bio-impedance [$Z(f_i)$] is discovered in EIS as the SUT's transfer function. To compute [$Z(f_i)$], voltage data [$V(f_i)$] measurement is divided by the applied current [$I(f_i)$] (Barsoukov and Macdonald 2005).

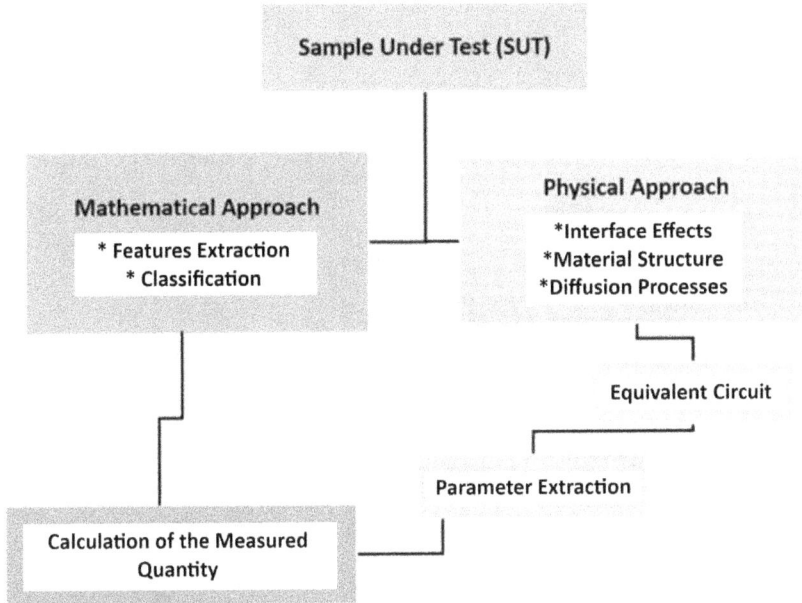

FIGURE 6.1 Layout for extracting data from impedance spectra.

$$Z(f_i) = \frac{V(f_i)}{I(f_i)}$$

In BIS, a weak AC signal having frequency less than 1 MHz is used to assess the impedance levels within the tissue. An electrical equivalent circuit made up of capacitors and resistors can be used to represent biological tissue (Dean et al. 2008). It is concluded that the membrane structure of the cell and bodily water (comprising extra- and intracellular fluid), respectively, are the sources of capacitance and resistance (Kamat, Bagul, and Patil 2014). Semipermeable cell membranes distinguish between intracellular and external regions. The most basic electrical representation of biological tissue is a parallel pairing of a capacitor and a conductor, while more accurate electrical circuit-based representations of tissue have also been developed to more accurately model the tissue (Dean et al. 2008; Chinen et al. 2015; Hernández-Balaguera, López-Dolado, and Polo 2016). Due to its label-free observations, impedance spectroscopy has lately grown significantly in popularity for the biosensing of a variety of chemicals. This allows it to achieve excellent sensitivity down to the femto- and attomolar levels (Grieshaber et al. 2008; Bahadir and Sezgintürk 2014; Muñoz, Montes, and Baeza 2017). It enables quick biocompatible surface screening, pathogenic bacteriological surveillance, and biological cell and tissue analysis. In this instance, a chemical/biochemical interaction among a specific biological receptor and the target analyte molecule, which is adsorbed from the solution, provide the primary basis for the impedance spectroscopy transducer signal. The kinetics

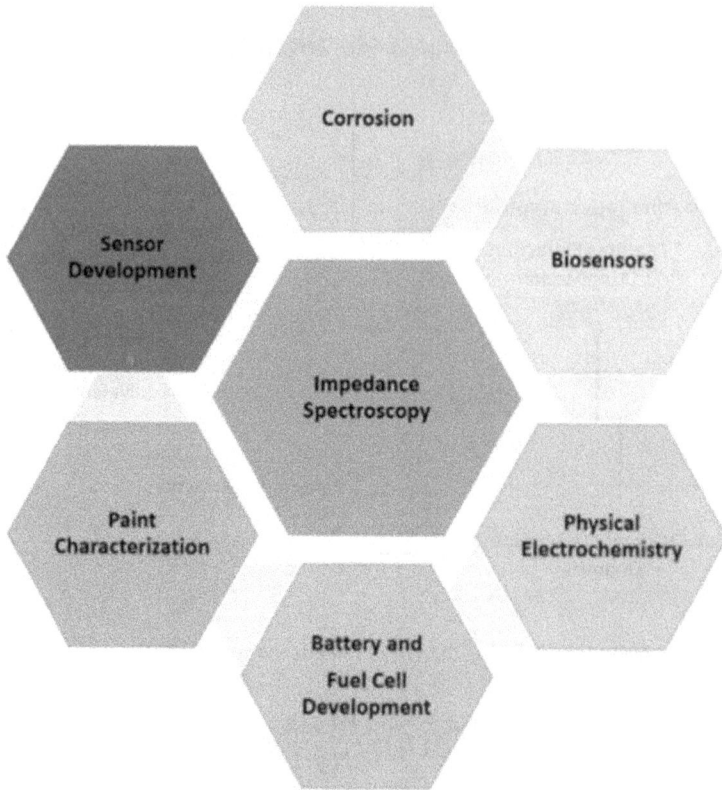

FIGURE 6.2 Application of impedance spectroscopy.

of interfacial electron transport between the conducting electrode and the analyte species are altered because of these interactions. Therefore, as the number of targets attached to the receptive surface grows, the charge transfer resistance also increases. As a result, the adoption of suitable materials and catalysts has a significant impact on the selectivity, detection limit, response time, and sensitivity (Figure 6.2).

6.2 PRINCIPLES OF NONINVASIVE SENSORS

Diagnostic methods based on integrated paper-based devices have been developed for a number of body fluid samples collected noninvasively. It means that sampling does not involve skin piercing, skin splitting, or entering a body cavity. Depending on the kind of sample, several sampling techniques are used. Schirmer paper strips, glass or plastic capillary tubes for sampling tears, or cotton swabs for swabbing can all be utilized as physical sample equipment. It is also possible to utilize chemical (oral antiperspirant medications) or electrical inducers (iontophoresis) to speed up sample processing. The sample material that is used most frequently is urine. Noninvasive samples of other body fluids, such as sweat, saliva, tears, breath, semen, and feces, is less common. Noninvasive (NI) impedance spectroscopy procedures for sensing are the main topic of this chapter. As a result, the fundamental ideas of modern NI sensing are examined in the sections that follow.

6.2.1 ANALYSIS OF SALIVA

Numerous biological elements found in saliva indicate the physiology and state of health (Zhao and Leung 2020). It has so far been widely utilized to diagnose drug misuse and human immunodeficiency virus (HIV) infections (Marley et al. 2014; Nunes, Mussavira, and Bindhu 2015; Kaufman and Lamster 2016). Therefore, saliva can be used in place of blood to signify bodily physiological processes (Caixeta et al. 2020).

6.2.2 ANALYSIS OF TEAR

Tears can also convey information and exhibit blood-like glucose levels (Geelhoed-Duijvestijn et al. 2021; Jones et al. 2021). Organic compounds seen in tears indicate one's wellness (Zhao and Leung 2020). Blood glucose levels are in the range of 90–140 mg/dL, whereas tear glucose levels are reasonably steady in the range of 0.9–90 mg/dL (0.05–5 mM) compared with saliva (Makaram, Owens, and Aceros 2014; S. Kim et al. 2020). As a result, tears have garnered a lot of interest for years. The widespread use of contact lenses is another important driver of tear glucose measurement (Baca, Finegold, and Asher 2007; Xiong et al. 2018).

6.2.3 ANALYSIS OF BREATH EXHALED

Exhaled breath is another intriguing biomarker, despite the presence of other markers like saliva and tears. It is generally known that there is a link between some illnesses and the scent of the exhaled breath (Dixit et al. 2021; Mule, Patil, and Kaur 2021). Examples include the "fruity scent" of acetone in the breath as a sign of diabetes and the "musty and fishy smell" as a sign of severe liver disease (Di Francesco et al. 2005). As a result, exhaled breath analysis may offer profound insights into the physiological and pathophysiological states of illnesses that are connected (Das, Pal, and Mitra 2016). Exhaled breath is often simple to acquire and gather, much like the other two media. In addition, compared with tear and saliva analysis, it is benign, kinder, and more satisfactory to patients (Dixit et al. 2021; Chen et al. 2021). Acetone, isoprene, carbon monoxide, isopropanol (IPA), and ethanol are some biomarkers found in human breath (Dixit et al. 2021).

6.2.4 DEPENDENCE OF IMPEDANCE SPECTROSCOPY ON SKIN STRUCTURE

An essential component of bio-impedance that may be utilized to assess human health is skin impedance. Skin impedance measurement can provide the functional data regarding the skin, such as the amount of moisture in skin, stratum corneum thickness, health of the skin's water channels, and more (Vanbever, Lecouturier, and Préat 1994; Kalia, Pirot, and Guy 1996; Dujardin et al. 2002; Birgersson et al. 2010; Davies, Chappell, and Melvin 2017). The use of additional bio-impedance factors, such as heart rate measurement, which significantly influences the desired quality of the signal, also considers electrode–skin impedance. A thorough investigation of skin impedance has been done using both bio-impedance measures and skin features (Clemente, Arpaia, and Manna 2013; Ghosh, Mahadevappa, and Mukhopadhyay 2017). The reaction of a particular area of skin to an externally applied electrical current is known as skin impedance (or voltage) (Clemente, Arpaia, and Manna 2013). The electrode is therefore essential for detecting skin impedance. The epidermis, dermis, and subcutaneous tissue of the human body are the three primary layers of skin. The outermost layer of our skin, or epidermis, acts as a barrier of defense. The stratum corneum is the epidermis's outermost layer. To maintain homeostasis, which allows for the measurement of glucose, electrolytes and water are continually exchanged between blood and other tissues. Because it alters stratum corneum hydration status, humidity has a significant impact on skin impedance (Jayaraman et al. 2007; Li et al. 2015). When utilizing dry electrodes, variables including skin roughness, level of moisture, and pressure applied to the sensor might change the electrical contact between the sensor and skin. Therefore, in the monitoring system, electrode materials or architectures as well as various kinds of electrical characteristics play a major role.

6.3 ROLE OF ELECTRODE MODIFICATION

An effective electrical method for analyzing the interfacial characteristics of bio-recognition processes that take place at the electrode surface, such as antibody–antigen recognition, substrate–enzyme interaction, or whole-cell capture, is EIS. From a synthetic perspective, many methods (physical, chemical, biological, or mixing processes) might be applied to the production of nanomaterials. The synthesis method is chosen based on the material of interest, the kind of nanomaterials (e.g., 0D, 1D, or 2D), the sizes, or the required quantities (Wang and Xia 2004). Due to benefits such as a high surface area, excellent connection, and electrical conductivity, nanomaterials are commonly used as sensors for physiological monitoring. Metallic and carbon-based nanoparticles are two of the most frequently utilized nanomaterials (Suni 2008; Lee et al. 2014; Bhalla et al. 2018). In recent work, a sensor with a low detection limit (LOD: 0.29 M) for uric acid was created utilizing ultra-small iron oxide nanoparticles coated with nitrogen-doped carbon (Tang et al. 2022). Additionally, Khalilzadeh and Borzoo (2016) described the environmentally friendly production of AgNPs for the impedance spectroscopy-based electrochemical detection of ascorbic acid.

6.3.1 Sensors with Molecular Imprints in Polymers

Due to their great selectivity, chemical and thermal durability, and ease of customization as compared to receptors from biological sources, molecular imprinted polymers (MIPs) are artificial recognition components utilized in the production of sensors. In the presence of the analyte template, a functional monomer is polymerized to produce MIPs (Crapnell et al. 2019). A very selective interaction with the target analyte is made possible when the template is removed because cavities with specified shapes are created (Saylan et al. 2019). MIPs attach to the target molecules, causing changes in the mass, absorbance, or electron transfer (ET) rate at the sensor surface. Through EIS measurements, it is possible to quantify the particular interaction that frequently hinders the ET between the electrode and the redox probe in the solution in the case of electrochemical sensors. MIPs generally interact with the analyte in solution while immobilized on the sensor surface. It was feasible to create a MIP by electro-polymerizing uric acid as a molecular template. The charge transfer resistance from the uric acid molecule was increased by the nanoimprinted polymer (NIP), but it was decreased by the MIP film at the polymer interface (Trevizan et al. 2021).

6.3.2 Metal-based composite sensors

Metal composites are made up of two different metals or metals plus another material type, such as a polymer. Due to their mesoporous architecture, metal–polymer composites have a huge surface area and improved electrical conductivity. A family of substances known as metal-organic frameworks (MOFs) is made up of metal ions or clusters that are coordinated with organic ligands to create one-, two-, or three-dimensional structures. MOFs offer a lot of potential in electrochemical sensing applications because of their adaptable structure and functionality, high porosity, and vast interior surface area (Kumar, Deep, and Kim 2015). Glassy carbon electrodes' (GCE) electrocatalytic capacity to oxidize uric acid (UA) is weak. Polyaniline (PANI) based GCE and Fe/GCE have demonstrated improved electrocatalysis for UA when compared with the unmodified electrode (Govindasamy et al. 2016).

6.3.3 Sensors based on graphene and carbon nanotubes

An allotrope of carbon-containing 2D layers of sp^2-hybridized carbon is called graphene. Due to its strong electrical conductivity and sizable surface area, which is susceptible to functionalization with biomolecules, it is utilized to modify sensors. Due to their high ET rate, surface area, reduced surface fouling, and stability, single-walled carbon nanotubes (SWCNT) and multiwall carbon nanotubes (MWCNT) were employed to alter the electrochemical transducers (Sireesha et al. 2018). For the detection of sweat metabolites, functionalization is done on the SWCNT-based flexible electrode arrays. For the enhancement of sensitivity and conductivity, platinum (Pt) nanoparticles were deposited. Further for the improvement of ion-electron transduction and catalytic performance for Na^+/K^+, glucose, and lactate, respectively, the modified layer was further electrodeposited with Prussian blue or poly(3,4-ethylenedioxythiopen) (PEDOT) (Hao et al. 2022) was used.

6.4 APPLICATION OF EIS AND BIS

6.4.1 Biological tissues analysis

EIS (Macdonald 1992; Wu, Ben, and Chang 2005; Houssin et al. 2010; Scrymgeour et al. 2010), BIS (Hoffer, Meador, and Simpson 1969; Lukaski et al. 1985; Hong-Yi and Kato 1995; Jakicic, Wing, and Lang 1998; Kyle et al. 2004; Parrinello et al. 2008), impedance cardiography (ICG) (Griffiths et al. 1981), electrical impedance plethysmography (IPG) (Hill, Jansen, and Fling 1967), and electrical impedance tomography (EIT) ("*Electrical Impedance Tomography: Methods, History and Applications* - Google Books" n.d.) are examples of electrical impedance-based non-invasive tissue characterizing techniques that are utilized to examine the frequency-based response of biological tissue's electrical impedance.

However, because EIS offers impedance fluctuations throughout frequencies, it is more widely used than BIS, IPG, and ICG in a number of application sectors. In contrast to BIS, IPG, and ICG, which are only employed in biological regimes, EIS has been researched for the noninvasive evaluation of biological and nonbiological materials in the frequency domain. The bioelectrical EIS examines and estimates electrical impedance data at various frequency ranges, to obtain lumped estimation of the impedance values of biological tissue in an appropriate frequency (typically 50 kHz) along with the knowledge necessary to comprehend a number of intricate bioelectrical phenomena, such as dielectric relaxation and dielectric dispersions (Schwan 1957, 2002). It may be summed up as an electrical technique that uses an electric current to circulate inside the body to detect voltage and then impedance before determining body fat and muscle mass. The measurements derived from these observations are then converted into useful data to determine an individual's body composition. Specialized biological cells with important electrical characteristics, such as impedance, give rise to biological tissues. Bio-impedance, which is a function of tissue architecture, composition, and signal frequency, kicks in as soon as an alternating current excites these tissues. The majority of the time, biological tissue is a complex heterogeneous medium made up of various absorbers like blood, water, and scatterers (collagen, keratin, etc.). It is possible to distinguish between healthy and malignant tissues in a number of organs using tissue electrical impedance, which is dependent on the structure of the tissue. When measuring electrical impedance, a bodily tissue's bioelectrical impedance is determined by passing an array of surface electrodes connected to the tissue's surface with a low-amplitude, low-frequency alternating current (often sinusoidal) (Bera 2014).

Further, EIS had been used for the noninvasive and rapid analysis of skin electroporation, and it is found that an increase in skin permeability due to the electroporation caused a higher transfer rate of drugs for the gene- and protein-based drugs (Pliquett and Prausnitz 2000). It is found that the degree of necrosis seen histologically in the cell population can be connected to the change in electrical impedance of a volume of tumor tissue that takes place during and/or after hyperthermia therapy (McRae, Esrick, and Mueller 1999). In another work, the authors (Osterman et al. 2004) investigated the capability of EIS to noninvasively assess and quantify the damage response in soft tissue following high dose rate irradiation, which is distinguished by largely localized dose distributions with steep spatial gradients. Both conductivity and permittivity are better metrics for tissue differentiation since they are both greater in healthy prostate tissues than in malignant tissues (Halter et al. 2007). Another research (Skourou et al. 2004) determined the early-stage growth and presence of tumors in rats with the help of EIS. Further, EIS is also used for the analysis of breast cancer (Estrela Da Silva, Marques De Sá, and Jossinet 2000; Kerner et al. 2002). Impedance spectroscopy is a prominent tool for the determination of the state of organs and biological tissues (Gersing 1998; Bera, Jampana, and Lubineau

2016). EIS is the most widely used and effective approach for multifrequency impedance analysis. A pulsed signal-based EIS apparatus is presently being used to study EIS approaches for tissue characterization and single-cell analysis (Chen et al. 2007; Gawad et al. 2007).

6.4.2   GLUCOSE ANALYSIS

The traditional approaches to measuring blood glucose levels are intrusive, uncomfortable, and inappropriate for long-term monitoring. Thus, the fight to create painless and bloodless blood glucose monitors has started over the past 30 years. One potential method for the creation of such monitors has been suggested: electrical BIS. Electrical bio-impedance (EBS) is defined as the estimate of a biological tissue's electrical characteristics (which include dielectric properties and electrical impedance) over a wide frequency range. Caduff's group started the research for the use of the NICBGM approach utilizing EBS. They created a noninvasive blood glucose monitor known as Pendra based on the findings of their research (Caduff et al. 2003). These days, various scientific and commercial noninvasive solutions are available in the market for the determination of glucose (Xue et al. 2022). Due to individual variations in skin thickness and underlying tissues, Pendra has a number of drawbacks, including the need for a two-point calibration process (Wentholt et al. 2005). Several research groups have discovered that changes in blood impedance are caused both directly and indirectly by variations in blood glucose levels. Blood glucose changes set off metabolic processes that alter the electrolytic balance across the erythrocyte membrane and modify the impedance of the subject's epidermis and underlying tissues (Caduff et al. 2003).

According to some reports, changes in blood's dielectric properties are directly caused by variations in blood glucose levels (Tura et al. 2007). By identifying the glucose-dependent electrical impedance characteristics of aqueous solution samples, researchers Satish, Sen, and Anand (2018) illustrated the change in impedance along the frequency spectrum. Sankhala et al. (2021) demonstrated a chemi-impedance biosensor for the noninvasive detection of glucose (Figure 6.3). EIS-based glucose sensors have further been integrated with humidity, temperature, and optical sensors to enhance their performance capability (Geng et al. 2017). It is important to note that taking too many extra sensors into account might cause misunderstandings and an increase in noise. Therefore, just the most important elements should be assessed. Another work (Ito et al. 2019) demonstrated third-generation impedimetric biosensors by conducting EIS for direct ET type flavin adenine dinucleotide-dependent glucose dehydrogenase modified gold disc electrodes without redox mediators. Further, a paper-based impedance spectroscopy sensor was fabricated for the detection of monosaccharides by using responsive gel (Daikuzono et al. 2017). Based on Bruggeman's effective medium theory, Pedro et al. (2020) suggested a simpler physical function for relating glucose concentration and electrical conductivity of blood.

6.4.3   URIC ACID ANALYSIS

UA is the main byproduct of purine metabolism in humans. It is crucial to monitor UA levels in the blood, urine, or both because they can serve as early warning

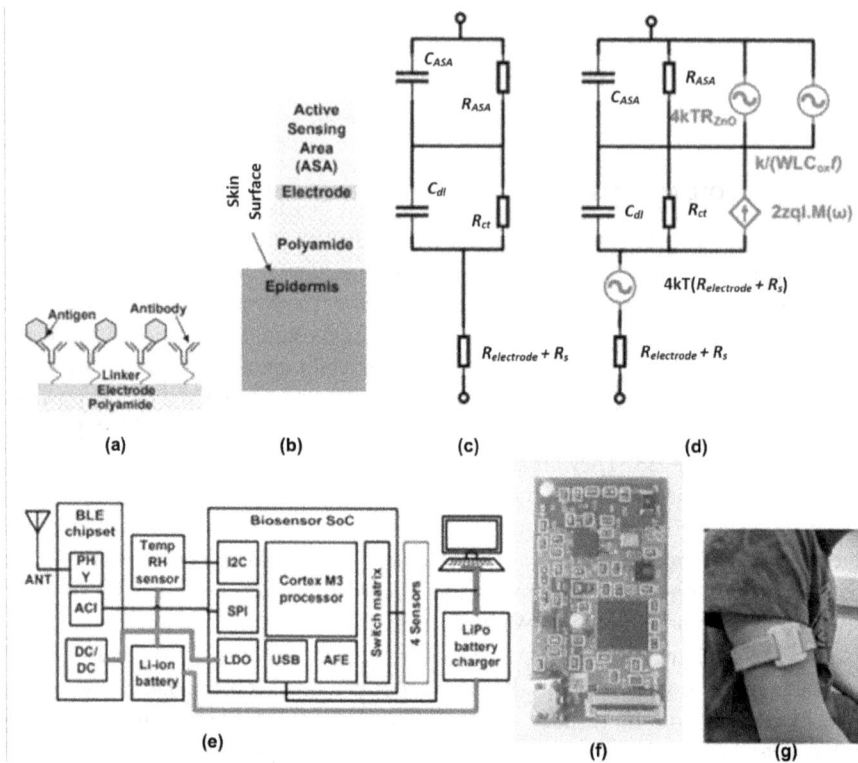

FIGURE 6.3 Immunoassay development on the functionalized surface as well as the cross-sectional view of the chemi-impedance biosensor. [Reprinted with permission from Sankhala et al. (2021). Copyright (2021) Arxiv.]

indicators of renal and metabolic diseases. One of the most effective methods for examining the characteristics of surface-modified electrodes is EIS (Hua et al. 2010). Using the direct transmission of electrons from the immobilized enzyme onto the graphene quantum dots (GQDs)-based GCE, a unique third-generation UA biosensor is created (Yu et al. 2018). Yan et al. (2020) reported a new biosensor with improved biocatalytic activity on the electrode surface for the sensitive and specific detection of UA. It was noticed that the constructed biosensor had a relatively low detection limit of 0.0596 μM, a lower K_m value of 34.7351 μM, and could successfully detect UA over a wide concentration range of 0.1–1,000 μM.

6.4.4 CANCER ANALYSIS

Serial mutations that happen as a result of genetic instability or environmental influences are what give rise to cancer (Aubele and Werner 1999; Ryu et al. 2003). Breast cancer is the term for the body's uncontrolled creation of breast cells (Shokoufi and Golnaraghi 2016). One of the medical methods for detecting cancer in its early stages is multifrequency EIS. Based on measuring the conductivity of bodily tissue, this method is used. The EIS approach works by identifying variations in the tissue's electrical impedance. We can calculate the variations in electrical impedances of diverse tissues by measuring the impedances of those tissues.

In addition to mammography and ultrasonography, the EIS-probe and the EIS-hand breast techniques can be used to identify different types of breast cancer in their early stages (Haeri et al. 2016). Shah et al. (2016) published a protocol for the use of bio-impedance in the treatment of breast cancer by concentrating on the early detection and treatment of lymphedema caused by breast cancer. Lee et al. (2019) concluded that in patients with metastatic disease, the extracellular-to-intracellular fluid volume (E/I) ratio might be utilized to forecast survival. Skin cancer detection also uses the differential impedances between pathological and healthy tissues. These diagnostic procedures involve measuring the in vivo impedance of two skin areas using a multi-electrode sensor to analyze the area of interest, which may contain a tumor, and the healthy nearby region. For instance, it is feasible to probe contralateral areas of skin on the arm that had a cancer test. Impedance-based skin cancer diagnosis, as previously mentioned, distinguishes between healthy and pathological skin areas, opening up new possibilities for oncology surveillance (Braun et al. 2017; Moqadam et al. 2018).

6.4.5 ANALYSIS OF BODY COMPOSITIONS

Human bodies are made up of a variety of chemicals, but among those that are now recognized, oxygen, carbon, and hydrogen have significant compositions. A healthy human body has the right amount of lipids, minerals, proteins, water, and other substances. The two additional subgroups of total body water, intracellular body water, and extracellular body water are in addition to the compositions mentioned above. Because a healthy body has less fat and hence exhibits less resistance when subjected to an electrical current, we may use the BIS technique to assess if a body is performing better or not. Additionally, tissues with little body fat have a higher heat conductivity than those with high body fat, hence it is important to discriminate between the two using bioelectrical impedance. Body composition (BC) estimate procedures frequently use bio-impedance methodologies. They are easy, safe, and noninvasive, and they give more accurate estimates than anthropometric methods without the limitations of those solutions (Moissl et al. 2006; Roa et al. 2013; Naranjo-Hernández, Reina-Tosina, and Min 2019). Naranjo-Hernández et al. (2020) estimated the BC through the smart BIS device and measured the data in multiple frequencies. The processed data was transmitted wirelessly for gathering information through a new algorithm for the identification of Cole model parameters. Despite having a normal body mass index (BMI), patients with the primary neuromuscular illness have proportionately more fat and less muscle mass than the general population. In these individuals, muscle mass can be determined by bio-impedance, although device performance and bias will vary. Because phase angle via bio-impedance corresponds to muscle mass, it may be utilized as a stand-in for muscle mass during follow-up (Ellegård et al. 2019).

A sophisticated bio-impedance analysis technique was created by Peppa et al. (2017) to assess the degree of osteosarcopenia and obesity in postmenopausal women with normal or depleted bone density. The total body water content of a pregnant woman can be quickly, easily, and noninvasively assessed using bioelectrical impedance; according to Piuri et al. (2016), bio-impedance may also be used to identify patients early in gestation who are at risk of developing various clinical phenotypes of hypertensive disease of pregnancy and SGA fetuses. According to Zając-Gawlak et al. (2017), an increasing visceral fat area is a significant risk factor for the emergence of metabolic syndrome. According to research by Kim et al. (2017), chronic hemodialysis patients with high extracellular fluid expansion and intracellular fluid ratios are not only fluid-overloaded. Also, the malnourished have stiff arteries with increased inflammation and have a high extracellular fluid and expansion ratio. According to Demirci et al. (2016), bio-impedance analysis is a reliable independent predictor of cardiovascular health which is a prominent cause of death in hemodialysis patients. According to Redondo-del-Río et al. (2016), body compartment changes in institutionalized elderly people are detected by bioelectrical impedance vector analysis and are not picked up by most common clinical practice nutritional indicators, such as BMI, waist circumference, and bio-impedance analysis estimated BC.

6.4.6 APPLICATION IN DIFFERENT BIOFIELDS

Impedance spectroscopy has a broad field of application in the bio/medical field. Suresh et al. (2018) demonstrated a paper-based microfluidic device for the detection of urea up to 1 pM concentration through a two-electrode assembly. In another work, the authors (Hao et al. 2022) disclosed functionalized wearable sensor arrays for the detection of electrolytes (Na$^+$, K$^+$) and sweat metabolites (glucose, lactate), which are ideal for sensitive and selective monitoring of a variety of sweat biomarkers. Figure 6.4 shows the fabrication of SWCNT-based flexible electrodes and their utilization for sweat analysis applications.

Wei et al. (2022) fabricated a portable sensor that concurrently detects Na+, ascorbic acid, and neuropeptide Y (NPY) in sweat as indicators of cardiovascular health. It has outstanding sensitivity, high selectivity, and good stability. Figure 6.5A shows the

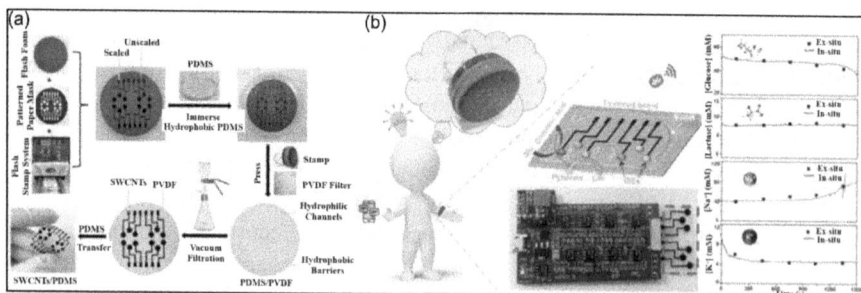

FIGURE 6.4 (A) Fabrication of flexible electrode arrays based on SWCNT; (B) images and schematics of an intelligent wearable sensor for noninvasive real-time sweat analysis that was inspired by photosensitive stamps. [Reprinted with permission from Hao et al. (2022). Copyright (2022) American Chemical Society.]

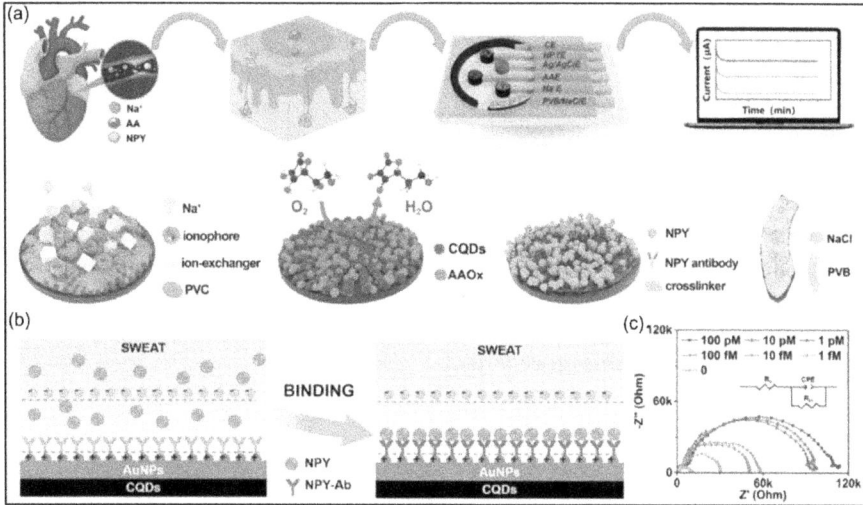

FIGURE 6.5   (A) Schematic of cardiovascular health sensor based on CQD; (B) schematic of NPY sensor's sensing system; (C) Nyquist plots showing the EIS readings for the 1 fM to 100 pM NPY concentration range. [Reprinted with permission from Wei et al. (2022). Copyright (2022) American Chemical Society.]

schematic of the cardiovascular health sensor; Figure 6.5B shows the schematic of the NPY sensor, and Figure 6.5C shows when the concentration of NPY is increased from 1 fM to 100 pM. It is shown that the impedance at low frequencies rises. This phenomenon is the consequence of alterations in capacitance brought on by the development of immunological complexes including NPY and NPY-Ab and results depict the NPY sensing electrode's real and hypothetical impedance as a function of frequency.

Biswas et al. (2021) provided a proof of concept for the development of paper-based impedimetric urine pH sensors with an average sensing accuracy of 95.53 ± 5.84% for urine pH. Chacón et al. (2022) showed that cell membrane capacitance could be incorporated into the current testing strategies for RhCE test methods, both as in-process quality control to assess the maturation of the epithelium prior to the testing procedure and as a novel criterion for the assessment of corneal irritation that would supplement the current endpoints based on cell viability.

6.4.7   Wearable Devices

Noninvasive wearable technology is widely used to track human health and identify disorders as soon as feasible. A noninvasive signal called bio-impedance is used in various clinical settings to diagnose and monitor conditions including congestive heart failure, BC, as well as hydration (Matthie 2008; Bera 2014). Villa et al. (2016) created a wearable multifrequency and multisegment BIS for covertly monitoring changes in bodily fluid levels caused by physical exercise. Bioresistance impedance and reactance play a crucial role in the analysis of biological tissues and the extraction of their properties (Dean et al. 2008). Ibrahim et al. (2018) presented a device for capturing wide signals for wearable applications. Using flexible, wearable biosensors based on gold (Sankhala, Muthukumar, and Prasad 2018), a four-channel EIS sensor analyzer module that could detect cortisol concentrations in both artificial and human sweat was reported. Figure 6.6 shows the fabricated sensor and its assembly along with the extracted incremental circuit model for the measuring procedure.

In order to effectively measure bioelectrical impedance even with extremely tiny sizes of electrodes, Jung et al. designed a unique wrist-worn bioelectrical impedance analyzer featuring a contact resistance compensation mechanism (Jung et al. 2021). Figure 6.7 shows the various formats (using single finger and watch style) of measurements as utilized in this procedure.

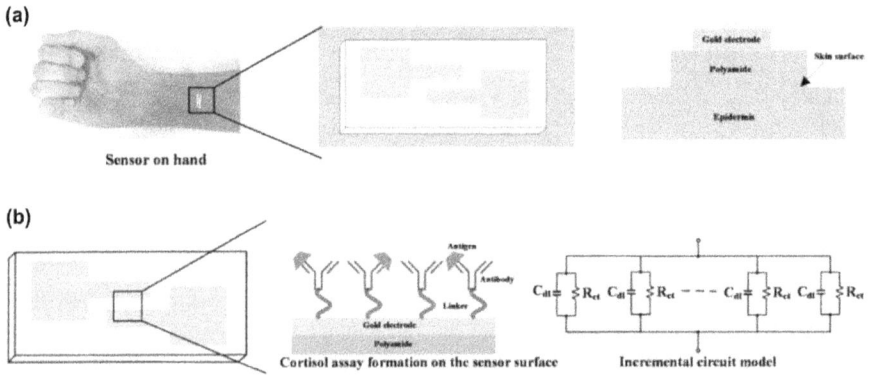

FIGURE 6.6 (A) Chemi-impedance biosensor's form factor; (B) magnified image of the functionalized cortisol test with equivalent electrical circuit modeling. [Reprinted with permission from Sankhala, Muthukumar, and Prasad (2018). Copyright (2018) SLAS.]

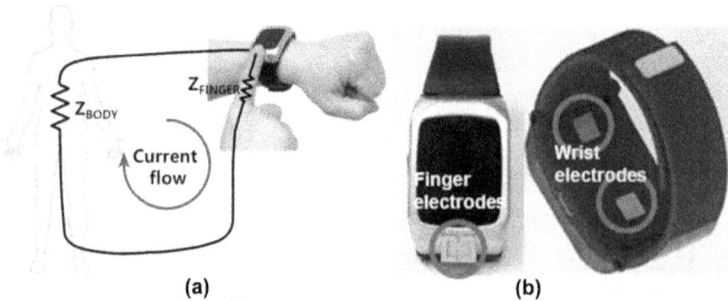

FIGURE 6.7 Measurement of bioelectrical impedance using (A) single finger; (B) watch-style BIA device. [Reprinted with permission from Jung et al. (2021). Copyright (2021) Nature.]

6.5 CONCLUSION AND FUTURE PERSPECTIVES

Impedance spectroscopy is a flexible technique that is becoming more and more important because it offers intriguing opportunities for analyzing materials, systems, and sensors. The use of high-speed and inexpensive electronic circuits in sensors now allows for more sophisticated production techniques and signal processing (Mondal and Sharma 2016). Similar features may currently be seen in impedance spectroscopy. The approach may be applied in a greater variety of applications, thanks to measurement instruments that make use of cutting-edge technology and offer solutions that are evermore effective.

For many years, impedance spectroscopy was thought of as an electrochemical laboratory technique. Today, a cost-effective electronic implementation is becoming more and more feasible, thanks to superior microelectronics breakthroughs. The medical industry has the widest range of embedded solutions, many of which are used in medical implants. The trend toward several options for sensors and diagnostic tools is now accelerating. In most situations, the goal is to discover solutions that are often constrained by the needs of a specific application, rather than to run a vast measurement range.

One of the crucial prerequisites for impedance spectroscopy is that the system remains unchanged throughout the whole impedance spectrum measurement. To achieve quasi-stable circumstances when the system has a greater dynamic, impedance testing should be done quickly. That is why shortening the measurement period is essential to extending the impedance spectroscopy's range of applications to more dynamic ones.

REFERENCES

Aubele, M., and M. Werner. 1999. "Heterogeneity in Breast Cancer and the Problem of Relevance of Findings." *Analytical Cellular Pathology : The Journal of the European Society for Analytical Cellular Pathology* 19 (2): 53–58. https://doi.org/10.1155/1999/960923.

Baca, Justin T., David N. Finegold, and Sandford A. Asher. 2007. "Tear Glucose Analysis for the Noninvasive Detection and Monitoring of Diabetes Mellitus." *The Ocular Surface* 5 (4): 280–93. https://doi.org/10.1016/S1542-0124(12)70094-0.

Bahadir, Elif Burcu, and Mustafa Kemal Sezgintürk. 2014. "A Review on Impedimetric Biosensors." *Artificial Cells, Nanomedicine, and Biotechnology* 44 (1): 248–62. https://doi.org/10.3109/21691401.2014.942456.

Barsoukov, Evgenij, and J. Ross Macdonald. 2005. "Impedance Spectroscopy: Theory, Experiment, and Applications." *Impedance Spectroscopy: Theory, Experiment, and Applications* January, 1–595. John Wiley & Sons. https://doi.org/10.1002/0471716243.

Bera, Tushar Kanti. 2014. "Bioelectrical Impedance Methods for Noninvasive Health Monitoring: A Review." *Journal of Medical Engineering* 2014 (June): 1–28. https://doi.org/10.1155/2014/381251.

Bera, Tushar Kanti, Nagaraju Jampana, and Gilles Lubineau. 2016. "A LabVIEW-Based Electrical Bioimpedance Spectroscopic Data Interpreter (LEBISDI) for Biological Tissue Impedance Analysis and Equivalent Circuit Modelling." *Journal of Electrical Bioimpedance* 7 (1): 35–54. https://doi.org/10.5617/JEB.2978.

Bhalla, Nikhil, Shivani Sathish, Abhishek Sinha, and Amy Q. Shen. 2018. "Large-Scale Nanophotonic Structures for Long-Term Monitoring of Cell Proliferation." *Advanced Biosystems* 2 (4): 1700258. https://doi.org/10.1002/ADBI.201700258.

Birgersson, Ulrik, Erik Birgersson, Peter Åberg, Ingrid Nicander, and Stig Ollmar. 2010. "Non-Invasive Bioimpedance of Intact Skin: Mathematical Modeling and Experiments." *Physiological Measurement* 32 (1): 1. https://doi.org/10.1088/0967-3334/32/1/001.

Biswas, Souvik, Arijit Pal, Koel Chaudhury, and Soumen Das. 2021. "Polyaniline Functionalized Impedimetric Paper Sensor for Urine PH Measurement." *IEEE Sensors Journal* 21 (13): 14474–82. https://doi.org/10.1109/JSEN.2020.3013405.

Braun, Ralph P., Johanna Mangana, Simone Goldinger, Lars French, Reinhard Dummer, and Ashfaq A. Marghoob. 2017. "Electrical Impedance Spectroscopy in Skin Cancer Diagnosis." *Dermatologic Clinics* 35 (4): 489–93. https://doi.org/10.1016/J.DET.2017.06.009.

Caduff, A., E. Hirt, Yu Feldman, Z. Ali, and L. Heinemann. 2003. "First Human Experiments with a Novel Non-Invasive, Non-Optical Continuous Glucose Monitoring System." *Biosensors and Bioelectronics* 19 (3): 209–17. https://doi.org/10.1016/S0956-5663(03)00196-9.

Caixeta, Douglas C., Emília M. G. Aguiar, Léia Cardoso-Sousa, Líris M. D. Coelho, Stephanie W. Oliveira, Foued S. Espindola, and Leandro Raniero et al. 2020. "Salivary Molecular Spectroscopy: A Sustainable, Rapid and Non-Invasive Monitoring Tool for Diabetes Mellitus during Insulin Treatment." *PLOS ONE* 15 (3): e0223461. https://doi.org/10.1371/JOURNAL.PONE.0223461.

Chacón, Manuel, Manuel Sánchez, Natalia Vázquez, Mairobi Persinal-Medina, Sergio Alonso-Alonso, Begoña Baamonde, Jose F. Alfonso, Luis Fernández-Vega-Cueto, Jesús Merayo-Lloves, and Álvaro Meana. 2022. "Impedance-Based Non-Invasive Assay for Ocular Damage Prediction on in Vitro 3D Reconstructed Human Corneal Epithelium." *Bioelectrochemistry* 146 (August): 108129. https://doi.org/10.1016/J.BIOELECHEM.2022.108129.

Chen, Ting, Tiannan Liu, Ting Li, Hang Zhao, and Qianming Chen. 2021. "Exhaled Breath Analysis in Disease Detection." *Clinica Chimica Acta* 515 (April): 61–72. https://doi.org/10.1016/J.CCA.2020.12.036.

Chen, Wan, Jiahui Fu, Qun Wu, Jan Schmidt, Abhishek Biswas, Tao Sun, Shady Gawad, Catia Bernabini, Nicolas G. Green, and Hywel Morgan. 2007. "Broadband Single Cell Impedance Spectroscopy Using Maximum Length Sequences: Theoretical Analysis and Practical Considerations." *Measurement Science and Technology* 18 (9): 2859. https://doi.org/10.1088/0957-0233/18/9/015.

Chinen, Koyu, Ichiko Kinjo, Aki Zamami, Kotoyo Irei, and Kanako Nagayama. 2015. "New Equivalent-Electrical Circuit Model and a Practical Measurement Method for Human Body Impedance." *Bio-Medical Materials and Engineering* 26 (s1): S779–86. https://doi.org/10.3233/BME-151369.

Clemente, Fabrizio, Pasquale Arpaia, and Carlo Manna. 2013. "Characterization of Human Skin Impedance after Electrical Treatment for Transdermal Drug Delivery." *Measurement* 46 (9): 3494–501. https://doi.org/10.1016/J.MEASUREMENT.2013.06.033.

Crapnell, Robert D., Alexander Hudson, Christopher W. Foster, Kasper Eersels, Bart van Grinsven, Thomas J. Cleij, Craig E. Banks, and Marloes Peeters. 2019. "Recent Advances in Electrosynthesized Molecularly Imprinted Polymer Sensing Platforms for Bioanalyte Detection." *Sensors* 19 (5): 1204. https://doi.org/10.3390/S19051204.

Daikuzono, C. M., C. Delaney, H. Tesfay, L. Florea, O. N. Oliveira, A. Morrin, and D. Diamond. 2017. "Impedance Spectroscopy for Monosaccharides Detection Using Responsive Hydrogel Modified Paper-Based Electrodes." *Analyst* 142 (7): 1133–39. https://doi.org/10.1039/C6AN02571D.

Das, Souvik, Saurabh Pal, and Madhuchhanda Mitra. 2016. "Significance of Exhaled Breath Test in Clinical Diagnosis: A Special Focus on the Detection of Diabetes Mellitus." *Journal of Medical and Biological Engineering* 36 (5): 605–24. https://doi.org/10.1007/S40846-016-0164-6.

Davies, L., P. Chappell, and T. Melvin. 2017. "Modelling the Effect of Hydration on Skin Conductivity." *Skin Research and Technology* 23 (3): 363–68. https://doi.org/10.1111/SRT.12344.

Dean, D. A., T. Ramanathan, D. Machado, and R. Sundararajan. 2008. "Electrical Impedance Spectroscopy Study of Biological Tissues." *Journal of Electrostatics* 66 (3–4): 165–77. https://doi.org/10.1016/J.ELSTAT.2007.11.005.

Demirci, C., G. Aşcı, M. S. Demirci, M. Özkahya, H. Töz, S. Duman, S. Sipahi, S. Erten, M. Tanrısev, and E. Ok. 2016. "Impedance Ratio: A Novel Marker and a Powerful Predictor of Mortality in Hemodialysis Patients." *International Urology and Nephrology* 48 (7): 1155–62. https://doi.org/10.1007/S11255-016-1292-1/FIGURES/2.

Dixit, Kaushiki, Somayeh Fardindoost, Adithya Ravishankara, Nishat Tasnim, and Mina Hoorfar. 2021. "Exhaled Breath Analysis for Diabetes Diagnosis and Monitoring: Relevance, Challenges and Possibilities." *Biosensors* 11 (12): 476. https://doi.org/10.3390/BIOS11120476.

Dujardin, Nathalie, Edith Staes, Yogeshvar Kalia, Peter Clarys, Richard Guy, and Véronique Préat. 2002. "In Vivo Assessment of Skin Electroporation Using Square Wave Pulses." *Journal of Controlled Release* 79 (1–3): 219–27. https://doi.org/10.1016/S0168-3659(01)00548-X.

"Electrical Impedance Tomography: Methods, History and Applications - Google Books." n.d. Accessed October 6, 2022. https://books.google.co.in/books?hl=en&lr=&id=5bBZEA AAQBAJ&oi=fnd&pg=PP1&dq=info:llNckCrnE6kJ:scholar.google.com&ots=9tXI6 bfB81&sig=YcWN2PBQdDp4k7uqSC_Nc-mRQqA&redir_esc=y#v=onepage&q&f=false.

Ellegård, L., A. Aldenbratt, M. K. Svensson, and C. Lindberg. 2019. "Body Composition in Patients with Primary Neuromuscular Disease Assessed by Dual Energy X-Ray Absorptiometry (DXA) and Three Different Bioimpedance Devices." *Clinical Nutrition ESPEN* 29 (February): 142–48. https://doi.org/10.1016/j.clnesp.2018.11.004.

Estrela Da Silva, J., J. P. Marques De Sá, and J. Jossinet. 2000. "Classification of Breast Tissue by Electrical Impedance Spectroscopy." *Medical & Biological Engineering & Computing* 38 (1): 26–30. https://doi.org/10.1007/BF02344684.

Francesco, F. Di, R. Fuoco, M. G. Trivella, and A. Ceccarini. 2005. "Breath Analysis: Trends in Techniques and Clinical Applications." *Microchemical Journal* 79 (1–2): 405–10. https://doi.org/10.1016/J.MICROC.2004.10.008.

Gawad, Shady, Tao Sun, Nicolas G. Green, and Hywel Morgan. 2007. "Impedance Spectroscopy Using Maximum Length Sequences: Application to Single Cell Analysis." *Review of Scientific Instruments* 78 (5): 054301. https://doi.org/10.1063/1.2737751.

Geelhoed-Duijvestijn, Petronella, Dovile Vegelyte, Alicja Kownacka, Nicoleta Anton, Maurits Joosse, and Christopher Wilson. 2021. "Performance of the Prototype NovioSense Noninvasive Biosensor for Tear Glucose in Type 1 Diabetes." *Journal of Diabetes Science and Technology* 15 (6): 1320–25. https://doi.org/10.1177/1932296820964844.

Geng, Zhanxiao, Fei Tang, Yadong DIng, Shuzhe Li, and Xiaohao Wang. 2017. "Noninvasive Continuous Glucose Monitoring Using a Multisensor-Based Glucometer and Time Series Analysis." *Scientific Reports* 7 (1): 1–10. https://doi.org/10.1038/s41598-017-13018-7.

Gersing, Eberhard. 1998. "Impedance Spectroscopy on Living Tissue for Determination of the State of Organs." *Bioelectrochemistry and Bioenergetics* 45 (2): 145–49. https://doi.org/10.1016/S0302-4598(98)00079-8.

Ghosh, Sudipta, M. Mahadevappa, and Jayanta Mukhopadhyay. 2017. "A 2D Electrode-Skin Model for Electrical & Contact Impedance Characterization of Bio Impedance." *IEEE Region 10 Annual International Conference, Proceedings/TENCON*, February, 2292–95. https://doi.org/10.1109/TENCON.2016.7848437.

Govindasamy, Mani, Veerappan Mani, Shen-Ming Chen, Anandaraj Sathiyan, Johnson Princy Merlin, and Vinoth Kumar Ponnusamy. 2016. "ELECTROCHEMICAL SCIENCE Sensitive and Selective Determination of Uric Acid Using Polyaniline and Iron Composite Film Modified Electrode." *International Journal Electrochemical Science* 11: 8730–37. https://doi.org/10.20964/2016.10.63.

Grieshaber, Dorothee, Robert MacKenzie, Janos Vörös, and Erik Reimhult. 2008. "Electrochemical Biosensors - Sensor Principles and Architectures." *Sensors (Basel, Switzerland)* 8 (3): 1400. https://doi.org/10.3390/S80314000.

Griffiths, R. W., M. E. Philpot, B. J. Chapman, and K. A. Munday. 1981. "Impedance Cardiography: Non-Invasive Cardiac Output Measurement after Burn Injury." *International Journal of Tissue Reactions* 3 (1): 47–55. https://europepmc.org/article/med/7287057.

Haeri, Z., M. Shokoufi, M. Jenab, R. Janzen, and F. Golnaraghi. 2016. "Electrical Impedance Spectroscopy for Breast Cancer Diagnosis: Clinical Study." *Integrative Cancer Science and Therapeutics* 3 (6): 1–6. https://doi.org/10.15761/ICST.1000212.

Halter, Ryan J., Alex Hartov, John A. Heaney, Keith D. Paulsen, and Alan R. Schned. 2007. "Electrical Impedance Spectroscopy of the Human Prostate." *IEEE Transactions on Bio-Medical Engineering* 54 (7): 1321–27. https://doi.org/10.1109/TBME.2007.897331.

Hao, Junxing, Zeqiang Zhu, Chengguo Hu, and Zhihong Liu. 2022. "Photosensitive-Stamp-Inspired Scalable Fabrication Strategy of Wearable Sensing Arrays for Noninvasive Real-Time Sweat Analysis." *Analytical Chemistry* 94 (10): 4547–55. https://doi.org/10.1021/ACS.ANALCHEM.2C00593/ASSET/IMAGES/LARGE/AC2C00593_0006.JPEG.

Hernández-Balaguera, E., E. López-Dolado, and J. L. Polo. 2016. "Obtaining Electrical Equivalent Circuits of Biological Tissues Using the Current Interruption Method, Circuit Theory and Fractional Calculus." *RSC Advances* 6 (27): 22312–19. https://doi.org/10.1039/C5RA24535D.

Hill, R. V., J. C. Jansen, and J. L. Fling. 1967. "Electrical Impedance Plethysmography: A Critical Analysis." *Journal of Applied Physiology* 22 (1): 161–68. https://doi.org/10.1152/JAPPL.1967.22.1.161.

Hoffer, E. C., C. K. Meador, and D. C. Simpson. 1969. "Correlation of Whole-Body Impedance with Total Body Water Volume." *Journal of Applied Physiology* 27 (4): 531–34. https://doi.org/10.1152/JAPPL.1969.27.4.531.

Hong-Yi, Hu, and Yuzuru Kato. 1995. "Body Composition Assessed by Bioelectrical Impedance Analysis (BIA) in Patients with Graves' Disease before and after Treatment." *Endocrine Journal* 42 (4): 545–50. https://doi.org/10.1507/ENDOCRJ.42.545.

Houssin, T., J. Follet, A. Follet, E. Dei-Cas, and V. Senez. 2010. "Label-Free Analysis of Water-Polluting Parasite by Electrochemical Impedance Spectroscopy." *Biosensors and Bioelectronics* 25 (5): 1122–29. https://doi.org/10.1016/J.BIOS.2009.09.039.

Hua, Mei, Manlan Tao, Ping Wang, Yinfeng Zhang, Zaisheng Wu, Yanbing Chang, and Yunhui Yang. 2010. "Label-Free Electrochemical Cocaine Aptasensor Based on a Target-Inducing Aptamer Switching Conformation." *Analytical Sciences : The International Journal of the Japan Society for Analytical Chemistry* 26 (12): 1265–70. https://doi.org/10.2116/ANALSCI.26.1265.

Ibrahim, Bassem, Drew A. Hall, and Roozbeh Jafari. 2018. "Bio-Impedance Spectroscopy (BIS) Measurement System for Wearable Devices." *2017 IEEE Biomedical Circuits and Systems Conference, BioCAS 2017- Proceedings* 2018-January (March): 1–4. https://doi.org/10.1109/BIOCAS.2017.8325138.

Ito, Yuka, Junko Okuda-Shimazaki, Wakako Tsugawa, Noya Loew, Isao Shitanda, Chi En Lin, Jeffrey La Belle, and Koji Sode. 2019. "Third Generation Impedimetric Sensor Employing Direct Electron Transfer Type Glucose Dehydrogenase." *Biosensors and Bioelectronics* 129 (March): 189–97. https://doi.org/10.1016/J.BIOS.2019.01.018.

Iwakura, C., H. Inoue, and S. Nohara. 2001. "Hydrogen–Metal Systems: Electrochemical Reactions (Fundamentals and Applications)." *Encyclopedia of Materials: Science and Technology* (January): 3923–41. https://doi.org/10.1016/B0-08-043152-6/00695-1.

Jakicic, J. M., R. R. Wing, and W. Lang. 1998. "Bioelectrical Impedance Analysis to Assess Body Composition in Obese Adult Women: The Effect of Ethnicity." *International Journal of Obesity and Related Metabolic Disorders : Journal of the International Association for the Study of Obesity* 22 (3): 243–49. https://doi.org/10.1038/SJ.IJO.0800576.

Jayaraman, Anandnayan, Kurt A. Kaczmarek, Mitchell E. Tyler, and Uchechukwu O. Okpara. 2007. "Effect of Localized Ambient Humidity on Electrotactile Skin Resistance." *Proceedings of the IEEE Annual Northeast Bioengineering Conference, NEBEC*, 110–11. https://doi.org/10.1109/NEBC.2007.4413303.

Jones, Lyndon, Alex Hui, Chau Minh Phan, Michael L. Read, Dimitri Azar, John Buch, Joseph B. Ciolino, et al. 2021. "BCLA CLEAR – Contact Lens Technologies of the Future." *Contact Lens and Anterior Eye* 44 (2): 398–430. https://doi.org/10.1016/J. CLAE.2021.02.007.

Jung, Myoung Hoon, Kak Namkoong, Yeolho Lee, Young Jun Koh, Kunsun Eom, Hyeongseok Jang, Wonjong Jung, Jungmok Bae, and Jongae Park. 2021. "Wrist-Wearable Bioelectrical Impedance Analyzer with Miniature Electrodes for Daily Obesity Management." *Scientific Reports* 11 (1): 1–10. https://doi.org/10.1038/s41598-020-79667-3.

Kalia, Yogeshvar N., Fabrice Pirot, and Richard H. Guy. 1996. "Homogeneous Transport in a Heterogeneous Membrane: Water Diffusion across Human Stratum Corneum in Vivo." *Biophysical Journal* 71 (5): 2692–2700. https://doi.org/10.1016/S0006-3495(96)79460-2.

Kamat, D. K., Dhanashri Bagul, and P. M. Patil. 2014. "Blood Glucose Measurement Using Bioimpedance Technique." *Advances in Electronics* 2014 (December): 1–5. https://doi. org/10.1155/2014/406257.

Kaufman, Eliaz, and Ira B. Lamster. 2016. "The Diagnostic Applications of Saliva— A Review." *Critical Reviews in Oral Biology & Medicine* 13 (2): 197–212. https://doi. org/10.1177/154411130201300209.

Kerner, Todd E., Keith D. Paulsen, Alex Hartov, Sandra K. Soho, and Steven P. Poplack. 2002. "Electrical Impedance Spectroscopy of the Breast: Clinical Imaging Results in 26 Subjects." *IEEE Transactions on Medical Imaging* 21 (6): 638–45. https://doi. org/10.1109/TMI.2002.800606.

Khalilzadeh, Mohammad A., and Mina Borzoo. 2016. "Green Synthesis of Silver Nanoparticles Using Onion Extract and Their Application for the Preparation of a Modified Electrode for Determination of Ascorbic Acid." *Journal of Food and Drug Analysis* 24 (4): 796–803. https://doi.org/10.1016/J.JFDA.2016.05.004.

Kim, Eun Jung, Myung Jin Choi, Jeoung Hwan Lee, Ji Eun Oh, Jang Won Seo, Young Ki Lee, Jong Woo Yoon, Hyung Jik Kim, Jung Woo Noh, and Ja Ryong Koo. 2017. "Extracellular Fluid/Intracellular Fluid Volume Ratio as a Novel Risk Indicator for All-Cause Mortality and Cardiovascular Disease in Hemodialysis Patients." *PLOS ONE* 12 (1): e0170272. https://doi.org/10.1371/JOURNAL.PONE.0170272.

Kim, Sooyeon, Hee Jae Jeon, Sijin Park, Dong Yun Lee, and Euiheon Chung. 2020. "Tear Glucose Measurement by Reflectance Spectrum of a Nanoparticle Embedded Contact Lens." *Scientific Reports* 10 (1): 1–8. https://doi.org/10.1038/s41598-020-65103-z.

Kumar, Pawan, Akash Deep, and Ki Hyun Kim. 2015. "Metal Organic Frameworks for Sensing Applications." *TrAC Trends in Analytical Chemistry* 73 (November): 39–53. https://doi. org/10.1016/J.TRAC.2015.04.009.

Kyle, Ursula G., Ingvar Bosaeus, Antonio D. De Lorenzo, Paul Deurenberg, Marinos Elia, José Manuel Gómez, Berit Lilienthal Heitmann et al. 2004. "Bioelectrical Impedance Analysis - Part I: Review of Principles and Methods." *Clinical Nutrition* 23 (5): 1226–43. https://doi.org/10.1016/j.clnu.2004.06.004.

Lee, Jee Young, Han Sung Ryu, Sung Soo Yoon, Eun Hye Kim, and Seong Woo Yoon. 2019. "Extracellular-to-Intracellular Fluid Volume Ratio as a Prognostic Factor for Survival in Patients With Metastatic Cancer." *Integrative Cancer Therapies* 18 (April): 1534735419847285.   https://doi.org/10.1177/1534735419847285/ASSET/IMAGES/ LARGE/10.1177_1534735419847285-FIG2.JPEG.

Lee, Seung Min, Hang Jin Byeon, Joong Hoon Lee, Dong Hyun Baek, Kwang Ho Lee, Joung Sook Hong, and Sang Hoon Lee. 2014. "Self-Adhesive Epidermal Carbon Nanotube Electronics for Tether-Free Long-Term Continuous Recording of Biosignals." *Scientific Reports* 4 (1): 1–9. https://doi.org/10.1038/srep06074.

Li, Dachao, Zhihua Pu, Wenshuai Liang, Tongkun Liu, Ridong Wang, Haixia Yu, and Kexin Xu. 2015. "Non-Invasive Measurement of Normal Skin Impedance for Determining the Volume of the Transdermally Extracted Interstitial Fluid." *Measurement* 62 (February): 215–21. https://doi.org/10.1016/J.MEASUREMENT.2014.11.015.

Lukaski, H. C., P. E. Johnson, W. W. Bolonchuk, and G. I. Lykken. 1985. "Assessment of Fat-Free Mass Using Bioelectrical Impedance Measurements of the Human Body." *The American Journal of Clinical Nutrition* 41 (4): 810–17. https://doi.org/10.1093/AJCN/41.4.810.

Macdonald, J. Ross. 1992. "Impedance Spectroscopy." *Annals of Biomedical Engineering* 20 (3): 289–305. https://doi.org/10.1007/BF02368532.

Makaram, Prashanth, Dawn Owens, and Juan Aceros. 2014. "Trends in Nanomaterial-Based Non-Invasive Diabetes Sensing Technologies." *Diagnostics* 4 (2): 27–46. https://doi.org/10.3390/DIAGNOSTICS4020027.

Marley, Gifty, Dianmin Kang, Erin C. Wilson, Tao Huang, Yuesheng Qian, Xiufang Li, Xiaorun Tao, Guoyong Wang, Huanmiao Xun, and Wei Ma. 2014. "Introducing Rapid Oral-Fluid HIV Testing among High Risk Populations in Shandong, China: Feasibility and Challenges." *BMC Public Health* 14 (1): 1–7. https://doi.org/10.1186/1471-2458-14-422/ TABLES/4.

Matthie, James R. 2008. "Bioimpedance Measurements of Human Body Composition: Critical Analysis and Outlook." *Expert Review of Medical Devices* 5 (2): 239–61. https://doi.org/10.1586/17434440.5.2.239.

McRae, Donald A., Mark A. Esrick, and Susette C. Mueller. 1999. "Changes in the Noninvasive, in Vivo Electrical Impedance of Three Xenografts during the Necrotic Cell-Response Sequence." *International Journal of Radiation Oncology*Biology*Physics* 43 (4): 849–57. https://doi.org/10.1016/S0360-3016(98)00487-8.

Moissl, Ulrich M., Peter Wabel, Paul W. Chamney, Ingvar Bosaeus, Nathan W. Levin, Anja Bosy-Westphal, Oliver Korth, et al. 2006. "Body Fluid Volume Determination via Body Composition Spectroscopy in Health and Disease." *Physiological Measurement* 27 (9): 921. https://doi.org/10.1088/0967-3334/27/9/012.

Mondal, Kunal, and Ashutosh Sharma. 2016. "Recent Advances in Electrospun Metal-Oxide Nanofiber Based Interfaces for Electrochemical Biosensing." *RSC Advances* 6 (97): 94595–94616. Royal Society of Chemistry. https://doi.org/10.1039/c6ra21477k.

Moqadam, Sepideh Mohammadi, Parvind Kaur Grewal, Zahra Haeri, Paris Ann Ingledew, Kirpal Kohli, and Farid Golnaraghi. 2018. "Cancer Detection Based on Electrical Impedance Spectroscopy: A Clinical Study." *Journal of Electrical Bioimpedance* 9 (1): 17–23. https://doi.org/10.2478/JOEB-2018-0004.

Mule, Nilakshi Maruti, Dipti D. Patil, and Mandeep Kaur. 2021. "A Comprehensive Survey on Investigation Techniques of Exhaled Breath (EB) for Diagnosis of Diseases in Human Body." *Informatics in Medicine Unlocked* 26 (January): 100715. https://doi.org/10.1016/J.IMU.2021.100715.

Muñoz, Jose, Raquel Montes, and Mireia Baeza. 2017. "Trends in Electrochemical Impedance Spectroscopy Involving Nanocomposite Transducers: Characterization, Architecture Surface and Bio-Sensing." *TrAC Trends in Analytical Chemistry* 97 (December): 201–15. https://doi.org/10.1016/J.TRAC.2017.08.012.

Naranjo-Hernández, David, Javier Reina-Tosina, and Mart Min. 2019. "Fundamentals, Recent Advances, and Future Challenges in Bioimpedance Devices for Healthcare Applications." *Journal of Sensors* 2019: 1–42. https://doi.org/10.1155/2019/9210258.

Naranjo-Hernández, David, Javier Reina-Tosina, Laura M. Roa, Gerardo Barbarov-Rostán, Nuria Aresté-Fosalba, Alfonso Lara-Ruiz, Pilar Cejudo-Ramos, and Francisco Ortega-Ruiz. 2020. "Smart Bioimpedance Spectroscopy Device for Body Composition Estimation." *Sensors (Basel, Switzerland)* 20 (1): 70. https://doi.org/10.3390/S20010070.

Nunes, Lazaro Alessandro Soares, Sayeeda Mussavira, and Omana Sukumaran Bindhu. 2015. "Clinical and Diagnostic Utility of Saliva as a Non-Invasive Diagnostic Fluid: A Systematic Review." *Biochemia Medica* 25 (2): 177–92. https://doi.org/10.11613/BM.2015.018/FULLARTICLE.

Osterman, K. Sunshine, P. Jack Hoopes, Christine DeLorenzo, David J. Gladstone, and Keith D. Paulsen. 2004. "Non-Invasive Assessment of Radiation Injury with Electrical Impedance Spectroscopy." *Physics in Medicine & Biology* 49 (5): 665. https://doi.org/10.1088/0031-9155/49/5/002.

Parrinello, Gaspare, Salvatore Paterna, Pietro Di Pasquale, Daniele Torres, Antonio Fatta, Manuela Mezzero, Rosario Scaglione, and Giuseppe Licata. 2008. "The Usefulness of Bioelectrical Impedance Analysis in Differentiating Dyspnea Due to Decompensated Heart Failure." *Journal of Cardiac Failure* 14 (8): 676–86. https://doi.org/10.1016/J.CARDFAIL.2008.04.005.

Pedro, Bruna Gabriela, David William Cordeiro Marcôndes, and Pedro Bertemes-Filho. 2020. "Analytical Model for Blood Glucose Detection Using Electrical Impedance Spectroscopy." *Sensors (Basel, Switzerland)* 20 (23): 1–11. https://doi.org/10.3390/S20236928.

Peppa, Melpomeni, Charikleia Stefanaki, Athanasios Papaefstathiou, Dario Boschiero, George Dimitriadis, and George P. Chrousos. 2017. "Bioimpedance Analysis vs. DEXA as a Screening Tool for Osteosarcopenia in Lean, Overweight and Obese Caucasian Postmenopausal Females." *Hormones (Athens, Greece)* 16 (2): 181–93. https://doi.org/10.14310/HORM.2002.1732.

Piuri, Gabriele, Enrico Ferrazzi, Camilla Bulfoni, Luciana Mastricci, Daniela Di Martino, and Attilio Francesco Speciani. 2016. "Longitudinal Changes and Correlations of Bioimpedance and Anthropometric Measurements in Pregnancy: Simple Possible Bed-Side Tools to Assess Pregnancy Evolution." *The Journal of Maternal-Fetal & Neonatal Medicine* 30 (23): 2824–30. https://doi.org/10.1080/14767058.2016.1265929.

Pliquett, Uwe, and Mark R. Prausnitz. 2000. "Electrical Impedance Spectroscopy for Rapid and Noninvasive Analysis of Skin Electroporation." *Electrochemotherapy, Electrogenetherapy, and Transdermal Drug Delivery* November, 37: 377–406. https://doi.org/10.1385/1-59259-080-2:377.

Redondo-del-Río, Ma Paz, Ma Alicia Camina-Martín, Laura Moya-Gago, Sandra de-la-Cruz-Marcos, Vincenzo Malafarina, and Beatriz de-Mateo-Silleras. 2016. "Vector Bioimpedance Detects Situations of Malnutrition Not Identified by the Indicators Commonly Used in Geriatric Nutritional Assessment: A Pilot Study." *Experimental Gerontology* 85 (December): 108–11. https://doi.org/10.1016/J.EXGER.2016.10.002.

Roa, Laura M., David Naranjo, Javier Reina-Tosina, Alfonso Lara, José A. Milán, Miguel A. Estudillo, and J. Sergio Oliva. 2013. "Applications of Bioimpedance to End Stage Renal Disease (ESRD)." *Studies in Computational Intelligence* 404: 689–769. https://doi.org/10.1007/978-3-642-27458-9_14/COVER.

Ryu, Hoon, Junghee Lee, Beatrix A. Olofsson, Aziza Mwidau, Alpaslan Deodoglu, Maria Escudero, Erik Flemington, Jane Azizkhan-Clifford, Robert J. Ferrante, and Rajiv R. Ratan. 2003. "Histone Deacetylase Inhibitors Prevent Oxidative Neuronal Death Independent of Expanded Polyglutamine Repeats via an Sp1-Dependent Pathway." *Proceedings of the National Academy of Sciences of the United States of America* 100 (7): 4281–86. https://doi.org/10.1073/PNAS.0737363100.

Sankhala, Devangsingh, Sriram Muthukumar, and Shalini Prasad. 2018. "A Four-Channel Electrical Impedance Spectroscopy Module for Cortisol Biosensing in Sweat-Based Wearable Applications." *SLAS Technology* 23 (6): 529–39. https://doi.org/10.1177/2472630318759257.

Sankhala, Devangsingh, Madhavi Pali, Kai-Chun Lin, Badrinath Jagannath, Sriram Muthukumar, and Shalini Prasad. 2021. "Analysis of Bio-Electro-Chemical Signals from Passive Sweat-Based Wearable Electro-Impedance Spectroscopy (EIS) Towards Assessing Blood Glucose Modulations." *arXiv preprint arXiv*: 2104.01793. https://doi.org/10.48550/arxiv.2104.01793.

Satish, Kushal Sen, and Sneh Anand. 2018. "Impedance Spectroscopy of Aqueous Solution Samples of Different Glucose Concentrations for the Exploration of Non-Invasive-Continuous-Blood-Glucose-Monitoring." *Mapan - Journal of Metrology Society of India* 33 (2): 185–90. https://doi.org/10.1007/S12647-017-0242-4/FIGURES/6.

Saylan, Yeşeren, Semra Akgönüllü, Handan Yavuz, Serhat Ünal, and Adil Denizli. 2019. "Molecularly Imprinted Polymer Based Sensors for Medical Applications." *Sensors* 19 (6): 1279. https://doi.org/10.3390/S19061279.

Schwan, H. P. 1957. "Electrical Properties of Tissue and Cell Suspensions." *Advances in Biological and Medical Physics* 5 (January): 147–209. https://doi.org/10.1016/B978-1-4832-3111-2.50008-0.

Schwan, Herman P. 2002. "Interface Phenomena and Dielectric Properties of Biological Tissue." *Encyclopedia of Surface and Colloid Science* 20: 2643–2653.

Scrymgeour, David A., Clark Highstrete, Yun Ju Lee, Julia W. P. Hsu, and Mark Lee. 2010. "High Frequency Impedance Spectroscopy on ZnO Nanorod Arrays." *Journal of Applied Physics* 107 (6): 064312. https://doi.org/10.1063/1.3319555.

Shah, Chirag, Frank A. Vicini, and Douglas Arthur. 2016. "Bioimpedance Spectroscopy for Breast Cancer Related Lymphedema Assessment: Clinical Practice Guidelines." *The Breast Journal* 22 (6): 645–50. https://doi.org/10.1111/TBJ.12647.

Shokoufi, Majid, and Farid Golnaraghi. 2016. "Development of a Handheld Diffuse Optical Breast Cancer Assessment Probe." *Journal of Innovative Optical Health Sciences* 9 (2): 1650007. https://doi.org/10.1142/S1793545816500073.

Sireesha, Merum, Veluru Jagadeesh Babu, A. Sandeep Kranthi Kiran, and Seeram Ramakrishna. 2018. "A Review on Carbon Nanotubes in Biosensor Devices and Their Applications in Medicine." *Nanocomposites* 4 (2): 36–57. https://doi.org/10.1080/20550324.2018.1478765.

Skourou, Christina, P. Jack Hoopes, Rendall R. Strawbridge, and Keith D. Paulsen. 2004. "Feasibility Studies of Electrical Impedance Spectroscopy for Early Tumor Detection in Rats." *Physiological Measurement* 25 (1): 335. https://doi.org/10.1088/0967-3334/25/1/037.

Suni, Ian I. 2008. "Impedance Methods for Electrochemical Sensors Using Nanomaterials." *TrAC Trends in Analytical Chemistry* 27 (7): 604–11. https://doi.org/10.1016/J.TRAC.2008.03.012.

Suresh, Vignesh, Ong Qunya, Bera Lakshmi Kanta, Lee Yeong Yuh, and Karen S.L. Chong. 2018. "Non-Invasive Paper-Based Microfluidic Device for Ultra-Low Detection of Urea through Enzyme Catalysis." *Royal Society Open Science* 5 (3): 171980. https://doi.org/10.1098/rsos.171980.

Tang, Tingfan, Menglin Zhou, Jiapei Lv, Hao Cheng, Huaisheng Wang, Danfeng Qin, Guangzhi Hu, and Xiaoyan Liu. 2022. "Sensitive and Selective Electrochemical Determination of Uric Acid in Urine Based on Ultrasmall Iron Oxide Nanoparticles Decorated Urchin-like Nitrogen-Doped Carbon." *Colloids and Surfaces B: Biointerfaces* 216 (August): 112538. https://doi.org/10.1016/J.COLSURFB.2022.112538.

Trevizan, Heitor Furlan, André Olean-Oliveira, Celso Xavier Cardoso, and Marcos F. S. Teixeira. 2021. "Development of a Molecularly Imprinted Polymer for Uric Acid Sensing Based on a Conductive Azopolymer: Unusual Approaches Using

Electrochemical Impedance/Capacitance Spectroscopy without a Soluble Redox Probe." *Sensors and Actuators B: Chemical* 343 (September): 130141. https://doi.org/10.1016/J.SNB.2021.130141.

Tura, A., S. Sbrignadello, S. Barison, S. Conti, and G. Pacini. 2007. "Impedance Spectroscopy of Solutions at Physiological Glucose Concentrations." *Biophysical Chemistry* 129 (2–3): 235–41. https://doi.org/10.1016/J.BPC.2007.06.001.

Vanbever, Rita, Nathalie Lecouturier, and Véronique Préat. 1994. "Transdermal Delivery of Metoprolol by Electroporation." *Pharmaceutical Research* 11 (11): 1657–62. https://doi.org/10.1023/A:1018930425591.

Villa, Federica, Alessandro Magnani, Martina A. Maggioni, Alexander Stahn, Susanna Rampichini, Giampiero Merati, and Paolo Castiglioni. 2016. "Wearable Multi-Frequency and Multi-Segment Bioelectrical Impedance Spectroscopy for Unobtrusively Tracking Body Fluid Shifts during Physical Activity in Real-Field Applications: A Preliminary Study." *Sensors* 16 (5): 673. https://doi.org/10.3390/S16050673.

Wang, Yuliang, and Younan Xia. 2004. "Bottom-up and Top-down Approaches to the Synthesis of Monodispersed Spherical Colloids of Low Melting-Point Metals." *Nano Letters* 4 (10): 2047–50. https://doi.org/10.1021/NL048689J/ASSET/IMAGES/MEDIUM/NL04 8689JN00001.GIF.

Wei, Jingwei, Xieli Zhang, Samuel M. Mugo, and Qiang Zhang. 2022. "A Portable Sweat Sensor Based on Carbon Quantum Dots for Multiplex Detection of Cardiovascular Health Biomarkers." *Analytical Chemistry* 94 (37): 12772–12780. https://doi.org/10.1021/ACS. ANALCHEM.2C02587/ASSET/IMAGES/LARGE/AC2C02587_0007.JPEG.

Wentholt, I. M. E., J. B. L. Hoekstra, A. Zwart, and J. H. DeVries. 2005. "Pendra Goes Dutch: Lessons for the CE Mark in Europe." *Diabetologia* 48 (6): 1055–58. https://doi.org/10.1007/S00125-005-1754-Y/FIGURES/2.

Wu, Jie, Yuxing Ben, and Hsueh Chia Chang. 2005. "Particle Detection by Electrical Impedance Spectroscopy with Asymmetric-Polarization AC Electroosmotic Trapping." *Microfluidics and Nanofluidics* 1 (2): 161–67. https://doi.org/10.1007/S10404-004-0024-5/FIGURES/8.

Xiong, Can, Tengfei Zhang, Weiyu Kong, Zhixiang Zhang, Hao Qu, Wei Chen, Yanbo Wang, Linbao Luo, and Lei Zheng. 2018. "ZIF-67 Derived Porous Co3O4 Hollow Nanopolyhedron Functionalized Solution-Gated Graphene Transistors for Simultaneous Detection of Glucose and Uric Acid in Tears." *Biosensors and Bioelectronics* 101 (March): 21–28. https://doi.org/10.1016/J.BIOS.2017.10.004.

Xue, Yirui, Angelika S. Thalmayer, Samuel Zeising, Georg Fischer, and Maximilian Lübke. 2022. "Commercial and Scientific Solutions for Blood Glucose Monitoring—A Review." *Sensors (Basel, Switzerland)* 22 (2): 425. https://doi.org/10.3390/S22020425.

Yan, Qinghua, Na Zhi, Li Yang, Guangri Xu, Qigao Feng, Qiqing Zhang, and Shujuan Sun. 2020. "A Highly Sensitive Uric Acid Electrochemical Biosensor Based on a Nano-Cube Cuprous Oxide/Ferrocene/Uricase Modified Glassy Carbon Electrode." *Scientific Reports* 10 (1): 1–10. https://doi.org/10.1038/s41598-020-67394-8.

Yu, Hong Wei, Ze Zhang, Tao Shen, Jing Hui Jiang, Dong Chang, and Hong Zhi Pan. 2018. "Sensitive Determination of Uric Acid by Using Graphene Quantum Dots as a New Substrate for Immobilisation of Uric Oxidase." *IET Nanobiotechnology* 12 (2): 191. https://doi.org/10.1049/IET-NBT.2016.0221.

Zając-Gawlak, Izabela, Barbara Kłapcińska, Aleksandra Kroemeke, Dariusz Pośpiech, Jana Pelclová, and Miroslava Přidalová. 2017. "Associations of Visceral Fat Area and Physical Activity Levels with the Risk of Metabolic Syndrome in Postmenopausal Women." *Biogerontology* 18 (3): 357–66. https://doi.org/10.1007/S10522-017-9693-9/FIGURES/3.

Zhao, Murong, and Po Sing Leung. 2020. "Revisiting the Use of Biological Fluids for Noninvasive Glucose Detection." *Future Medicinal Chemistry* 12 (8): 645–47. https://doi.org/10.4155/FMC-2020-0019.

7 Recent progress in biomedical devices and importance of impedance spectroscopy

Vibhas Chugh, Abhishek Naskar, Manshu Dhillon, Adreeja Basu, and Aviru Kumar Basu

CONTENTS

DOI: 10.1201/9781003358091-9

7.1 INTRODUCTION

Equipment, instruments, implants, in vitro reagents, materials, or other similar items used for the safe and effective detection, therapy, and recovery of disease and illness in humans are referred to as biomedical devices. These tools can be used to accomplish a variety of goals, including the diagnosis of illness and damage, life supports, sample analysis, treatment assessment, and body part replacement. Thus, biomedical devices cover a wide range of goods with various degrees of complexity and functionality, from straightforward instruments like thermometers, scalpels, syringes, lenses, tongue depressors, and blood sugar meters to more intricate ones like X-ray machines, heart valves, neuroprosthetics, medical robots for surgeries, artificial intelligence (AI)-based programmable pacemakers, or micro/nanochip implants (Lam and Chen 2019).

To progress global health scope, deal with medical emergencies and encourage healthier communities, access to high-quality, reasonably priced, and adequate medical products is essential. In the absence of medical devices, routine medical procedures like bandaging a sprained ankle, detection of diseases, artificial implanting procedures, or performing any type of surgery will be impossible to perform. These medical technologies are employed in the diagnosis and treatment of acute and chronic illnesses, as well as in the support of persons with disabilities (Khan et al. 2014).

Scientists and engineers have explored biomedical production through a new body of information. This information includes product planning and optimization, material selection and properties, manufacturing processes, quality assurance, the interplay between the body and the materials used, legal commitments, and an understanding of physiology and biochemistry. To produce novel biomedical products successfully, understanding biomedical production and legislation is a crucial hurdle. Once they have this information, biomedical engineers create products that are well-balanced in terms of costs, materials, and machining processes (Manickam et al. 2022).

The biggest sectors of the biomedical device market at the moment are imaging diagnostics, implantable medical gadgets, home test apparatus, common surgical tools, and all types of consumables. In recent years, the COVID-19 pandemic has impacted almost every facet of life that has brought about many changes in our lifestyle and healthcare (Singh et al. 2020). We witnessed numerous changes, particularly in how the global population interacted with one another. Doctors, researchers, businesses, and nations worked together in novel ways to share technology and medical developments that could aid in the treatment of the virus or limit its transmission.

This movement towards globalization, exchange of knowledge, and shared objectives are expected to persist in a number of fields in the coming years, particularly medical technology. All areas of technology, including the Internet of things (IoT), AI, manufacturing materials, sensing technologies, wireless and wearable capabilities, and others, continue to experience substantial advancements (Rahaman et al. 2019). Medical devices develop along with broader trends in technology as a result of these developments and the availability of better components. This chapter covers the latest technology that is being used to develop biomedical devices. It also covers the materials that are being used or have the capability to be incorporated into such devices to increase their efficiency. Further, the chapter discusses various devices for diversified biomedical applications.

7.2   TYPES OF BIOMEDICAL DEVICES

Utilizing a device for health reasons carries a large risk of concerns; hence, medical devices must be demonstrated to be safe and effective with a fair degree of assurance before the public sale of the device is permitted. The quantity of testing needed to verify a device's effectiveness and security typically grows along with the associated risk of the device. Additionally, if related risk rises, the patient's potential benefit must rise as well. Many implantable medical devices have been created in the past few decades for the treatment of diseases and disorders as well as the monitoring of patients' biological parameters. In order to monitor physiological processes, provide medications, or maintain the operations of particular organs or tissues, implantable medical devices are inserted into the human body either permanently or temporarily. Certain implants are prostheses used to replace broken physical parts (Jung et al. 2020). Nonimplantable items are frequently utilized to prevent infection, absorb blood and exudates, and accelerate healing. Nonimplantable refers to surface wound treatments for various body sections that cannot be implanted. Materials that are applied externally to the body that may or may not come into contact with skin are not implantable. This covers products used for eye protection and wound treatment, and orthopaedic belts and plasters (Wang, Shi, and Lee 2022).

7.2.1   Implantable Devices

Implantable medical devices are inserted into the human body either temporarily or permanently to detect physiological processes, distribute drugs, or support the functions of particular tissues or organs. A medical device that is "designed to be wholly or partially inserted, surgically or therapeutically, into the human body or by surgical assistance into a natural orifice, and which is planned to remain after the surgery" is known as an active implanted medical device. Artificial joints, breast implants, contraceptive intrauterine devices (IUDs), and hardware for bone, muscle, and joint fusion are among the most frequently utilized implantable medical devices. An implanted vascular access device is another popular form of implantable medical device that is utilized in patients with poor peripheral venous access or who need regular vein access for therapies like chemotherapy (Khan et al. 2014). Examples of some biomedical implants are discussed in further sections.

7.2.1.1 Medical textiles

Development in the textiles industry, whether it is for natural or synthetic materials, is frequently focused on how the users' comfort is improved. One such invention that is essentially intended to turn patients' uncomfortable days into comfortable days is the development of medical textiles. General medical textile requirements include strength, biodegradability, nontoxicity, biological compatibility, dimensional stability, resistance to cancer and allergies, increased bodily comfort, and antifungal and antibacterial efficiency. In general, implantable materials are designed to heal bodily sections that have been compromised (Guru et al. 2022). They are widely utilized as wound sutures, surgical time replacement, and other segmental replacements such as artificial ligaments and vascular grafts, including soft tissue implants, orthopaedic implants, and cardiovascular implants. For exterior applications, such as bandages, plasters, wound dressing, and wound healing products, nonimplantable materials are used.

7.2.1.2 Cardiac stimulator device

Cardiac stimulator devices are active implantable medical devices that are routinely required to treat heart problems. A pacemaker, a tiny, battery-operated wireless implantable medical device that helps the heart beat in a regular rhythm, or an implantable cardioverter defibrillator (ICD), a device put under the skin that monitors the heart rate, may be required by a patient to help control the heart's rhythm (Tarakji et al. 2010). A patient may require a left ventricular assist device (LVAD), a battery-operated mechanical pump-type device to sustain the pumping ability of a heart that cannot properly perform on its own, to sustain the structure of the heart and circulation (Ausra et al. 2022).

7.2.1.3 Sleep apnoea devices

Sleep apnoea is another condition for which an implantable medical device may be used. An implanted sleep apnoea device may be used to treat patients who struggle with obstructive sleep apnoea or those who are unable to utilize a continuous positive airway pressure machine. In addition to regulating breathing, this medical device prevents the tongue from impeding breathing during sleep by activating the nerve that causes the tongue to move.

7.2.1.4 Artificial joints

Metal, ceramic, or robust plastic can be used to make the artificial joint (prosthesis). The artificial joint behaves similarly to the native joint and has a comparable appearance. Any joint in the body can be replaced by a surgeon; however, hip and knee replacements are the two most popular arthroplasty procedures (Mamidi et al. 2019). The primary polymer utilized in joints is ultra-high-molecular-weight polyethylene (UHMWPE), which is also a popular material for acetabular cups, because of its improved biocompatibility, high impact strength, and lack of toxicity. Despite the fact that metal or plastic couplings produce relatively modest frictional torques, the UHMWPE's high wear rate has become the main issue. The volume, size, and shape of the wear particles that are created, more frequently, trigger a biological response in the bone that surrounds the prosthesis, rather than that most joints wear down (Patil,

Njuguna, and Kandasubramanian 2020). This reaction causes osteolysis or bone loss, and the prosthesis subsequently becomes looser.

7.2.1.5   Wearable artificial throat

Stretchable, sensitive strain sensors with conformability to the skin are needed for wearable sound detectors, and 2D materials show tremendous promise in this regard. The majority of mute patients are unable to communicate owing to vocal-chord lesions. A wearable skin-like ultrasensitive artificial graphene throat (WAGT) that combined sound/motion detection and sound production into one device was presented to help the deaf and hard of hearing to "talk." The WAGT system, which combines a thermoacoustic sound generator with a sound/motion detector in a single device, is suggested. First, the WAGT gadget exhibits a maximum 150% strain range as a sound detector, exceptional skin adherence, and an ultrathin thickness. With the accumulated particular waveforms for each word, it can identify sound signals. The WAGT motion-detected system is able to convert various human motions to a variety of noises when acting as a motion detector. Overall, the WAGT is able to exactly speak on behalf of mute people by sensibly detecting their throat movement signals. This is very significant for assisting mute people in the future in "speaking" for themselves (Wei et al. 2019). An MXene-based sound detector is successfully constructed with better recognition and sensitive reaction to pressure and vibration using the combination of deep learning (DL) and 2D MXenes, which makes it possible to create a sound detector with high recognition and resolution. The long vowels and short vowels of human pronunciation are successfully identified by training and testing a DL network model with a significant amount of data collected by an MXene-based sound detector (Jin et al. 2020).

7.2.1.6   Contraceptive devices

Small contraceptive devices called IUDs placed inside the uterus are the safest and most efficient reversible contraceptives in use today (Esther et al. 2019). Additionally, it is the most efficient emergency contraception and is especially well suited for use by women in developing nations because it is accessible, inexpensive, and does not necessitate regular clinic visits. The copper IUD (Cu-IUD) and the hormonal IUD are the two varieties that are offered. Widely used Cu-IUDs, containing copper exhibit common side effects, including bleeding and pain right after insertion. Cu material replacement is a must requirement in these devices. Since 1969, zinc and its alloys, an emerging class of biodegradable materials, have also demonstrated contraceptive properties demonstrating it as an active component in IUDs (Bao et al. 2022).

7.2.2   Nonimplantable devices

The significance of in vitro biomedical applications, including biological entity placement, isolation, detection, stimulation, and cultivation, has increased significantly in recent times. Such implementations help to elucidate the behaviours and features of the associated entities, including their stimulus–response, receptor function, and various other properties (Ahmad et al. 2021). Instead of reproducing the full cell microenvironment, new approaches in the in vitro modelling condense the range of human organ architecture to the basic cellular microanatomy allowing for a thorough evaluation of complex metabolic activities and structural inquiry. Due to its potential application in the creation of microfluidic devices for health monitoring, detail analysis, and clinical research, organ-on-a-chip (OoC) materials constitute an innovative study area that has gained major interest (Basu, Basu, and Bhattacharya 2019). Ongoing animal rights resurgence and awareness, specific concerns, and related process failures have arisen along with the monetary constraints and messy nature of using animal for studies. Consequently, to accelerate and increase the favourable outcomes of clinical trials, improved tissue sample models for drug screening in developmental investigations are desired. Systems built on microfluidic chips, also known as OoC systems, can be used to solve this problem by integrating 3D biocompatible tissue architectures as well as a dynamic milieu into chip-based systems to mimic in vitro physiological processes at the organ level.

Structural architecture, sensors, diagnostic investigation, clinical acceptance, and programming employed on OoC materials provide valuable insights into human health and illness, and the present knowledge of disease development is based on these factors. In the coming years, widespread adoption will be made feasible at a low cost, thanks to the incorporation of various technological inputs to varying degrees, including emerging materials, automation, AI, machine learning (ML), and improved genomics (Nahak et al. 2022). Enhanced comprehension of the fundamental processes underlying neuronal activity, even at the single-cell level, is necessary to comprehend its behaviour, under both normal and pathological circumstances that have prompted major work mostly on the creation of tiny sensor gadgets to track neural activity with great spatial and transmission precision.

A very novel and powerful self-aligned functionalization method that enables unmatched accuracy in organizing neural networks at the single-cell level is developed to completely separate microfabrication from bioassays and offers superior temporal and physical reliabilities. This technique produced wide algorithms at sparse populations with extremely excellent sensitivity (low parasitic cell connection). This method enables executing the cell culture on packed chips compatible with complementary metal-oxide-semiconductor line technology at a wafer size thus giving 75% of unicellular nodes controlled at the micron scale at the soma. In vitro versions of the nervous system have been created using the influence of tetrahedral DNA nanostructures (TDNs) on neuro-ectodermal stem cells, allowing them to additionally grow and develop into a neuronal lineage. This administration of TDNs induced the characteristic reestablishment of neuroectodermal (NE-4C) stem cells by promoting the activation of the Wnt/β-catenin channel that is necessary for the control of the cell cycle and neural development by suppressing the Notch signalling pathway. In the absence of any transfection agents, TDNs were further introduced into dental pulp stem cells produced by human dental pulp.

The lithography method used in a highly efficient way quickly coordinated the parallel assembling of multiple solid-phase ligands over length scales. By displaying diverse inputs spatially on cell surfaces, such DNA-directed approaches provide fascinating insights into how tissues encode regulatory signals (Hivare et al. 2021). A younger generation of fluorescent sensors and biosensors that make use of nanostructures like quantum dots or carbon nanomaterials are being developed for the measurement of neurohormones in order to identify abnormalities in neurohormone levels brought on by long-term psychological stress and develop effective stress-reduction strategies.

For sustaining health, both mentally and physically, hormones including adrenaline, dopamine, and cortisol are essential. Small variations in their concentrations in bodily secretions can generate neurodegenerative and cardiovascular issues, indicative of the existence of a variety of illnesses. Smartphones integrated biosensors, including cameras, light sources, image processing, and communication capabilities, can minimize prices, facilitate the large-scale arrangement, and spread cutting down on processing time. Microfluidic integrated chips often 3D printed may be used to wirelessly transfer the findings for analysis, evaluate data using customized apps, and monitor signals directly through clinical specimens. Expenses might be decreased by adopting microfluidic-based biodevices because they demand tiny volumes of reagents and reduced materials (Halicka et al. 2022). The goal of in vitro science is to provide a collection of straightforward, affordable, transportable, and durable instruments that might be employed in a variety of healthcare engineering domains, including drug development, diagnostics, and therapeutic methods for regenerative medicine. The recent trends of integrated programmable circuitry based on developing technologies such as microfluidics and 3D printing make it an area of interest from the application, research, and commercial point of view.

7.2.3 SENSORS FOR BIOMEDICAL APPLICATIONS

Specialized electronic instruments with either a biological component or a biomimetic (such as antibodies or nucleic acids) that are intimately correlated to or integrated into a physicochemical transducer or transducing microsystems are called biomedical sensors, which convert biological impulses into quantifiable electrical signals. The bio-receptor, the transducing element, the excitation element, and the readout modality are the components that make up a biosensor (Vasan et al. 2013). Several factors, including the biosensing technique (label-based or label-free), transduction process, and kind of receptor employed, may be used to categorize biomedical sensors. The transducer, which helps the sensor produce an electrical signal proportionate to a single analyte or a collection of related analytes used and the mode of operation it is built upon, is one of the crucial components of how biosensors work (Basu, Basu, and Bhattacharya 2020). As shown in Figure 7.1, biosensors can be categorized into certain specific groups, such as thermal, mechanical/piezoelectric, electrical/electrochemical, optical, and chemical. Though other categories may be included in the group depending on the many factors, such as its type of operation, and material utilized, each group has a certain operating principle that it adheres to. The following sections further detail these sensors.

7.2.3.1 Piezoelectric sensors

Certainly, if we discuss piezoelectric biosensors, we know that their working is based on the basic principle of oscillations generated from the deformation of the anisotropic crystal, which in turn is associated with the generation of alternating voltage. The fact that biological molecules may exhibit piezoelectric effect opens up new

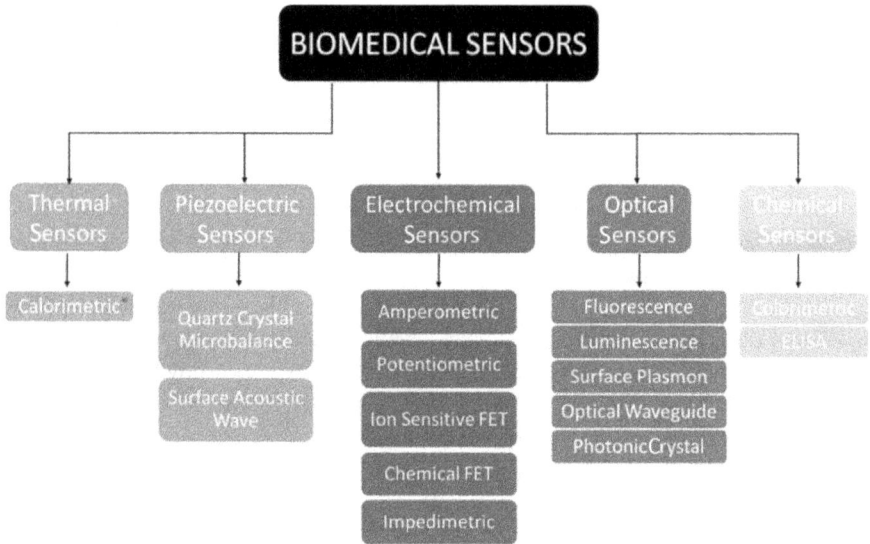

FIGURE 7.1 Classification of biosensors based on transduction mechanisms.

possibilities for the design of biosensors. The interesting class of biological entities known as protein ion channels, with anisotropic characteristics merits additional
study (Pohanka 2017). This class of piezoelectric biosensors is further characterized
into quartz crystal microbalance (QCM) and surface acoustic wave (SAW). In QCM,
a thin quartz disc with electrodes acts as a sensing surface on both faces of the crystal.
Mechanical stresses are created in quartz when an external electric field is applied.
As a result, the crystal oscillates perpendicular to the surface of the plate because of
the applied alternating voltage characterized by a resonant frequency depending on
the mass per unit area. For disease biosensing, the mass-based detection approach of
QCM offers numerous intrinsic benefits (Lim et al. 2020).

In the case of SAW, on the surface of the piezoelectric substrate, interdigital transducers are constructed to effectively transform the electrical signal to an acoustic
one. Vapour and biomolecule detection by observing frequency shifts can be done
using SAW. Also, it is more efficient than the traditional ultrasound methods in the
case of cell lysis. It is also beneficial in cell adhesion monitoring and sorting of cells
and particles (Huang, Das, and Bhethanabotla 2021).

7.2.3.2 Electrochemical sensors

Electrochemical biosensors convert biological interactions into electrical impulses.
A major element of these systems is an electrode, which serves as a stable platform
for the immobilization of biomolecules and the passage of electrons (Cho, Kim, and
Park 2020). In this line, amperometric, voltametric, conductometric, and impedimetric sensors help extensively in the determination of various components in aqueous
samples (Soldatkin et al. 2021).

Biosensors are created using the direct electrochemical processes of oxidore-
ductases for quick, inexpensive, and interactive analyte measurement. The analyte

array includes a wide variety of pertinent analytes, including phenolic compounds, saccharides, and tiny molecules like sulphite or alcohol compounds that are crucial for tracking environmental changes, food sector chemicals, and by-products of biomimetic operations (Kumar Basu et al. 2021). While there are numerous such enzymes with adequate electrochemical characteristics that may be used to identify some sources, numerous research papers focused on glucose detection because of the significant clinical interest in blood glucose estimation. The several electrode materials/shapes integrated with various electrode modifications that are capable of DET can be efficiently utilized for third-generation amperometric sensors (Schachinger et al. 2021).

Given its potential for miniaturization, parallel sensing, quick response times, and easy integration with digital production processes, including CMOS semiconductors, detecting utilizing field-effect transistors has gained a lot of attention between many different potentiometric techniques. Despite the difficulties, a number of businesses hope to provide point-of-care diagnostics soon. For instance, DNA is now working on an ISFET-based sepsis diagnostic that is produced using standard CMOS manufacturing procedures wherein the method of detection relies on the detection of hydrogen ions generated during DNA amplification (Kaisti 2017).

7.2.3.3 Thermal sensors

Dopamine is a molecule that is physiologically active and serves a variety of important roles as a hormone and neurotransmitter and it is used for various treatments such as Parkinson's disease, schizophrenia, and depression. Hence, it is crucial for biomedical research to determine the body's dopamine content and various related medicinal research. The majority of enzyme-catalysed reactions have an exothermic character. When an enzyme reacts with an analyte solution, calorimetric biosensors assess the change in the solution's temperature and translate it into the concentration of the analyte. A tiny fixed bed section made up of immobilized enzymes is used to filter the analyte solution. A different thermistor is used to monitor the temperature of the solution just as it enters the column and just as it exits the column.

Recently, thermal biosensors are being used for the detection of chemical oxygen demand, which is a parameter for the organic pollutants in wastewater. Heat is generated during oxidizing the organic contaminants in wastewater. Heat generated during this oxidation process is measured by the biosensor. The flow injection analysis method is utilized in conjunction with the test. This continuous analytical method allows for the control of test water samples and carrier solution amounts. Batch manufacture at cheap costs and low-cost integration of miniature devices are some of the advantages showcased by the MEMS (Vasuki et al. 2019).

7.2.3.4 Optical sensors

With a strong track record of identifying biological mechanisms, and having significantly advanced clinical diagnostics, drug development, food process improvement, and environmental control, optical biosensors prove to be worthy of their use. Without the pre-treatment complication and the possible effect on the composition of target molecules, they provide additional advantages, such as high sensitivity, durability, dependability, and the capacity to be integrated into a single chip. Optical biosensing class has seen the emergence of enhanced approaches based on surface plasmon resonance (SPR), such as localized SPR, SPR imaging, and long-range surface plasmon. As a biosensor, optical waveguides with the widely used shapes of rib waveguide, slot waveguide, and photonic waveguide are encouraged by an evanescent-based mechanism.

The interferometric waveguide was created as a new type of biosensor by cleverly combining the waveguiding and interferometry methods (Chen and Wang 2020). For point-of-care detections of dangerous viruses to stop a potential pandemic outbreak, colorimetric biosensors are extremely desirable. These sensors may be utilized to identify a specific analyte through colour changes. Rapid discovery, better efficiency, lower cost, and more dependability with highly consistent test findings make the loop-mediated isothermal amplification (LAMP) approach more advantageous.

7.2.3.5 Chemical sensors

For highly accurate on-site viral identification, nanoparticle-based colorimetric biosensors may be linked with mobile PCR or LAMP equipment. This is essential for prompt virus infection diagnosis and rapid and managed epidemic prevention (Zhao et al. 2020).

With the help of their precise binding to immobilized antibodies or antigens, cells, viruses, and molecular antigens may be detected and quantified using the commonly used traditional approach known as ELISA. For POC diagnosis and many other bioanalyses, a multi-colorimetric ELISA biosensor on a paper/polymer hybrid analytical device is extremely attractive as it is mobile, inexpensive, and enables quick quasi- and quantitative analysis without the necessity of expensive specialized equipment (Ma, Abugalyon, and Li 2021).

7.3   MODERN-DAY MATERIALS FOR BIOMEDICAL DEVICES

Recent advancements in cutting-edge nanotechnologies have made it possible for scientists to make great progress in the investigation of novel materials with the potential for use in medical biotechnologies (Pankhi Singh et al. 2023). Many intriguing 2D substances have received more attention, some of which are discussed in the next section.

7.3.1   GRAPHENE

A single layer of atoms organized in a 2D honeycomb lattice nanostructure makes up the carbon allotrope known as graphene. A new family of broad-spectrum antibacterial substances called graphene materials is quickly evolving. Examples include graphene quantum dots, graphene oxide, and reduced graphene oxide (Tabish, Abbas, and Narayan 2021). Their antibacterial qualities may be beneficial in all medical specialties where antiseptics are desired. Because of their excellent electrical and mechanical characteristics, these materials are advantageous with improved stretchability and flexibility for precise incorporation with the curvilinear, soft faces of human organs or tissues while keeping their distinguished functionality intact (Kumar et al. 2019; Basu et al. 2020). Even if CNTs and graphene have exceptional physical qualities, they cannot absorb into the body of humans, if their clear biocompatibility has not been proved. Graphene and CNTs have specific characteristics of having concentration-dependent biocompatibility, functionalization, production process, and type of cell lines for which they can be utilized.

Even though the majority of research with carbon nanomaterials in mammals has demonstrated biocompatibility, serious analysis of the types of applications is still required as some specific studies have been found harmful to specific tissues, such as the skin and lungs. Standard microelectrodes with a surface coating of CNTs or CNT composites and various other conductive materials, such as gold and PEDOT have a significantly higher surface roughness than usual, which leads to a high SNR (Karahan et al. 2018). The capacity to measure analytes and physiological status as well as the special material properties of graphene make it possible to identify important biomarkers that are suggestive of human disease. Many patients could benefit from the creation of user-friendly, cost-effective, and antibacterial transdermal biosensing equipment for continuous and customized monitoring of target molecules. The main areas of discussion are the usage of graphene-based transcutaneous biosensors for health monitoring, assessment of these devices for glucose and hydrogen peroxide sensing using in vitro, in vivo, and ex vivo research, recent technological advancements, and potential difficulties (Kim, Cho, and Yu 2018). A hybrid material made of graphene nanoribbons and sheets offers increased surface area. As a result, it can be employed to increase loading capacity in the medication delivery industry (Johnson, Gangadharappa, and Pramod 2020).

7.3.2 Borophene

Borophene (BO) is made of boron atoms; it is organized in a hexagonal shape with typical covalent interactions binding each boron atom. It is a 2D allotrope of boron, and because of its crystalline atomic monolayer, it is also referred to as a boron sheet. BO is an active and intriguing material with electrical, optical, and thermal properties (Tatullo et al. 2019). BO, both in nanostrip form and when combined with other metals, has the potential to improve the functionality or extend the lifecycle of biomedical devices. Recently, a 2D triangular structure named BO was produced by the synthesis of thin 2D boron sheets on the Ag (111) substrates. BO has repeatedly displayed graphene-like anisotropic behaviour.

An intelligent and highly sensitive method of converting temperature, pH, and other stimuli into electrical, optical, and mechanical information is the combination of BO with hydrogel-based substrates (Kumar Sharma et al. 2022). 2D-BO-enabled biosensors are intended to have the ultrahigh sensitivity necessary for low LOD, a quick respond, and a combination with microelectronics to create inexpensive diagnostics. A highly anisotropic material with good electrical conductivity has been explored, and it is called 2D-BO (Shao et al. 2021). BO nanosheets display potently inhibitive effects on a number of bacterial and fungal pathogens that are medically dangerous (Taşaltın, Güllülü, and Karakuş 2022).

7.3.3 MXENE

MXenes are a group of 2D inorganic chemicals. These substances have layers of transition metal carbides, nitrides, or carbonitrides that are only a few atoms thick. MXenes were first introduced in 2011, and they integrate the hydrophilic nature because of oxygen-terminated surfaces or their hydroxyl group with the metallic conductivity of transition metal carbides (Taşaltın, Güllülü, and Karakuş 2022). Because of their outstanding mechanical, optical, electrical, and tunable capabilities, MXenes quickly gained popularity. These remarkable qualities made them desirable materials for biomedical and biosensing purposes, including drug delivery systems, antibacterial treatments, tissue engineering, sensor probes, and supplementary agents for photothermal therapy and hyperthermia uses, among others. When compared with hydrophobic nanoparticles that can need laborious surface functionalization, MXenes hydrophobicity and abundance of surface functional groups are helpful for medicinal applications (Koyappayil et al. 2022). Although the impact of surface termination on MXene performance has been extensively studied, not knowing the surface structure and how it changes with the environment has proven to be a major obstacle from both a theoretical and experimental standpoint (Wang et al. 2021). The search for new prospective biomedical applications of MXenes in combination with imaging agents can produce useful outcomes in cancer theranostics, which pave substantial strides in oncology research, thanks to their fascinating biocompatibility and tunable electrical and optical properties (Sivasankarapillai et al. 2020). MXene can be used to create face shields, PPE kits, efficient biosensors with stronger antiviral properties, and face masks, among other biomedical devices. In addition to treating comorbidities in COVID-19 patients, MXenes also have a high drug loading and precise release capacity. Additionally, MXene's outstanding biocompatibility is a plus point for biomedical applications (Panda et al. 2022).

7.3.4 TRANSITION METAL DICHALCOGENIDES (TMDs)

A possible substitute is TMDs, which are semiconductors of type MX2, in which M is a transition metal atom (like Mo or W) and X is a chalcogen atom (like S, Se, or Te). MoS_2 is the most researched material in this group due to its robustness. TMDs are intriguing for fundamental research as well as applications in advance electronics, spintronics, optoelectronics, fuel cells, flexible electronics, DNA sequencing, and personalized medicine because TMDs display a peculiar pairing of atomic-scale thickness, direct bandgap, strong spin-orbit coupling, and desirable electronic and mechanical properties (Manzeli et al. 2017). Because of their unique atomic-scale thickness, direct bandgap, strong spin-orbit coupling, and excellent mechanical, chemical, physical, optical, and electrical properties, 2D TMDs have an enormous amount of potential to substitute graphene. TMDs are effective for dual-model cancer therapy due to their comparatively high photothermal coefficient (Cheng et al. 2020).

The huge surface area to volume ratio of TMD nanosheets shows that they are optimal for improved drug loading and optimized sustainable drug release. TMD compounds demonstrated improved mechanical strength in bone regeneration scaffolds, making them ideal materials for 3D printing and for creating tissue engineering scaffolds (Gong and Gu 2019). Surface modification of 2D TMDCs by inorganic or organic nanoparticles may modify the charge-transfer properties and modify the Fermi level. The electrical and optoelectronic characteristics of 2D TMDs may be improved as a result of surface modifications. The charge flow based on the Fermi energy difference primarily addresses the performance improvement, but less attention is paid to the electrical and chemical causes (Azadmanjiri et al. 2020).

7.4 SMART TECHNOLOGIES FOR BIOMEDICAL DEVICES

7.4.1 DEVICES BASED ON ARTIFICIAL INTELLIGENCE

Artificial intelligence (AI) enables a computer to carry out tasks or make judgements that closely resemble and emulate human behaviour and intelligence. A contemporary technique based on computer science called AI creates programs and algorithms to make machines smart and effective at carrying out tasks that often demand expert human intellect. ML, DL, traditional neural networks, fuzzy logic, and voice recognition are only a few of the subdivisions of AI that have distinctive capabilities and functionalities that can enhance the performance of contemporary medical sciences (Hee Lee and Yoon 2021). The first humanoid robot in history, called "WABOT-1," was created by Japanese scientists in 1972. It is capable of communicating with humans in Japanese and measuring the coordinates of points and directions.

However, until the late 1990s, AI research was at a standstill because of financial and computer capacity constraints. Large IT firms like IBM began developing AI-based models in the late 1990s. In the middle of the 2000s, firms that process a lot of data, such as social network platforms, electronic mail services, search engines, and many others, prospered from AI models and programs. The development of CPUs' processing capabilities and the use of GPUs for computations are two main factors that have resulted in the growth of AI. The implementation of an AI-based system is also motivated by the huge data generated by user desire for better analytics (Car et al. 2019). The network of physical items, or "things" that are implanted with sensors, software, and other innovations for the purpose of communicating and transferring data with other devices and systems through the Internet is referred to as the IoT. The real-time position of medical devices like wheelchairs, defibrillators, nebulizers, oxygen pumps, and other monitoring equipment is tracked using IoT devices tagged with sensors. By using a linked network, a healthcare system with IoT capabilities was useful for properly monitoring COVID-19 patients (Rahaman et al. 2019). The readmission rate to hospitals was decreased and patient satisfaction was increased because of this technology. To combat the COVID-19 epidemic, IoT offered a vast, integrated network for healthcare. Since every medical gadget is connected online, whenever a critical scenario arises, a message is automatically sent to the medical team. With reliable teledevices, infected cases could be addressed effectively in a remote location (Singh et al. 2020).

7.4.2 QUANTUM TECHNOLOGY

The next major transformation in the medical industry will be led by quantum computing. This technology has numerous benefits across various industries, particularly in those that have an impact on the health sector (Bakhshandeh 2022). The blending of quantum physics in quantum tech allows for the processing of massive amounts of data, the development of specialized drugs, the creation of simulations, and the molecular manipulation of organs, among other things, using AI nanotechnology, particularly qubit computing. The leaders of the latest revolution in healthcare are quantum technologies, which are expected to have a benign effect on this sector (Lavroff et al. 2021; Shimazoe et al. 2021).

While healthcare devices have long used quantum technologies like laser beams and magnetic resonance imaging, new qubit technologies, particularly those found in desktops and other systems, would then have a greater influence. These technologies are making it possible to create customized devices, and modify organs safely at the cellular scale, and other medications (Efros et al. 2018). The initial and foremost crucial step is to understand the actual nature of quantum technology. Quantum entanglement and aggregation of matter serve as the foundation for this field. Because of them, it creates a computation that differs from the conventional one, functioning with significantly more effective arithmetic algorithms and equipped to store a great deal greater states per unit of data. The restrictions of conventional computing are removed by utilizing quantum mechanics, the branch of science that examines subatomic and atomic entities (Halicka et al. 2022). Devices capable of doing some computations and simulations much more quickly and nimbly than current computers are known as quantum processors or simulators. The study of all-optical phenomena, including the absorbance and release of photons by particles, is known as quantum optics. It makes use of gains from tiny world traits and qualities that do not exist in the macrocosm but have shown a lot of promise. Quantum computing, quantum optics, and quantum simulation stand out among the newly rising quantum technologies for the development of healthcare and wellness (Kjaergaard et al. 2020).

7.4.3 REMOTE PATIENT MONITORING

Today, treating and caring for patients often involves using remote patient monitoring. By utilizing the UCI Repository dataset and the medical sensors for predicting the individuals who have been severely impacted by diabetes, a novel systematic approach is employed for diabetes and associated diseases and the related medical data is obtained. In addition, a brand-new classification technique known as the fuzzy rule-based neural classifier is suggested for determining the severity of illnesses. These technologies do, however, also provide significant privacy hazards and security issues with regard to data transfer and transaction logging. Therefore, using a blockchain to securely manage and analyse large amounts of healthcare data is crucial. Blockchains are not entirely compatible with the majority of resource-constrained IoT devices intended for smart cities because they are computationally expensive, require high bandwidth, and require additional computational power. IoT devices that depend on their distributed nature and other extra privacy and security features of the network can use a fresh framework of modified blockchain models. The model's extra privacy and security features are built on sophisticated cryptographic primitives. It improves the security and anonymity of IoT application data and transactions across a blockchain-based network (Dwivedi et al. 2019).

7.4.4 SMARTPHONE-BASED HEALTH MONITORING SYSTEMS

Smartphones have a vast array of features, and they are increasingly used in a variety of healthcare applications with specific attachments. It will aid in the regular management of health, preserving lives from opponents that frequently go unnoticed because they have nonsymptomatic or delayed beginnings. Additionally, users can use smartphone-based analytical imaging technologies to identify the different food components and identify any pollutants in food while they are dining out, lowering the likelihood of food allergies. A mobile voice health monitoring system is introduced by using a smartphone with an accelerometer sensor. In the device under examination, a tiny accelerometer serves as a voice sensor, and the smartphone serves as the platform for data collection. This system is placed around the neck of the patient. Although the frame-based voice parameters were employed in this system, monitoring can also be done with the raw accelerometer data (Hunt, Ruiz, and Pogue 2021).

7.4.5 ROBOTIC SURGERIES

A revolution in the surgical sector has been brought about by AI technology in the form of collaborating robots. When making little incisions, the revolution may be noticed in terms of depth and cutting rate. In general, the skill of the surgeon can alter the outcome of surgery, especially one involving a novel or complicated procedure. Even the most talented surgeons can operate more efficiently when AI is used to reduce case-to-case variances. AI robots reduce the possibility of tremors or other unintentional movements during surgery because they are precise. For instance, AI-controlled robots can conduct fundamental tasks like precise cutting and sewing while functioning with more accuracy and miniaturization. AI systems can use information from previous surgeries to create novel surgical techniques (Collins et al. 2022).

7.4.6 HAND-HYGIENE MONITORING

One of the best ways to stop the spread of illnesses and diseases is to comply with hand-hygiene rules. According to the Centers for Disease Control and Prevention (CDC), washing hands for at least 20 seconds is advised. But at the moment, either direct observation or sophisticated electronic compliance monitoring technologies are used to check hand-hygiene compliance: a "smart ring," which is a straightforward, affordable, small, wearable fluid-sensing device, for tracking hand-hygiene compliance in real time. The smart ring has an internal electrical board, an LED indication, and an electrochemical fluid sensor that was prototyped using 3D printing technology. Without the use of an expensive electronic monitoring device or trained employees, the ring is capable of recognizing hand-washing events and tracking hand-washing time (Zhang et al. 2018). Various hand-washing products, including soaps, sanitizers, and antimicrobial compounds, as well as tap water, have all been successfully detected. Such a straightforward, small, wearable gadget could pave the way for increased hand-hygiene compliance and a decline in infection rates in daily life, particularly in healthcare facilities, restaurants, and educational institutions.

7.4.7 Virtual nursing assistants

AI systems relieve the need for on-call virtual nursing aides. Virtual nursing assistants could help the healthcare sector significantly by communicating with patients and guiding them to the most suitable settings for their care. They can keep an eye on patients, respond to their inquiries, and so provide prompt real-time responses. The majority of virtual nursing assistant applications available today make it possible for patients and healthcare professionals to communicate frequently and consistently. There are fewer risks of unnecessary hospital trips or readmission to the hospital as this occurs in between patient visits to their doctors' offices. In addition to scheduling doctor visits and keeping track of patients' health, AI-powered virtual assistants offer individualized experiences to patients and assist them in identifying their ailments based on their symptoms. You are guided through the procedure by a virtual nurse assistant. This use of AI may be employed in order to improve patient participation and improve their self-management abilities in order to stop serious conditions from getting worse. The world's first virtual nurse assistant, Care Angel, can do wellness checks using voice and AI (Hee Lee and Yoon 2021).

7.4.8 Cardiac monitoring

It has been determined that applications based on the IoT combined with AI for cardiac care can serve as virtual assistants for cardiac patients. The diagnosis of the patient is made simpler and more quickly with real-time monitoring of a specific patient. It is acknowledged that using data analysis and ML techniques to compose the result interpretation is extremely effective and improves clinical decision-making (Manickam et al. 2022). Such criteria are the essential building blocks for creating compact and composite smartphone-based products. When laboratory facilities are scarce or nonexistent, the goal of point-of-care diagnostics is to quickly start the treatment or prognosis. In rural and less developed areas, the IoT minimizes or eliminates active human intervention (Cao et al. 2021).

7.5 MICROFLUIDICS AND 3D PRINTING-BASED BIOMEDICAL APPLICATIONS AND DEVICES

One of the priceless treasures of the modern industrial period is additive manufacturing, sometimes known as 3D printing. Since the beginning, it has greatly simplified our life. Its applications span a wide range of cutting-edge fields, from aerospace and defence to nanotechnology and space technology. We now have greater technical comfort in disease modelling, developing individualized implants and prostheses, printing organs, practicing veterinary medicine, and tissue engineering, thanks to the use of 3D printing for the production of various biomedical devices, related parts, and other related biomedical applications. At a reasonable price, gadgets that fit the patient's anatomical structure can be produced (Basu, Basu, and Bhattacharya 2021). That is why this method is used to create the majority of hearing devices that are made specifically for the patient that needs to fit the patient's ear structure thus expanding its use and affordability towards eye lenses, stethoscopes, and glasses designed for visually impaired persons. Various drugs may be created and further simplification may be done using this technology in all dimensions and geometries thus proving to be a remedy for the issues with the existing production procedures and thus allowing for the creation of expensive and time-consuming medications with complicated formulas as well. Numerous desirable layouts may now be quickly prototyped and verified in a limited amount of time at low cost, thanks to the widespread commercialization of 3D printing in prosthetic developments.

A developing technique that can be used to create different functioning tissue architectures to replace tissues that have been damaged or are ill is 3D bioprinting, which is founded on the concept of layer-after-layer placement of biological components and live cells (Bozkurt and Karayel 2021). It assembles cells, developmental elements, and biomimetics to provide a microhabitat where cells may develop into tissue architectures.

Manufacturing 3D-printed biosensors might involve creating moulds for casting sensors, inserting commercial devices into printed structures, or manufacturing the whole system itself. It is indeed crucial to note that the strength and brittleness of 3D-printed biosensors will rely on the biological elements' makeup as well as the functionalization strategy used to insert or immobilize elements upon that electrode surface. Because of its ease of usage, broad applicability, and accuracy, extrusion-based 3D printing is one of the most popular printing techniques, and fused deposition modelling is ranked the most popular printing technique due to its capacity to produce complicated geometries, affordable cost, and ease of use. Although this method does have some limitations—the printing process is slow when compared with other alternatives like stereolithography (SLA), and the printing resolution is less precise on the whole than printing methods like binder jetting and vat photopolymerization but still it is very helpful for rapid prototyping. Among photopolymerization printing methods, SLA and digital light processing (DLP) have attracted attention for their suitability in the production of biosensors (Remaggi, Zaccarelli, and Elviri 2022).

For highly efficient biosensing, the sensor must pass through some tests and be competent based on the understated parameters:

(1) high sensitivity,
(2) high stability and repeatability,
(3) real-time analysis (quick response),
(4) lower sample volume consumption,
(5) operational ease,
(6) microfluidics is the phenomenon of systems that process or manipulate tiny volumes (10^{-9} to 10^{-18} litres) of fluids by employing channels within the widths of tens to hundreds of micrometres, thus giving radically new capacities for controlling molecule concentrations.

Being transparent and efficient in supporting the microfluidic components, polydimethyl siloxane, is the favoured one for microfluidics fabrication along with glass, polycarbonate, polyolefin, and silicon, and the intriguing ones for research comprising of clothes and wearables. Microfluidics supports lower sample usage and tiny structure formation along with lesser area consumption and hence its integration with the biosensing field may produce highly efficient sensors with higher sensitivities.

A microfluidic biosensor has understated advantages (Wang et al. 2020):

(1) To increase sensitivity, microfluidics offers a confined and consistent biosensing habitat.
(2) The sensing area may be effectively, precisely, and dramatically reduced using a microfluidic channel.
(3) Several functionalities can be integrated into a single system using microfluidic architecture without the use of additional techniques or apparatus.
(4) Capable of automation.
(5) Permits the mutual and independent treatment of several binding tests on unique or several specimens at once.

A microfluidic biosensor for rapid nucleic acid quantitation was developed based on hyperspectral interferometric amplicon-complex mapping wherein the chip resembling a circular saucer with a hollow centre is attached to the supporting structure having four recurring units located on the front side of the device to allow simultaneous detection of four different targets having two storage chambers and a reaction chamber in each unit (Fu et al. 2021).

Microfluidic chips may be created via 3D printing in a step process, negating the requirement to create an isolated channel. This is unique to microfluidics and provides the ability to produce items quickly with a variety of characteristics and functions, such as mechanical, electrical, chemical, or optical ones, and it overcomes a key industrial barrier. Microfluidics-related 3D printing processes include material extrusion, vat photopolymerization, material jetting, powder bed fusion, and laminated item production processes, all of which use a variety of fabrication methods, some of which are stated in the preceding paragraphs. FFF printers provide additional advantages for the 3D printing of micro- and milli-fluidic systems in spite of lower channel size limitations as they provide accessibility to different materials and ease in multimaterial printing (Balakrishnan et al. 2020).

SLS is a 3D printing method that belongs to the powder bed fusion category in which a beam is frequently used to create layered objects by melting and fusing the powder. This is done by a variety of particle-binding processes, such as chemical reactions, solid-state sintering, total or partial melting, or a combination of these (Prabhakar et al. 2021). This technique has been successfully used for the construction of 3D microfluidic chips using cell-laden hydrogel and a Herringbone mixing channel. With DLP, multidirectional channels are printed with interface overlap distances ranging from 20 to 200 micrometres (Rupal et al. 2019). The stereolithography technique was used for 3D-printing a microfluidic system to use in the natural spotting of cancer biomarker proteins (Prabhakar et al. 2021). The use of 3D-printed microfluidic interfaces is becoming increasingly common and may eventually displace automated and conventional fluid-handling equipment.

7.6 IMPORTANCE OF IMPEDANCE-BASED BIOMEDICAL DEVICES

Scientists have been looking for novel platforms, approaches, and methodologies to overcome recurrent issues with the present healthcare detection technologies for the past decade (Sawhney and Conlan 2019). Cost-effectiveness, point-of-care detection, quick outcomes, and primary illness identification are some of the very important concerns that require additional research in this area. Point-of-care medical equipment has the capacity to detect and report biomarkers of health issues quickly, often without the need for offsite diagnostic testing resources (Lasserre et al. 2022). Due to its cheaper cost compared with current healthcare technologies, high sensitivity built into the methodology, quick signal analysis, and the ability to be shrunk, electrochemical systems and approaches are more enticing for disease detection (Diáz-Cartagena et al. 2019).

The label-free potential and comparatively simple measuring procedures of electrochemical detection platforms have drawn notice across various research channels. Although there are many variants of electrochemical measurements as stated, electrochemical impedance spectroscopy (EIS) has been enormously emerging these days due to the various advantages it offers. One is the low stimulation voltage requirement to get the detection done as it does not cause any change in analyte constituent and can be worked with very less volumes, even in the picolitre range. The various other advantages it offers are its modest energy requirement, customizable by calibration, and allowing detection of small electrochemical alterations with limited complexity (Yen, Chao, and Yeh 2020; Lorenzo et al. 2021). EIS can also be integrated with various other technologies involving the optical, electrical, magnetic, and mechanical ones to enhance measurement capability.

In terms of the biomedical microdevices that have been discussed in this chapter, EIS can be integrated to most of the technologies in implantable (Priti Singh et al. 2016) and nonimplantable devices (Bera 2014) with various substrate materials and detection mechanisms, which make this technique more adaptable to general diagnostic usage.

7.7 CONCLUSION AND FUTURE PERSPECTIVES

Medical device design is a highly collaborative process that frequently combines the expertise and abilities of manufacturing engineers, biologists, product engineers, packaging experts, physicians, quality control experts, and many more consultants. The bigger design problem must be divided into smaller, easier-to-manage subtasks as part of the design phase for these kinds of devices. These smaller, less controllable problems can then be assessed, resolved, and merged into a full device by various specialist teams. As a still-emerging field of technology, medical device design requires successful collaboration between more conventional goods engineers, focusing on factors like device physics and medical professionals interested in the clinical consequences of a device's activities. The discipline of designing medical devices has merged with software engineering, and both are subjected to similar design regulations. Despite having independent knowledge bases in the past, these professions have merged as a result of the development of biomedical engineering, necessitating future collaboration among researchers to address unmet medical requirements. The biomedical device sector has been expanding quickly in recent years to include cutting-edge technologies including point-of-care gadgets, telemedicine, chemical diagnosis, medical robotics, and imaging diagnosis. A large investment is needed to conduct research, develop, and manufacture new biomedical goods and innovations in order to meet the enormous and growing need for biomedical devices. The design, manufacture, process optimization, material characterization and processing, computer and data science, tissue engineering, medicine, and other traditional and developing fields are all included in the biomedical device business. As new technologies evolve quickly, product research is required with ever-shorter life cycles and speed to market. The function and compactness of biomedical devices are also clearly on the rise, as demonstrated by more sophisticated production techniques. Overall, the necessity for biomedical manufacturing is growing and will increasingly address a wider spectrum of healthcare challenges, ultimately enhancing the effectiveness and calibre of patient treatment as well as human life.

Additionally, micromagneto fluidics, electrical systems, multimaterial input–output devices, automated processes, and multifunctional programmable systems are all components of integrated 3D microfluidics that are progressively gaining practice. Developments in 3D microfluidics including modular printing, automated analysis employing 3D-printed 3D-microfluidic devices, and maker-space microfluidics can be seen as the focus of research and common use in the future (Rupal et al. 2019).

REFERENCES

Ahmad, Belal, Michaël Gauthier, Guillaume Laurent, Aude Bolopion, and Guillaume J Laurent. 2021. "Biomedical Applications: A Survey." *IEEE Transactions on Robotics* 1: 1.

Ausra, Jokubas, Micah Madrid, Rose T. Yin, Jessica Hanna, Suzanne Arnott, Jaclyn A. Brennan, Roberto Peralta et al. 2022. "Wireless, Fully Implantable Cardiac Stimulation and Recording with on-Device Computation for Closed-Loop Pacing and Defibrillation." *Science Advances* 8 (43): eabq7469. https://doi.org/10.1126/SCIADV.ABQ7469/SUPPL_FILE/SCIADV.ABQ7469_MOVIE_S1.ZIP.

Azadmanjiri, Jalal, Parshant Kumar, Vijay K. Srivastava, and Zdenek Sofer. 2020. "Surface Functionalization of 2D Transition Metal Oxides and Dichalcogenides via Covalent and Non-Covalent Bonding for Sustainable Energy and Biomedical Applications." *ACS Applied Nano Materials* 3 (4): 3116–43. https://doi.org/10.1021/ACSANM.0C00120/ASSET/IMAGES/LARGE/AN0C00120_0014.JPEG.

Bakhshandeh, Sadra. 2022. "Quantum Sensing Goes Bio." *Nature Reviews Materials* 7 (4): 254–54. https://doi.org/10.1038/s41578-022-00435-y.

Balakrishnan, Hari Kalathil, Faizan Badar, Egan H. Doeven, James I. Novak, Andrea Merenda, Ludovic F. Dumeé, Jennifer Loy, and Rosanne M. Guijt. 2020. "3D Printing: An Alternative Microfabrication Approach with Unprecedented Opportunities in Design." *ACS Publications* 93 (1): 350–66. https://doi.org/10.1021/acs.analchem.0c04672.

Bao, Guo, Kun Wang, Lijun Yang, Jialing He, Bin He, Xiaoxue Xu, and Yufeng Zheng. 2022. "Feasibility Evaluation of a Zn-Cu Alloy for Intrauterine Devices: In Vitro and in Vivo Studies." *Acta Biomaterialia* 142 (April): 374–87. https://doi.org/10.1016/J. ACTBIO.2022.01.053.

Basu, Aviru Kumar, Adreeja Basu, and Shantanu Bhattacharya. 2019. "Study of PH Induced Conformational Change of Papain Using Polymeric Nano-Cantilever." *AIP Conference Proceedings* 2083 (1): 030001. https://doi.org/10.1063/1.5094311.

———. 2020. "Micro/Nano Fabricated Cantilever Based Biosensor Platform: A Review and Recent Progress." *Enzyme and Microbial Technology* 139: 109558. https://doi. org/10.1016/j.enzmictec.2020.109558.

———. 2021. "Recent Trends and Progress in Mems-Based Bioinspired/Biomimetic Sensors." *MEMS Applications in Biology and Healthcare* December: 2–1. https://doi. org/10.1063/9780735423954_002.

Basu, Aviru Kumar, Amar Nath Sah, Mayank Manjul Dubey, Prabhat K. Dwivedi, Asima Pradhan, and Shantanu Bhattacharya. 2020. "MWCNT and α-Fe2O3 Embedded RGO-Nanosheets Based Hybrid Structure for Room Temperature Chloroform Detection Using Fast Response/Recovery Cantilever Based Sensors." *Sensors and Actuators B: Chemical* 305 (February): 127457. https://doi.org/10.1016/J.SNB.2019.127457.

Bera, Tushar Kanti. 2014. "Bioelectrical Impedance Methods for Noninvasive Health Monitoring: A Review." *Journal of Medical Engineering* 2014: 1–28. https://doi. org/10.1155/2014/381251.

Bozkurt, Yahya, and Elif Karayel. 2021. "3D Printing Technology; Methods, Biomedical Applications, Future Opportunities and Trends." *Journal of Materials Research and Technology* 14 (September): 1430–50. https://doi.org/10.1016/J.JMRT.2021.07.050.

Cao, Xiaole, Yao Xiong, Jia Sun, Xiaoxiao Zhu, Qijun Sun, and Zhong Lin Wang. 2021. "Piezoelectric Nanogenerators Derived Self-Powered Sensors for Multifunctional Applications and Artificial Intelligence." *Advanced Functional Materials* 31 (33): 2102983. https://doi.org/10.1002/ADFM.202102983.

Car, Josip, Aziz Sheikh, Paul Wicks, and Marc S. Williams. 2019. "Beyond the Hype of Big Data and Artificial Intelligence: Building Foundations for Knowledge and Wisdom." *BMC Medicine* 17 (1): 1–5. https://doi.org/10.1186/S12916-019-1382-X/METRICS.

Chen, Chen, and Junsheng Wang. 2020. "Optical Biosensors: An Exhaustive and Comprehensive Review." *Analyst* 145 (5): 1605–28. https://doi.org/10.1039/C9AN01998G.

Cheng, Liang, Xianwen Wang, Fei Gong, Teng Liu, Zhuang Liu, L Cheng, X Wang, F Gong, Z Liu, and T Liu. 2020. "2D Nanomaterials for Cancer Theranostic Applications." *Advanced Materials* 32 (13): 1902333. https://doi.org/10.1002/ADMA.201902333.

Cho, Il Hoon, Dong Hyung Kim, and Sangsoo Park. 2020. "Electrochemical Biosensors: Perspective on Functional Nanomaterials for on-Site Analysis." *Biomaterials Research* 24 (1): 1–12. https://doi.org/10.1186/S40824-019-0181-Y.

Collins, Justin W., Hani J. Marcus, Ahmed Ghazi, Ashwin Sridhar, Daniel Hashimoto, Gregory Hager, Alberto Arezzo et al. 2022. "Ethical Implications of AI in Robotic Surgical Training: A Delphi Consensus Statement." *European Urology Focus* 8 (2): 613–22. https://doi.org/10.1016/J.EUF.2021.04.006.

Diáz-Cartagena, Diana C., Griselle Hernández-Cancel, Dina P. Bracho-Rincón, José A. González-Feliciano, Lisandro Cunci, Carlos I. González, and Carlos R. Cabrera. 2019. "Label-Free Telomerase Activity Detection via Electrochemical Impedance Spectroscopy." *ACS Omega* 4 (16): 16724–32. https://doi.org/10.1021/ACSOMEGA.9B00783/ASSET/ IMAGES/LARGE/AO-2019-00783A_0007.JPEG.

Dwivedi, Ashutosh Dhar, Gautam Srivastava, Shalini Dhar, and Rajani Singh. 2019. "A Decentralized Privacy-Preserving Healthcare Blockchain for IoT." *Sensors* 19 (2): 326. https://doi.org/10.3390/S19020326.

Efros, Alexander L., James B. Delehanty, Alan L. Huston, Igor L. Medintz, Mladen Barbic, and Timothy D. Harris. 2018. "Evaluating the Potential of Using Quantum Dots for Monitoring Electrical Signals in Neurons." *Nature Nanotechnology* 13 (4): 278–88. https://doi.org/10.1038/S41565-018-0107-1.

Esther, Nonye-Enyidah, EjikemMazi EC, Nonye-Enyidah Esther, and EjikemMazi EC. 2019. "Profile of Intrauterine Contraceptive Device (IUCD) Acceptors at the Rivers State University Teaching Hospital, Southern Nigeria." *World Journal of Advanced Research and Reviews* 4 (2): 096–101. Https://Wjarr.Com/Sites/Default/Files/WJARR-2019-0099.Pdf https://doi.org/10.30574/WJARR.2019.4.2.0099.

Fu, Rongxin, Wenli Du, Xiangyu Jin, Ruliang Wang, Xue Lin, Ya Su, Han Yang et al. 2021. "Microfluidic Biosensor for Rapid Nucleic Acid Quantitation Based on Hyperspectral Interferometric Amplicon-Complex Analysis." *ACS Sensors* 6 (11): 4057–66. https://doi.org/10.1021/ACSSENSORS.1C01491/ASSET/IMAGES/LARGE/SE1C01491_0008.JPEG.

Gong, Linji, and Zhanjun Gu. 2019. "Transition Metal Dichalcogenides for Biomedical Applications." *Two Dimensional Transition Metal Dichalcogenides*: 241–92. https://doi.org/10.1007/978-981-13-9045-6_8.

Guru, Ramratan, Anupam Kumar, Deepika Grewal, Rohit Kumar, Ramratan Guru, Anupam Kumar, Deepika Grewal, and Rohit Kumar. 2022. "To Study the Implantable and Non-Implantable Application in Medical Textile." *Next-Generation Textiles [Working Title]*, May. https://doi.org/10.5772/INTECHOPEN.103122.

Halicka, Kinga, Francesca Meloni, Mateusz Czok, Kamila Spychalska, Sylwia Baluta, Karol Malecha, Maria I. Pilo, and Joanna Cabaj. 2022. "New Trends in Fluorescent Nanomaterials-Based Bio/Chemical Sensors for Neurohormones Detection A Review." *ACS Omega* 7 (38): 33749–68. https://doi.org/10.1021/ACSOMEGA.2C04134/ASSET/IMAGES/LARGE/AO2C04134_0016.JPEG.

Hee Lee, Don, and Seong No Yoon. 2021. "Application of Artificial Intelligence-Based Technologies in the Healthcare Industry: Opportunities and Challenges." *International Journal of Environmental Research and Public Health* 18 (1): 271. https://doi.org/10.3390/IJERPH18010271.

Hivare, Pravin, Chinmaya Panda, Sharad Gupta, and Dhiraj Bhatia. 2021. "Programmable DNA Nanodevices for Applications in Neuroscience." *ACS Chemical Neuroscience* 12 (3): 363–77. https://doi.org/10.1021/ACSCHEMNEURO.0C00723/ASSET/IMAGES/LARGE/CN0C00723_0008.JPEG.

Huang, Yuqi, Pradipta Kr Das, and Venkat R. Bhethanabotla. 2021. "Surface Acoustic Waves in Biosensing Applications." *Sensors and Actuators Reports* 3 (November): 100041. https://doi.org/10.1016/J.SNR.2021.100041.

Hunt, Brady, Alberto J. Ruiz, and Brian W. Pogue. 2021. "Smartphone-Based Imaging Systems for Medical Applications: A Critical Review." *Journal of Biomedical Optics* 26 (4): 040902. https://doi.org/10.1117/1.JBO.26.4.040902.

Jin, Yukun, Bo Wen, Zixiong Gu, Xiantao Jiang, Xiaolan Shu, Zhenping Zeng, Yupeng Zhang et al. 2020. "Deep-Learning-Enabled MXene-Based Artificial Throat: Toward Sound Detection and Speech Recognition." *Advanced Materials Technologies* 5 (9): 2000262. https://doi.org/10.1002/ADMT.202000262.

Johnson, Asha P., H. V. Gangadharappa, and K. Pramod. 2020. "Graphene Nanoribbons: A Promising Nanomaterial for Biomedical Applications." *Journal of Controlled Release* 325 (September): 141–62. https://doi.org/10.1016/J.JCONREL.2020.06.034.

Jung, Yei Hwan, Jong Uk Kim, Ju Seung Lee, Joo Hwan Shin, Woojin Jung, Jehyung Ok, and Tae il Kim. 2020. "Injectable Biomedical Devices for Sensing and Stimulating Internal Body Organs." *Advanced Materials* 32 (16): 1907478. https://doi.org/10.1002/ADMA.201907478.

Kaisti, Matti. 2017. "Detection Principles of Biological and Chemical FET Sensors." *Biosensors and Bioelectronics* 98 (December): 437–48. https://doi.org/10.1016/J.BIOS.2017.07.010.

Karahan, Hüseyin Enis, Christian Wiraja, Chenjie Xu, Jun Wei, Yilei Wang, Liang Wang, Fei Liu, and Yuan Chen. 2018. "Graphene Materials in Antimicrobial Nanomedicine: Current Status and Future Perspectives." *Advanced Healthcare Materials* 7 (13): 1701406. https://doi.org/10.1002/ADHM.201701406.

Khan, Wahid, Eameema Muntimadugu, Michael Jaffe, and Abraham J. Domb. 2014. "Implantable Medical Devices." *Focal Controlled Drug Delivery*, 33–59. https://doi.org/10.1007/978-1-4614-9434-8_2.

Kim, Taemin, Myeongki Cho, and Ki Jun Yu. 2018. "Flexible and Stretchable Bio-Integrated Electronics Based on Carbon Nanotube and Graphene." *Materials* 11 (7): 1163. https://doi.org/10.3390/MA11071163.

Kjaergaard, Morten, Mollie E. Schwartz, Jochen Braumüller, Philip Krantz, Joel I. J. Wang, Simon Gustavsson, and William D. Oliver. 2020. "Superconducting Qubits: Current State of Play." *Annual Review of Condensed Matter Physics* 11 (March): 369–95. https://doi.org/10.1146/ANNUREV-CONMATPHYS-031119-050605.

Koyappayil, Aneesh, Sachin Ganpat Chavan, Yun Gil Roh, and Min Ho Lee. 2022. "Advances of MXenes; Perspectives on Biomedical Research." *Biosensors* 12 (7): 454. https://doi.org/10.3390/BIOS12070454.

Kumar, Aviru, Pankaj Singh, Mohit Awasthi, and Shantanu Bhattacharya. 2019. "α -Fe2O3 Loaded RGO Nanosheets Based Fast Response / Recovery CO Gas Sensor at Room Temperature." *Applied Surface Science* 465 (September 2018): 56–66. https://doi.org/10.1016/j.apsusc.2018.09.123.

Kumar Basu, Aviru, Adreeja Basu, Sagnik Ghosh, Shantanu Bhattacharya, and Mems Applications in Biology. 2021. "Introduction to MEMS in Biology and Healthcare." *MEMS Applications in Biology and Healthcare*, 1–1. Melville, New York: AIP Publishing LLC. December. https://doi.org/10.1063/9780735423954_001.

Kumar Sharma, Parshant, Antonio Ruotolo, Raju Khan, Yogendra K. Mishra, Nagendra Kumar Kaushik, Nam Young Kim, and Ajeet Kumar Kaushik. 2022. "Perspectives on 2D-Borophene Flatland for Smart Bio-Sensing." *Materials Letters* 308 (February): 131089. https://doi.org/10.1016/J.MATLET.2021.131089.

Lam, Raymond H. W., and Weiqiang Chen. 2019. "Introduction to Biomedical Devices." *Biomedical Devices*, Springer: 1–30. https://doi.org/10.1007/978-3-030-24237-4_1.

Lasserre, Perrine, Banushan Balansethupathy, Vincent J. Vezza, Adrian Butterworth, Alexander Macdonald, Ewen O. Blair, and Liam McAteer et al. 2022. "SARS-CoV-2 Aptasensors Based on Electrochemical Impedance Spectroscopy and Low-Cost Gold Electrode Substrates." *Analytical Chemistry* 94 (4): 2126–33. https://doi.org/10.1021/ACS.ANALCHEM.1C04456/ASSET/IMAGES/LARGE/AC1C04456_0007.JPEG.

Lavroff, Robert H., Doran L. Pennington, Ash Sueh Hua, Barry Yangtao Li, Jillian A. Williams, and Anastassia N. Alexandrova. 2021. "Recent Innovations in Solid-State and Molecular Qubits for Quantum Information Applications." *Journal of Physical Chemistry A* 125 (44): 9567–70. https://doi.org/10.1021/ACS.JPCA.1C08677/ASSET/IMAGES/LARGE/JP1C08677_0001.JPEG.

Lim, Hui Jean, Tridib Saha, Beng Ti Tey, Wen Siang Tan, and Chien Wei Ooi. 2020. "Quartz Crystal Microbalance-Based Biosensors as Rapid Diagnostic Devices for Infectious Diseases." *Biosensors & Bioelectronics* 168 (November): 112513. https://doi.org/10.1016/J.BIOS.2020.112513.

Lorenzo, Melvin F., Suyashree P. Bhonsle, Christopher B. Arena, and Rafael V. Davalos. 2021. "Rapid Impedance Spectroscopy for Monitoring Tissue Impedance, Temperature, and Treatment Outcome during Electroporation-Based Therapies." *IEEE Transactions on Biomedical Engineering* 68 (5): 1536–46. https://doi.org/10.1109/TBME.2020.3036535.

Ma, Lei, Yousef Abugalyon, and Xiu Jun Li. 2021. "Multicolorimetric ELISA Biosensors on a Paper/Polymer Hybrid Analytical Device for Visual Point-of-Care Detection of Infection Diseases." *Analytical and Bioanalytical Chemistry* 413 (18): 4655–63. https://doi.org/10.1007/S00216-021-03359-8/TABLES/1.

Mamidi, Siva Kumar, Kristin Klutcharch, Shradha Rao, Julio C. M. Souza, Louis G. Mercuri, and Mathew T. Mathew. 2019. "Advancements in Temporomandibular Joint Total Joint Replacements (TMJR)." *Biomedical Engineering Letters* 9 (2): 169–79. https://doi.org/10.1007/S13534-019-00105-Z/FIGURES/5.

Manickam, Pandiaraj, Siva Ananth Mariappan, Sindhu Monica Murugesan, Shekhar Hansda, Ajeet Kaushik, Ravikumar Shinde, and S. P. Thipperudraswamy. 2022. "Artificial Intelligence (AI) and Internet of Medical Things (IoMT) Assisted Biomedical Systems for Intelligent Healthcare." *Biosensors* 12 (8): 562. https://doi.org/10.3390/BIOS12080562.

Manzeli, Sajedeh, Dmitry Ovchinnikov, Diego Pasquier, Oleg V. Yazyev, and Andras Kis. 2017. "2D Transition Metal Dichalcogenides." *Nature Reviews Materials* 2 (8): 1–15. https://doi.org/10.1038/natrevmats.2017.33.

Nahak, Bishal Kumar, Anshuman Mishra, Subham Preetam, and Ashutosh Tiwari. 2022. "Advances in Organ-on-a-Chip Materials and Devices." *ACS Applied Bio Materials* 5 (8): 3576–3607. https://doi.org/10.1021/ACSABM.2C00041/ASSET/IMAGES/LARGE/MT2C00041_0016.JPEG.

Panda, Subhasree, Kalim Deshmukh, Chaudhery Mustansar Hussain, and S. K. Khadheer Pasha. 2022. "2D MXenes for Combatting COVID-19 Pandemic: A Perspective on Latest Developments and Innovations." *FlatChem* 33 (May): 100377. https://doi.org/10.1016/J.FLATC.2022.100377.

Patil, Nikhil Avinash, James Njuguna, and Balasubramanian Kandasubramanian. 2020. "UHMWPE for Biomedical Applications: Performance and Functionalization." *European Polymer Journal* 125 (February): 109529. https://doi.org/10.1016/J.EURPOLYMJ.2020.109529.

Pohanka, Miroslav. 2017. "The Piezoelectric Biosensors: Principles and Applications, a Review." *International Journal of Electrochemical Science* 12 (1): 496–506. https://doi.org/10.20964/2017.01.44.

Prabhakar, Priyanka, Raj Kumar Sen, Neeraj Dwivedi, Raju Khan, Pratima R. Solanki, Avanish Kumar Srivastava, and Chetna Dhand. 2021. "3D-Printed Microfluidics and Potential Biomedical Applications." *Frontiers in Nanotechnology* 3: 609355. https://doi.org/10.3389/fnano.2021.609355.

Rahaman, Ashikur, Md Milon Islam, Md Rashedul Islam, Muhammad Sheikh Sadi, and Sheikh Nooruddin. 2019. "Developing Iot Based Smart Health Monitoring Systems: A Review." *Revue d'Intelligence Artificielle* 33 (6): 435–40. https://doi.org/10.18280/ria.330605.

Remaggi, Giulia, Alessandro Zaccarelli, and Lisa Elviri. 2022. "3D Printing Technologies in Biosensors Production: Recent Developments." *Chemosensors* 10 (2): 65. https://doi.org/10.3390/CHEMOSENSORS10020065/S1.

Rupal, Baltej Singh, Elisa Aznarte Garcia, Cagri Ayranci, and Ahmed Jawad Qureshi. 2019. "3D Printed 3D-Microfluidics: Recent Developments and Design Challenges." *Journal of Integrated Design and Process Science* 22 (1): 5–20. https://doi.org/10.3233/jid-2018-0001.

Sawhney, M. Anne, and R. S. Conlan. 2019. "POISED-5, a Portable on-Board Electrochemical Impedance Spectroscopy Biomarker Analysis Device." *Biomedical Microdevices* 21 (3): 1–14. https://doi.org/10.1007/S10544-019-0406-9/FIGURES/7.

Schachinger, Franziska, Hucheng Chang, Stefan Scheiblbrandner, and Roland Ludwig. 2021. "Amperometric Biosensors Based on Direct Electron Transfer Enzymes." *Molecules* 26 (15): 4525. https://doi.org/10.3390/MOLECULES26154525.

Shao, Wei, Guoan Tai, Chuang Hou, Zenghui Wu, Zitong Wu, and Xinchao Liang. 2021. "Borophene-Functionalized Magnetic Nanoparticles: Synthesis and Memory Device Application." *ACS Applied Electronic Materials* 3 (3): 1133–41. https://doi.org/10.1021/ACSAELM.0C01004/ASSET/IMAGES/MEDIUM/EL0C01004_M001.GIF.

Shimazoe, K., H. Tomita, D. Watts, P. Moskal, A. Kagawa, P. G. Thirolf, D. Budker, and C. S. Levin. 2021. "Quantum Sensing for Biomedical Applications." *2021 IEEE Nuclear Science Symposium and Medical Imaging Conference Record*, NSS/MIC 2021 and 28th International Symposium on Room-Temperature Semiconductor Detectors, RTSD 2022. https://doi.org/10.1109/NSS/MIC44867.2021.9875702.

Singh, Pankhi, Vibhas Chugh, Antara Banerjee, Surajit Pathak, Sudeep Bose, and Ranu Nayak. 2023. "Nanomaterials: Compatibility Towards Biological Interactions." *Practical Approach to Mammalian Cell and Organ Culture,* 1059–1089. Singapore: Springer Nature Singapore. https://doi.org/10.1007/978-981-19-1731-8_19-1.

Singh, Priti, Shailendra Kumar Pandey, Jyoti Singh, Sameer Srivastava, Sadhana Sachan, and Sunil Kumar Singh. 2016. "Biomedical Perspective of Electrochemical Nanobiosensor." *Nano-Micro Letters* 8 (3): 193–203. https://doi.org/10.1007/s40820-015-0077-x.

Singh, Ravi Pratap, Mohd Javaid, Abid Haleem, and Rajiv Suman. 2020. "Internet of Things (IoT) Applications to Fight against COVID-19 Pandemic." *Diabetes & Metabolic Syndrome: Clinical Research & Reviews* 14 (4): 521–24. https://doi.org/10.1016/J.DSX.2020.04.041.

Sivasankarapillai, Vishnu Sankar, Ajeesh Kumar Somakumar, Jithu Joseph, Sohrab Nikazar, Abbas Rahdar, and George Z. Kyzas. 2020. "Cancer Theranostic Applications of MXene Nanomaterials: Recent Updates." *Nano-Structures & Nano-Objects* 22 (April): 100457. https://doi.org/10.1016/J.NANOSO.2020.100457.

Soldatkin, O. O., I. S. Kucherenko, D. V. Siediuko, D. Yu Kucherenko, S. V. Dzyadevych, and A. P. Soldatkin. 2021. "Development of Enzyme Conductometric Biosensor for Dopamine Determination in Aqueous Samples." *Electroanalysis* 33 (10): 2187–95. https://doi.org/10.1002/ELAN.202100257.

Tabish, Tanveer A., Aumber Abbas, and Roger J. Narayan. 2021. "Graphene Nanocomposites for Transdermal Biosensing." *Wiley Interdisciplinary Reviews: Nanomedicine and Nanobiotechnology* 13 (4): Article E1699. https://doi.org/10.1002/WNAN.1699.

Tarakji, Khaldoun G., Eric J. Chan, Daniel J. Cantillon, Aaron L. Doonan, Tingfei Hu, Steven Schmitt, Thomas G. Fraser, Alice Kim, Steven M. Gordon, and Bruce L. Wilkoff. 2010. "Cardiac Implantable Electronic Device Infections: Presentation, Management, and Patient Outcomes." *Heart Rhythm* 7 (8): 1043–47. https://doi.org/10.1016/J.HRTHM.2010.05.016.

Taşaltın, Nevin, Selim Güllülü, and Selcan Karakuş. 2022. "Dual-Role of β Borophene Nanosheets as Highly Effective Antibacterial and Antifungal Agent." *Inorganic Chemistry Communications* 136 (February): 109150. https://doi.org/10.1016/J.INOCHE.2021.109150.

Tatullo, Marco, Barbara Zavan, Fabio Genovese, Bruna Codispoti, Irina Makeeva, Sandro Rengo, Leonzio Fortunato, and Gianrico Spagnuolo. 2019. "Borophene Is a Promising 2D Allotropic Material for Biomedical Devices." *Applied Sciences* 9 (17): 3446. https://doi.org/10.3390/APP9173446.

Vasan, Arvind Sai Sarathi, Dinesh Michael Mahadeo, Ravi Doraiswami, Yunhan Huang, and Michael Pecht. 2013. "Point-of-Care Biosensor System." *Frontiers in Bioscience - Scholar*. Front Biosci (Schol Ed) 5 (1): 39–71. https://doi.org/10.2741/s357.

Vasuki, S., V. Varsha, R. Mithra, R. A. Dharshni, S. Abinaya, R. Deva Dharshini, and N. Sivarajasekar. 2019. "Thermal Biosensors and Their Applications." *American International Journal of Research in Science, Technology, Engineering & Mathematics* 1: 262–64.

Wang, Chan, Qiongfeng Shi, and Chengkuo Lee. 2022. "Advanced Implantable Biomedical Devices Enabled by Triboelectric Nanogenerators." *Nanomaterials* 12 (8): 1366. https://doi.org/10.3390/NANO12081366.

Wang, Jing, Yong Ren, Bei Zhang, Jing Wang, Yong Ren, and Bei Zhang. 2020. "Application of Microfluidics in Biosensors." *Advances in Microfluidic Technologies for Energy and Environmental Applications* May. https://doi.org/10.5772/INTECHOPEN.91929.

Wang, Xuepeng, Gary M. C. Ong, Michael Naguib, and Jianzhong Wu. 2021. "Theoretical Insights into MXene Termination and Surface Charge Regulation." *Journal of Physical Chemistry C* 125 (39): 21771–79. https://doi.org/10.1021/ACS.JPCC.1C07076/ASSET/IMAGES/LARGE/JP1C07076_0009.JPEG.

Wei, Yuhong, Yancong Qiao, Guangya Jiang, Yunfan Wang, Fangwei Wang, Mingrui Li, Yunfei Zhao, et al. 2019. "A Wearable Skinlike Ultra-Sensitive Artificial Graphene Throat." *ACS Nano* 13 (8): 8639–47. https://doi.org/10.1021/ACSNANO.9B03218/SUPPL_FILE/NN9B03218_SI_004.MP4. Yen, Yi Kuang, Chen Hsiang Chao, and Ya Shin Yeh. 2020. "A Graphene-PEDOT:PSS Modified Paper-Based Aptasensor for Electrochemical Impedance Spectroscopy Detection of Tumor Marker." *Sensors* 20 (5): 1372. https://doi.org/10.3390/S20051372.

Zhang, Xin, Karteek Kadimisetty, Kun Yin, Carlos Ruiz, Michael G. Mauk, and Changchun Liu. 2018. "Smart Ring: A Wearable Device for Hand Hygiene Compliance Monitoring at the Point-of-Need." *Microsystem Technologies* 25 (8): 3105–10. https://doi.org/10.1007/S00542-018-4268-5.

Zhao, Victoria Xin Ting, Ten It Wong, Xin Ting Zheng, Yen Nee Tan, and Xiaodong Zhou. 2020. "Colorimetric Biosensors for Point-of-Care Virus Detections." *Materials Science for Energy Technologies* 3 (January): 237–49. https://doi.org/10.1016/J.MSET.2019.10.002.

8 Nonbiological applications of impedimetric sensors

Nitish Katiyar, Rishi Kant, and
Sagnik Sarma Choudhury

CONTENTS

8.1 INTRODUCTION

Electrochemical impedance spectroscopy (EIS) has proven to be a highly employed characterization technique for different nonbiological applications (Figure 8.1), such as corrosion monitoring (Ribeiro and Abrantes 2016), medicine, characterization of energy devices (Tröltzsch, Kanoun, and Tränkler 2006), predicting lubrication's life (Lvovich and Smiechowski 2006), electroactive polymer characterization (Hong, Almomani, and Montazami 2014), and synthesis of materials like chlorine and aluminum.

A wide variety of sensors based on EIS are heavily deployed by many industries, such as automobile and aerospace industries, for their assembly, painting, joining, welding, and other processes (Bhattacharya et al. 2019; Pandey Tatiya et al. 2019a,

DOI: 10.1201/9781003358091-10

2019b; Tatiya et al. 2019), medical and health-care industries for the development of continuous health-monitoring systems (Pandey, Shahare et al. 2019; Pandey, Srivastava et al. 2019; Pandey, Tatiya, and Bhattacharya 2021), and environmental monitoring (Chauhan, Pandey, and Bhattacharya 2019). The EIS technique is also utilized to recognize different interfaces that are involved within devices (i.e., solid/solid, electrolyte/solid) (Pandey et al. 2022). The detection of distinct interfaces is achieved through the frequency response of corresponding interfaces to attenuated potential and successive decoupling of capacitive and resistive components used in the specified circuit arrangement (Bhatt et al. 2019, 2021). The development of the impedance measurement technique started toward the end of the nineteenth century. Franceschetti et al. (1991) established basic science for the EIS technique and studied the mass transfer effect on impedance while Gerischer (Gerischer 1951) proposed an impedance model for heterogeneous reactions that were affected by mass transfer phenomena. For further development of EIS measurement, other techniques having different transfer functions have also been deployed within EIS, like thermos-electro-chemical impedance spectroscopy (Citti et al. 1997), electro-hydrodynamic impedance spectroscopy (Tribollet and Newman 1983), and electro-gravimetric impedance spectroscopy (Gabrielli, Garcia-Jareno, and Perrot 2001). Apart from this, advance manufacturing techniques, such as laser treatment, 3D printing (Kumar et al. 2019), grafting, and plasma processing (Sundriyal, Pandey, and Bhattacharya 2020) are amalgamated with different polymeric materials to realize newer development in EIS measurements (Choudhury, Pandey, and Bhattacharya 2021; Pandey, Rashiku, and Bhattacharya 2021). The correlation between physical properties (reaction rate and diffusion rate) and EIS data can be obtained from EIS that helps in predicting the behavior of a device or an electrochemical system. Nevertheless, there exists a certain level of vagueness in the EIS analysis during the interpretation of results, which requires efficient and intelligent modeling techniques and statistical tools to obtain an optimized correlation between data and properties. So far, EIS is widely used for the characterization of electrocatalysis and energy devices especially in batteries.

The latest development in impedance measurements has offered a simpler route to electrochemical measurement. This has happened because of significant technological developments in electronic instrumentation in recent years, which have helped

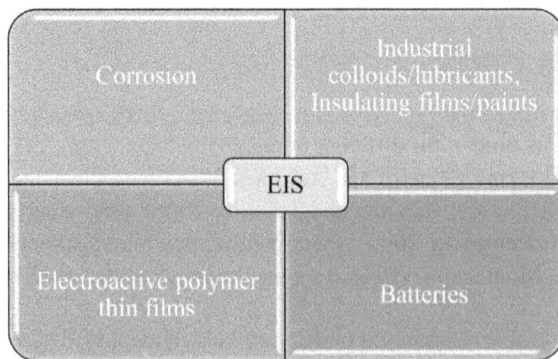

FIGURE 8.1   Potential nonbiological applications for EIS research and use.

the EIS technique. Now, specified objectives like localized electrochemical measurement can also be accomplished by utilizing coupled EIS with additional transfer function measurements (Bard Allen and Faulkner Larry 2001; Orazem and Tribollet 2008). EIS is being utilized widely in nonbiological application that includes internal and external corrosion monitoring, evaluation of battery life, prediction of other energy device behavior, and synthesis of materials like aluminum and electropolymers. Moreover, the recent nonbiological applications reported (Simon Araya et al. 2019) are different various faults diagnoses, such as H_2S poisoning, CO detection, drying condition, and reactant starvation, via the frequency spectrum of impedance spectroscopy.

This chapter provides an in-depth discussion of nonbiological applications of the EIS technique. It discusses EIS application in corrosion monitoring of concrete and other materials like steel for solar-thermal power plants and other applications, condition monitoring of lubricants used in heavy industries, characterization of engine oil (EO), battery life prediction, and electroactive polymers.

8.1.1 MODELING OF ELECTROCHEMICAL IMPEDANCE SPECTROSCOPY

Since the impedance spectroscopy technique does not yield direct measurements of physio-chemical phenomena, it requires an appropriate interpretation from acquired

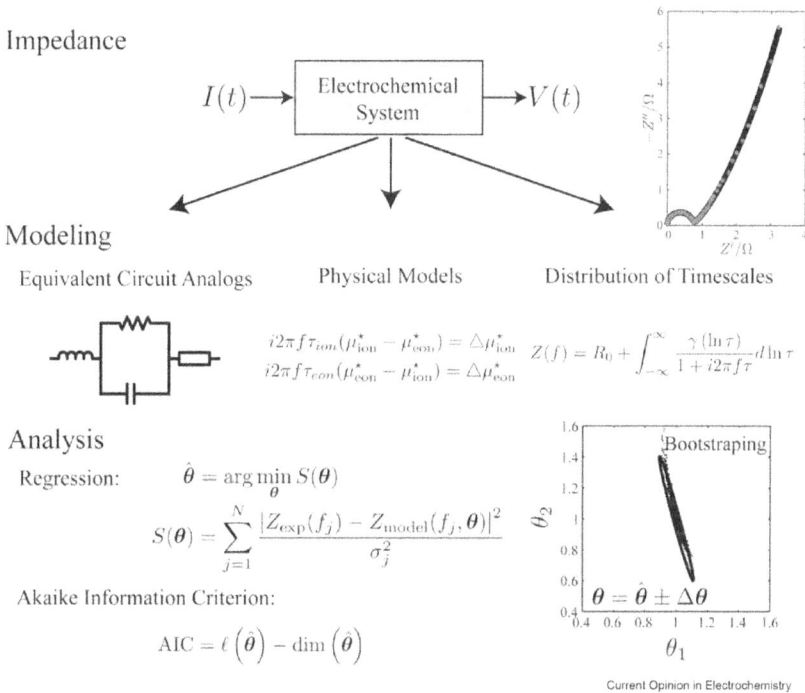

Current Opinion in Electrochemistry

FIGURE 8.2 Schematic representation of model based on the physical partial differential equation, distribution of timescales, regression analysis, and Akaike information criterion. [Reprinted with permission from Ciucci (2019). Copyright (2016) Elsevier].

TABLE 8.1

List of circuit components, their chemical formulae, and symbols used to model EIS data

Resistor (R)	$Z_R = R$	
Capacitor(C)	$Z_c = -j / \omega C$	
Inductor (L)	$Z_L = j\omega L$	
Constant phase elements (CPE)	$Z_{CPE} = \dfrac{1}{Q(j\omega)^{1/2}}$	
Warburg diffusion (W)	$Z_W = \dfrac{\sigma\sqrt{2}}{(j\omega)^{1/2}}$	$\boxed{Z_W}$

data collected from impedance measurements that needs a suitable model utilizing various transfer functions, regression analysis, and mathematical modeling. Figure 8.2 illustrates various tools and elements involved in EIS modeling.

Generally, an electrical circuit model contains resistors (R), capacitors (C), inductors (L), CPE (constant phase elements) and other distinct distributed elements. But an efficient model is always essential for the proper interpretation of the physioelectrochemical phenomena. The basic equivalent circuits made up of CPEs, diffusion elements, and basic elements as C, R, and L are summarized in Table 8.1.

8.1.1.1 Resistors (R)

The transfer of faradaic charge across interfacial layers causes an oxidation/reduction reaction that comprises basic elements like a resistor. This reaction phenomenon involves the electron transport from the electrolyte solution to the corresponding electrode and vice versa (Bhatt and Bhattacharya 2019). In this case, the impedance of a resistor, denoted by the symbol Z_R, is equivalent to its resistance (R).

8.1.1.2 Capacitors (C)

The solid/solid as well as solids/liquids interfaces carry a nonfaradaic charge that is generated during reaction, can be described as capacitance. The double-layer interfaces occur at the solid/solution interface, with contacts between nearby metal oxide particles or nanocrystals, and contacts between a metal oxide and a conductive substrate (Liu et al. 2008). Additionally, the observed capacitance is generally associated with the density of electronic states at the surface of the metal oxide. The capacitor's impedance represented by Z_C as expressed in Table 8.1 is inversely proportional to frequency, which means impedance will be higher at lower frequencies and vice versa.

8.1.1.3 Inductors (L)

Due to the presence of metallic connections, EIS models often need the inclusion of inductors. For illustration, in compact devices in which the cathodes and anodes are located close to one another, as is the case with heterojunction solar cells, in which the two electrodes are sandwiched together and only have a minimal amount of space between them. The impedance of an inductor is denoted by the symbol Z_L, and it is equal to $j\omega L$, which is opposite to that of the capacitor.

8.1.1.4 Constant phase elements

Inhomogeneities on the surface of the metal oxide electrodes are the root cause of the nonideal capacitance seen in the double layer that stays at the interface between the solid and the electrolyte. Constant phase elements (CPEs) are used almost exclusively nowadays rather than pure capacitors for modeling the interfacial layer because of their superior performance. The symbol Z_{CPE} represents the impedance of a constant phase element.

8.1.1.5 Warburg diffusion (W)

When analyzing EIS data, the diffusion of mobile charges through metal oxide electrodes and in solution is an issue that must not be overlooked. The impedance of diffusion (Warburg diffusion), Z_W is characterized using either finite or infinite diffusion models. When there are electroactive or electro-inactive ions present in a fluid solution (Kant et al. 2013), the diffusion layer is often rather substantial. The value "σ" is used to characterize the resistance associated with diffusion, and it does so by relating it to the concentration of charge carriers and the diffusion coefficients of those charge carriers. The expression for the impedance of Warburg diffusion is given in Table 8.1.

8.2 IMPEDIMETRIC SENSING FOR CORROSION MONITORING

Steel corrosion is one of the most expensive concerns that a wide variety of sectors deal with daily because of the harsh circumstances that existed during the process of manufacturing steel parts or stress corrosion cracking, which causes early tool failure in steel, as well as the deterioration of steel parts in machinery and equipment used for a specific purpose. Steel is also intended for use in the manufacturing of ships, heavy machinery, agricultural equipment, and pipelines for the transportation of fluids while using corrosion-resistant steel (Marcus 2002; Morcillo et al. 2011). Ahmad (2006) elaborated on the importance of considering the environment that metals will be exposed to for extended periods of time for defining corrosion. The following is a list of the environments that are corrosive to some degree: (a) the relative humidity of air, (b) saline water, fresh, distilled, (c) the natural atmospheres of urban, marine, and industrial environments, (d) gases and steams, (e) ammonia (NH_3) and hydrogen sulfide (H_2S), (f) sulfur dioxide (SO_2) oxides of nitrogen (NOx), and (g) soils, acids, alkalis, and fuel gases (Kant et al. 2019). EIS is regarded as a fruitful new electrochemical method that has undergone significant development in recent years and is now a crucial analytical tool for researching materials science (Mansfeld 1988; Shih and Mansfeld 1989). Multiple parameters like the properties of concrete, the existence of films on the surface, interfacial corrosion, and the occurrence of mass transfer processes can be studied by using EIS.

8.2.1 Corrosion Monitoring in Reinforced Concrete

Kim et al. (2019) reported the rate of corrosion of reinforcing steel that is embedded in concrete structures when placed in an environment that simulates a marine setting with a high chloride concentration. Cement mortars were subjected to 25 wet–dry cycles, each of which consisted of immersion for eight hours in a 3-wt% NaCl solution, followed by a drying time of 16 h at room temperature. This was done so that the corrosion of reinforced concrete in a marine environment could be simulated. Ribeiro and Abrantes (2016) reported a novel approach to analysis that is predicated on the relaxation angular frequency, which is unique to each phenomenon, as well as the correlation between typical capacitances and frequency ranges. The setup that is used to monitor corrosion in reinforced concrete is shown in Figure 8.3.

An electrode, seen as a reinforcing bar (rebar), is embedded in an electrolyte, which is represented by concrete, and is then exposed to an alternating signal with a relatively low amplitude (5–20 mV). By studying the phase shift as well as the amplitude of the voltage and current, a comparison is made between the applied initial disturbance and the response of the electrode. The interpretations of the measured EIS can be done by either using graphical representations or by drawing a correlation between the data on the impedance and an analogous circuit that represents the physical processes that are taking place in the system that is the subject of the inquiry. An equivalent circuit can also be used to do either of these two things. Figure 8.4 shows the correlation between each phenomenon's characteristic capacitances and frequencies. The characteristic frequencies are clearly defined, as can be seen.

8.2.2 Corrosion Monitoring in Metals

Corrosion that occurs at high temperatures is among the most important things that must be considered when choosing materials, making structures, and figuring out how long engineering parts that work in high temperatures will last. Any type of manufacturing industry that is linked with a high-temperature process, such as the industry for storing solar energy, which uses inorganic molten salts, has a critical role to play in the prevention of corrosive attacks on materials caused by high

FIGURE 8.3 Instrumentation for conducting EIS measurement for the corrosion process. [Reprinted with permission from Ribeiro and Abrantes (2016). Copyright (2016) Elsevier].

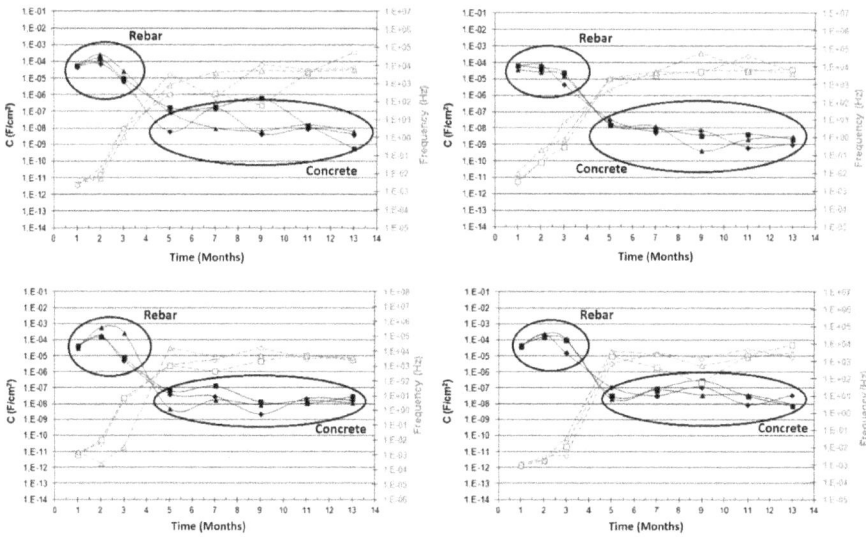

FIGURE 8.4   Relationship among each phenomenon's capacitance and its characteristic frequency, as determined by EIS. [Reprinted with permission from Ribeiro and Abrantes (2016). Copyright (2016) Elsevier].

temperatures. Aspects such as quality, profitability, reliability, and safety are directly affected by this prevention.

Fernandez and Mallco (2018) described a method for monitoring corrosion that is based on EIS, and he proposed a mechanism for corrosion that takes place during the test. At a temperature of 390°C, tests were conducted in a standard solar salt mixture (0.4 KNO_3 and 0.6 $NaNO_3$) and in contact with a low-Cr alloy steel (T22) and a Ni-base alloy (HR224). Both types of steel showed a low rate of corrosion, which meant that they might be considered for use as a container material in CSP facilities. However, this study showed the effective development of a corrosion-monitoring approach for control storage systems that focuses on thermal energy storage materials at high temperatures and is based on EIS. Encinas-Sánchez et al. (2019) studied the behavior of ferritic–martensitic steel (F/M steels) in concentrating solar-thermal power systems, where EIS was utilized in the presence of molten solar salt [$NaNO_3/KNO_3$ (3/2 wt%)]. The apparatus used in the experiment to detect corrosion is depicted in Figure 8.5a, where the corrosion tests were carried out for up to 1000 h at a temperature of 580°C. Figure 8.5b presents the P91 specimens in superficial macrographs before 5b-1, and after 5b-2, the test has been run for 1000 h.

The P91 sample's impedance spectra were taken at various times while they were immersed in molten binary salt (Figure 8.6). After the test's initial stages, the impedance spectra displayed a well-formed loop at high frequency and a straight line at low frequency. This behavior is typical in diffusion-controlled reactions, and it demonstrates that a porous layer mechanism was followed in the corrosion process. The variations in R_w (the finite resistance at low frequencies) and T_w (the time for diffusion) values were recorded after the experimentation, and these variations indicated

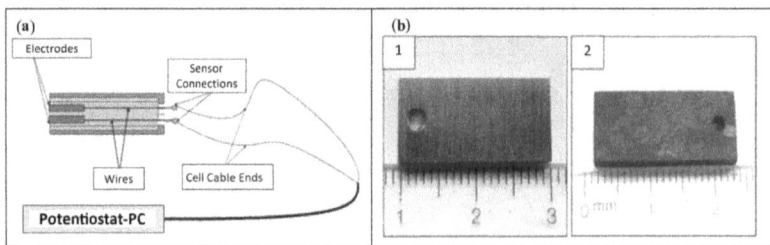

FIGURE 8.5 (a) Diagram of the apparatus used in the experiment to detect corrosion; (b) macrophotographs of the P91 specimens outside surfaces: (1) original sample (2) sample after 1,000 h of the test. [Reprinted with permission from Encinas-Sánchez et al. (2019). Copyright (2019) Elsevier].

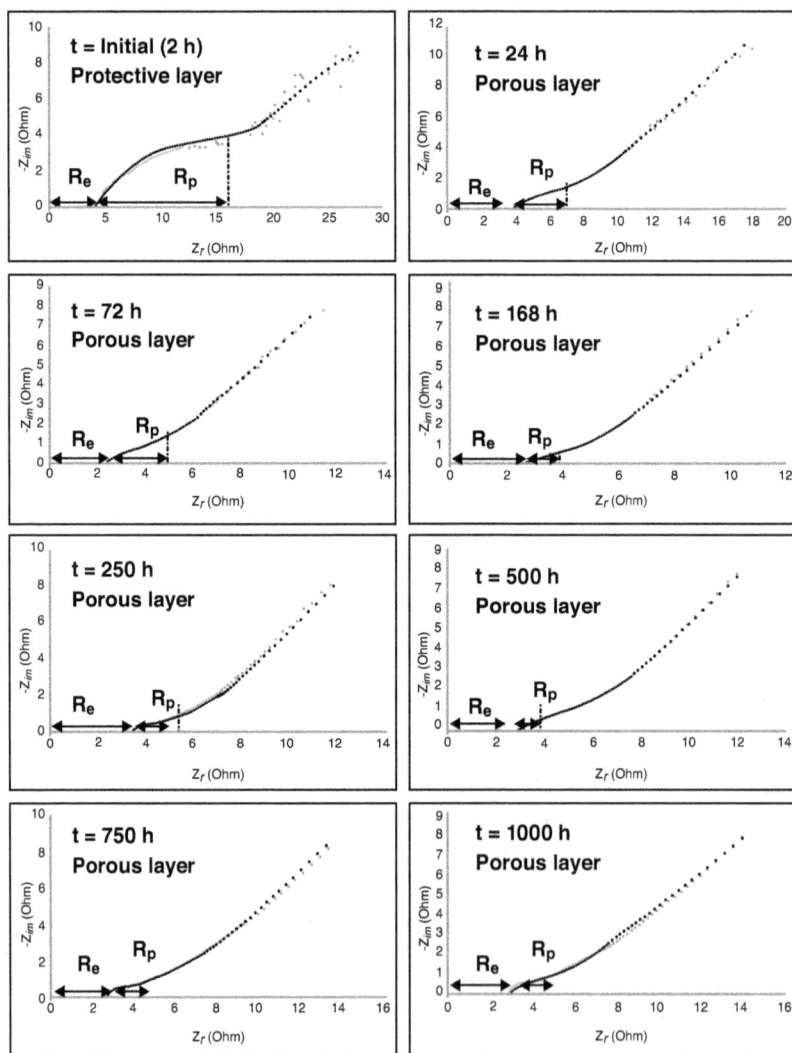

FIGURE 8.6 Spectra represent the impedance of P91 samples tested up to 1,000 h at 580°C in solar salt. [Reprinted with permission from Encinas-Sánchez et al. (2019). Copyright (2019) Elsevier].

TABLE 8.2

Corrosion rate calculated using the results of the EIS test. [Reprinted with permission from (Encinas-Sánchez et al. 2019). Copyright (2019) Elsevier]

Time (h)	$Z_r = (R_p + R_e)$	v_{corr} (µm year^{-1})
2	15.67	23
24	7.15	81
72	4.20	152
168	4.28	174
250	4.50	346
500	3.97	301
750	3.83	280
1000	3.91	305

that there were differences in the thickness of the corrosion layer. There is a correlation between the values of R_w and T_w and the effective diffusion thickness. In terms of the value of the exponent P_w, also known as the roughness of the diffusion media, it significantly deviated from 0.5, which suggested that a rough scale was forming on the P91 steel. Because of the instability of the salt throughout the experiment, the resistance of the electrolyte, denoted by the symbol "Re," as well as its conductivity, denoted by the symbol "K" both changed over the course of the test.

The findings of this experiment were backed up by a chemical analysis of the salt, during which its levels of nitrate and nitrite anions were measured. Table 8.2 provides a concise summary of the findings obtained from the EIS test in terms of the corrosion rate. The rate of corrosion of the P91 steel sample was found to be almost the same (300 µm/year) as before after 500 h of testing.

8.3 IMPEDIMETRIC SENSING IN VARIOUS INDUSTRIAL APPLICATIONS

Lubricants are essential in both automotive and industrial equipment for isolating moving parts, neutralizing corrosive acids, guarding wear surfaces, suspending contaminants, dissipating heat, and offering a variety of other performance-enhancing functionalities. The amount of time that EO can be used is impacted by several factors, some of which include the size of the engine, the amount and kind of oil additives used, the base oil's composition, and the driving circumstances. High-temperature deterioration and contamination of lubricating oils by water, ethylene glycol, fuel, soot, and wear metals through considerable chemical changes over the course of their lifetime has also been observed. These changes can be observed in the oil's composition. After a certain amount of time has passed since the last service or after a certain number of miles have been driven, lubricants are frequently changed without testing. The existing tests for determining the quality of lubricating oil include analyzing the oil's viscosity, total base number, fuel and water dilution, insoluble (such as soot) content, glycol contamination, total acid number (TAN), and metal content (Wang and

Lee 1994; Wang 2001). In general, electrochemical methods do not have the problems that come with the standard industry testing procedures, and they offer a quick, easy, and affordable method that does not have to worry about sample preparation or temperature restrictions. EOs (Smiechowski and Lvovich 2002) and nonaqueous colloidal dispersions (Smiechowski and Lvovich 2005) have both been characterized using EIS.

8.3.1 IMPEDIMETRIC MONITORING OF INDUSTRY-GRADE LUBRICANTS

Impedance spectra (Lvovich and Smiechowski 2006) for a variety of conditions including temperature, electrode geometry, potential, and degradation have been studied for analyzing the lubricant condition. Electrochemical information gathered from several different series of lubricant drains collected during engine field testing was carried out by the Lubrizol Corporation (Cleveland, Ohio). An impedance/dielectric analyzer was used during the tests that were carried out (Novocontrol GmbH, Hundsangen, Germany) at three frequency regimes, viz. low-frequency zone (1–100 mHz), medium-frequency zone (0.1–10 Hz), and high-frequency zone (10 Hz to 10 MHz). For the low-frequency zone, the impedance signature for a particular lubricant is depicted by one/two depressed capacitive semicircles (the first one requires 50 seconds of relaxation time, while the second one requires 200 seconds), or occasionally a configuration that combines a Warburg-type line with a depressed capacitive semicircle as shown in Figure 8.7. The depressed capacitive semicircle with a relaxation duration of about 1 s serves as a representation of the impedance characteristic for the medium-frequency band (0.1–10 Hz) as shown in Figure 8.7. At temperatures of 100–120°C, the resistance, capacitance, and α parameter of this trend for different industrial lubricants are measured and found to be 100 KΩ, 1 μF, and 0.5–0.9, respectively. An efficient way to express the medium-frequency characteristic is to combine a constant phase element and a resistor in parallel.

The high-frequency impedance response of an industrial lubricant is made up of one highly noticeable capacitive semicircular feature (Figure 8.7) that has a relaxation period that is on the order of 1 ms. The values of the resistance (1–10 MΩ), capacitance (approximately 100 pF), and α parameter (~1) of this trend for various industrial lubricants were recorded at temperatures 100–120°C.

For the purpose of elucidating the primary characteristics of the lubrication system, a detailed equivalent circuit diagram, Figure 8.8 was produced. The overall equivalent circuit model expressed four processes: diffusion from the bulk solution, charge transfer at the electrode, bulk relaxations, and adsorption on the electrode interface. It is possible to divide these four processes into two distinct regimes: the bulk solution regime and the electrode interface regime. The bulk solution was represented by two resistance/capacitance circuits that are linked in parallel and then connected in series. Charge transfer, adsorption, and diffusion processes are all part of the interfacial regime. Resistance R_{ADS} and constant phase element CPE_{ADS} in parallel combinations represented adsorption. The effects of pseudo capacitance, which were caused by charge build-up on the electrode surface, were accounted for by CPE_{ADS}. In low-frequency modeling, faradic impedance is often added in parallel to C_{DL} (replaced by CPE_{DL} because of dispersion in low-frequency capacitance)

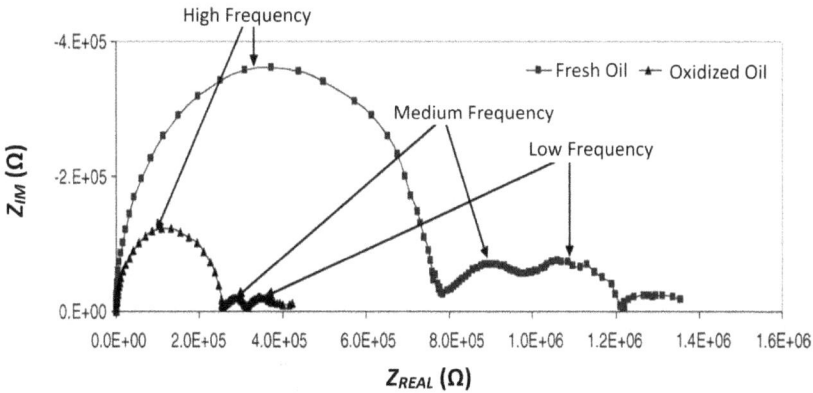

FIGURE 8.7   Impedance spectra at 120°C for fresh oils as well as oils that have been oxidized. [Reprinted with permission from Lvovich and Smiechowski (2006). Copyright (2006) Elsevier].

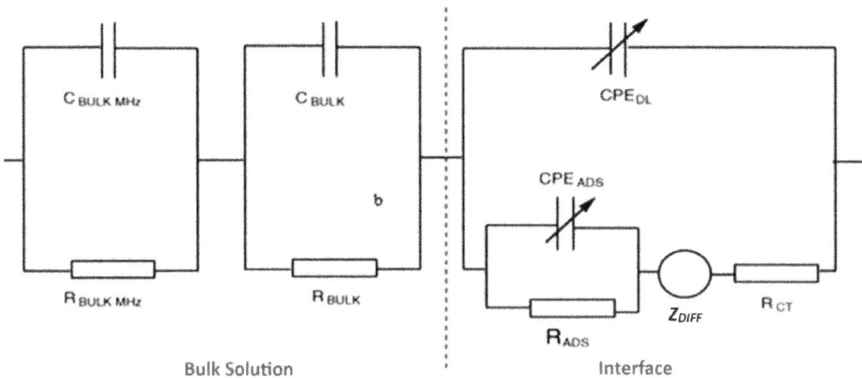

FIGURE 8.8   An equivalent circuit model was created for the investigation of lubricant impedance spectra for the bulk solution frequency ranges of 10 MHz to 10 Hz and for the interface frequency ranges of 10 Hz to 1 mHz. [Reprinted with permission from Lvovich and Smiechowski (2006). Copyright (2006) Elsevier].

to describe the electrode interface. The adsorption and diffusion components were arranged in series with the resistance to charge transfer R_{CT} on the electrode surface because the diffusion prevented the discharge of adsorbed species.

When trying to determine the acidity of a nonaqueous solution, there are a few different methods that can be used as possible options. Infrared spectroscopy, often known as IR spectroscopy, provides an all-encompassing perspective of the chemical system. However, it does suffer from a few drawbacks that make it unsuitable for use in the creation of individual sensors. The techniques used for analyzing IR data quickly become mathematically intensive, necessitating the use of computer systems for data analysis. Electrochemical approaches that are adaptable, sensitive, and affordable have shown a surprising potential for direct study and monitoring of chemistry in highly resistive nonaqueous environments, such as oil deterioration (Price and Clarke 1991;Farrington and Slater 1997). However, current research on the pH sensitivity, ion, and oxidation/reduction interference, operating pH range, and hysteresis effects of different potentiometric metal oxide sensors demonstrated that IrO_2-based sensors attained the most promising performance overall (Fog and Buck 1984; Yao, Wang, and Madou 2001). Industrial lubricants' acidity and basicity can also be assessed using iridium oxide sensors (Smiechowski and Lvovich 2003). The acidity and basicity levels in a nonaqueous medium are factors that are similar to many diverse types of industrial fluids and are intimately related to the rate at which the lubricant oxidation breakdown occurs. Different methods were used to fabricate macro/micro scale iridium oxide sensors (Yao, Wang, and Madou 2001). Case western university (the Electronics Design Center) fabricated a sputter iridium oxide sensor and micro-electromechanical systems (MEMS) based potentiometric iridium oxide sensor. Four different drained diesel oil samples were collected to study the acidity and basicity of industrial lubricants at a temperature of 80°C. Every single one of these sensors was submerged in a variety of diesel fuels. Figure 8.9a depicts the findings obtained from conducting potentiometric testing on aqueous solutions. As the basicity of the substance rose, the response from all the fabricated sensors was in the opposite direction, which was negative.

All the sensors were compared across the entire pH range, and they showed linear behavior. When comparing the responses of each sensor to the change in pH, the MEMS sensor exhibited a sub-Nernstian response, but the Melt IrOx sensor exhibited a super-Nernstian response (~80 mV per pH unit). Figure 8.9b depicts the findings of long-term stability studies. The long-term stability of all developed sensors showed that they failed long-term stability tests over a 24-h period in diesel oil. The fabricated sensor's instability is attributed to the creation of a nonuniform film thickness and adsorption of material (Bock and Birss 1999; Lassali, Boodts, and Bulhões 2000). In the future, a study in this field may entail refining and optimizing the

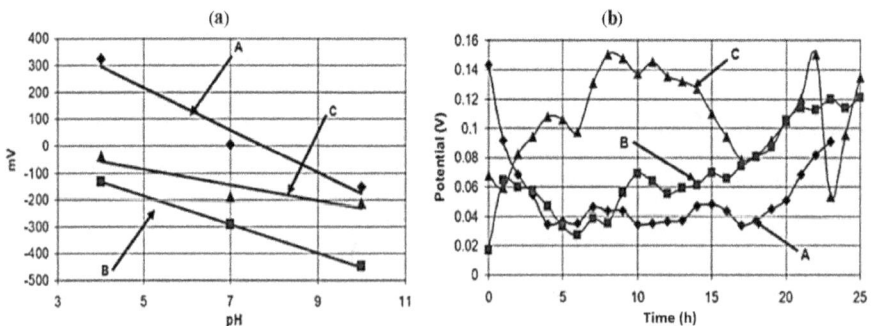

FIGURE 8.9 (a) Experimental data for the sensor response versus pH in aqueous standard solutions: (A) chronopotentiometric; (B) sputter iridium oxide sensor; (C) MEMS iridium oxide sensor; (b) the testing of the long-term stability of the fabricated sensors: (A) chronopotentiometric sensor; (B) sputter iridium oxide sensor; (C) MEMS iridium oxide sensor. [Reprinted with permission from Smiechowski and Lvovich (2003). Copyright (2003) Elsevier].

design of the sensor as well as the manufacturing process to deliver higher levels of stability and greater sensor performance.

Ethylene glycol is a key component of both antifreeze and coolant used in automobiles. These fluids work together to prevent the engine from overheating in the summer or freezing in the winter. A concentration of ethylene glycol as low as a few hundred parts per million (ppm) can have a negative impact on EO (Stehouwer and Hudgens 1987). Also, antifreeze leakage into oil causes heavy sludge deposits, which is prevented by antiwear additive zinc di-alkyl dithiophosphate (ZDP) that can be hydrolyzed by a mixture of glycol and water. By causing an increase in the creation of the acidic organic reaction products of the EO, glycol quickens the process of the oil's deterioration. Researchers (Wang and Lee 1997) have evaluated the bulk layer resistance of a mixture of glycol and EO at a temperature of 100°C by using an alternating current impedance technique at a broad spectrum of frequencies, from 0.001 to 1000 Hz. For testing lubricants, the authors utilized two different SAE 5W-30 API SG EOs, designated as A and B, in addition to commercial grade ethylene glycol. EOs A and B were offered by the same company, but they were blended in various sets. Before putting them together in a glass beaker to make test samples, glycol and EO were measured separately. Afterward, for the alternating current impedance measurement, the mixture was then vigorously mixed using a magnetic stirrer, and, after that, it was heated to 100°C. From Table 8.3, EO B has higher levels of calcium and sodium than EO A. Furthermore, the base number derived from B is 50% greater than the base number derived from EO A.

Figure 8.10a depicts the top view of the manufactured molybdenum interdigitated electrodes. The distance between the electrodes, which was held constant at 5 μm throughout this work, was the most important dimension. The completed interdigitated pattern covered an area that measured 0.7 cm × 0.7 cm. Figure 8.10b displays the resistance of the bulk layer (evaluated at 30 Hz) vs. glycol content (ppm), which shows that increasing the glycol content from 50 to 150 ppm reduces the resistance to 0.9 MΩ from 2.4 MΩ, because of which, it is possible to make use of the bulk layer resistance to find EO that has been contaminated with glycol.

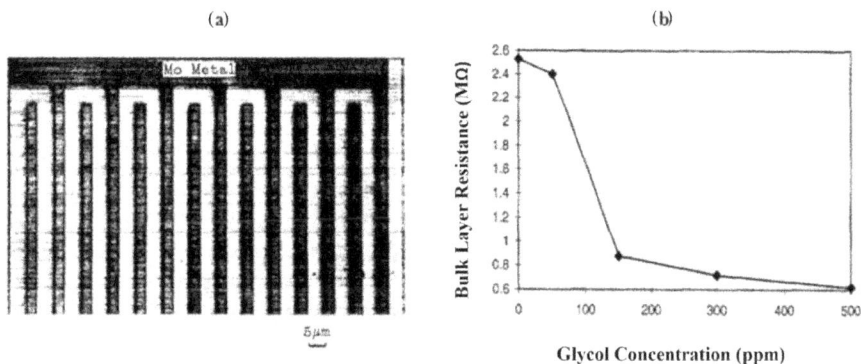

FIGURE 8.10 (a) Interdigitated molybdenum electrodes seen from the top; (b) the plots depict the bulk layer resistance of EO (A) versus the glycol concentration at temperature 100°C. [Reprinted with permission from Wang and Lee (1997). Copyright (1997) Elsevier].

TABLE 8.3

Results of the tests performed on engine oils
(EO) A and B. [Reprinted with permission from
Wang and Lee (1997). Copyright (1997) Elsevier]

Entities (%)	EO (A)	EO (B)
Aluminum	< 0.001	< 0.001
Acid number	2.80	2.8
Base number	8.0	12.0
Calcium	0.10	0.17
Chromium	< 0.00I	< 0.00I
Copper	< 0.001	0.020
Iron	< 0.001	< 0.001
Lead	< 0.01	< 0.01
Magnesium	0.09	0.11
Phosphorus	0.10	0.11
Sodium	< 0.01	0.05
Zinc	0.14	0.15

8.3.2 REAL-TIME MONITORING OF ENGINE OILS WITH EIS TECHNIQUE

The lifespan of EO can vary significantly depending on several factors. Hence, having an onboard sensor that can continuously monitor the oil's state would be quite helpful. Ulrich et al. (2007) reported using EIS and multivariate data analysis to forecast the quantities of soot and diesel in EO at the same time. For this reason, they used a precise measuring setup in order to reduce the amount of interference caused by background noise and to acquire a substantial quantity of data in a relatively short period of time. A preliminary experiment is conducted to evaluate the impact of oil temperature. This is done because the temperature within an engine might change. This research contributes to the progress of an electrochemical onboard sensor for continuous EO monitoring. The impedance is affected by soot over the full frequency range that is being used. Diesel, on the other hand, is mostly responsible for the impact seen at low frequencies in the impedance. Consequently, the use of multi-frequency impedance measurements has allowed for the capability of distinguishing between these pollutants. Wang and Lee (1994) reported in situ micro- and macro-oil conditioning sensors to measure the EOs' capacity to carry current. To test the sensor, two sets of EO samples A and B were collected. Set was collected from the engine-dynamometer test (heavy load and high speed), and the other set was taken from a vehicle while it was being driven in different settings. Macro and micro electrochemical sensors were fabricated; Figure 8.11 displays a diagrammatic representation of the microsensor in its most basic form.

Two different kinds of microsensors, sensors X and Y were built, with sensor X having smooth electrode surfaces and sensor Y having electrode surfaces that were rough because of excessive etching in zincate solution. Experiments were conducted through micro and macro electrochemical sensors and the obtained results were compared to find a correlation between the outputs of the sensors and important chemical and physical characteristics, such as TAN. The sensors had been evaluated by placing them in a furnace at a predetermined temperature and then submerging them in test oil samples. Figure 8.12 depicts a diagram in the form of a schematic. An alternating current of a triangular potential waveform had been implemented between the two gold electrodes while the operation was in progress.

In Figure 8.13, the TAN values of the microsensor Y for both the EOs A and B are plotted against the output current. As a measure of the oil's overall quality, TAN is useful as the degraded oil with a rising TAN has likely reached the end of its shelf life. Therefore, microsensor Y might very well take advantage of this to determine when the oil has reached the end of its useful lifespan.

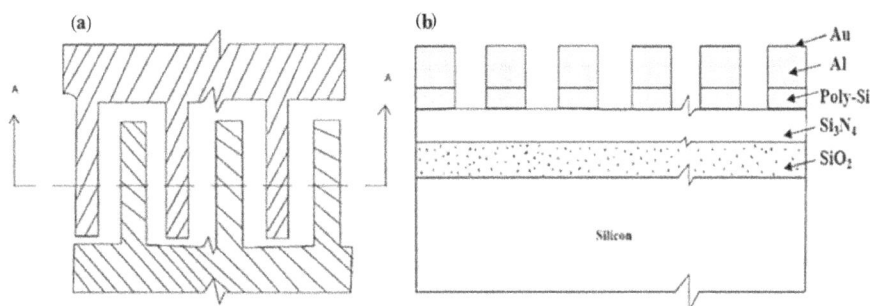

FIGURE 8.11 Fabricated microsensor (a) top view and (b) image of the structure through its cross-section from section AA. [Reprinted with permission from Wang and Lee (1994). Copyright (1994) Elsevier].

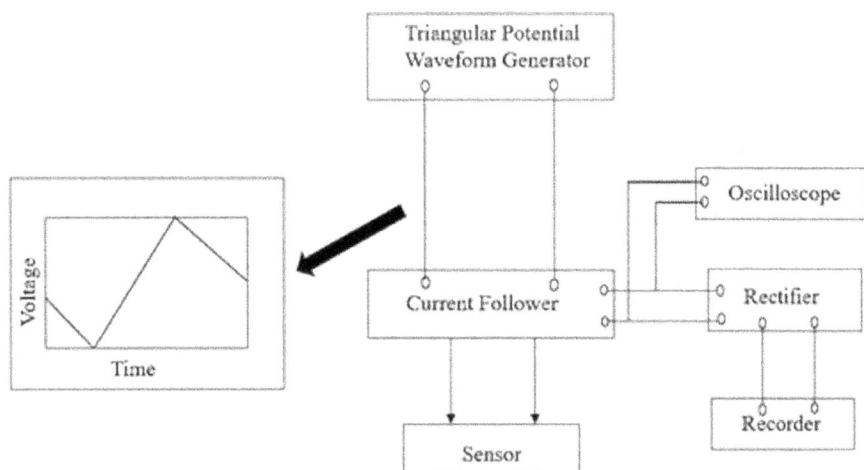

FIGURE 8.12 A diagrammatic representation of the system that was utilized for the sensor. [Reprinted with permission from Wang and Lee (1994). Copyright (1994) Elsevier].

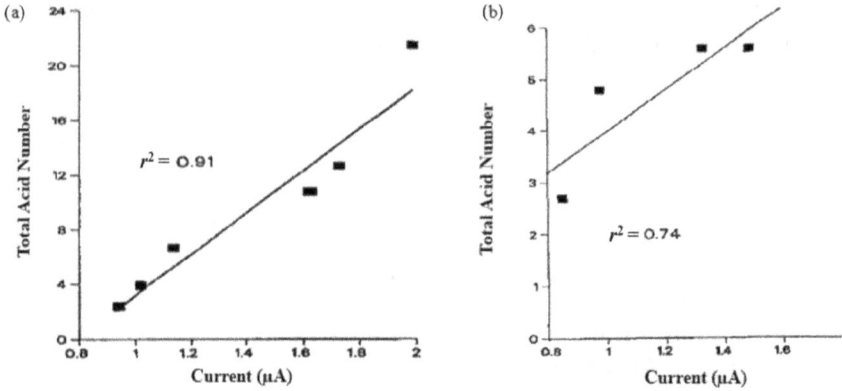

FIGURE 8.13 Link between the total number of acids and current being measured by microsensor Y using sample sets a and b, at 100°C. [Reprinted with permission from Wang and Lee (1994). Copyright (1994) Elsevier].

8.4 IMPEDANCE SPECTROSCOPY FOR ENERGY STORAGE DEVICES

The process of capturing energy generated at one time for using it later, with the goal of reducing imbalances between energy consumption-production, is known as energy storage. The term "accumulator" or "battery" is often used to refer to a device that accumulates energy. Some popular energy storage devices are Li-ion batteries (LIB) (Tröltzsch, Kanoun, and Tränkler 2006), fuel cells [alkaline cell, proton-exchange membrane cell (Santarelli and Torchio 2007), phosphoric acid cell], capacitors, compressed air, flywheel, hydrogen, and super magnets. Scientists, governments, and the general public all fully acknowledge fuel cells as ecologically benign power production devices as a special answer to some of the most pressing problems we face today, such as global warming, dwindling fossil fuel supply, and pollution. Due to increasing financing and research efforts over the past several years, fuel-cell technology has advanced significantly because of its high-power density, low operating temperature, and variety of applications. In this chapter, we mainly discuss Li-ion and fuel-cell monitoring using EIS.

8.4.1 LITHIUM-ION BATTERY LIFE PREDICTION USING IMPEDANCE SPECTROSCOPY

A LIB's calendar life is between two and four years (Scrosati 2002). The main cause of aging is the development of a surface layer at the negative carbon electrode known as the solid electrolyte interface during storage. If a battery is overcharged or undercharged for a long time, the electrolyte component starts to break down. Afterward, the system internal resistance of the LIBs increases because of the reduction in the conductivity of the electrolyte (Arora, White, and Doyle 1998). Every LIB cell has an anode, a cathode, a separator, an electrolyte, current collectors, and electrical connections. Every one of these components is responsible for the production of a variety of impedances, including inductive, capacitive, and resistive contributions that have their roots in electrical and electrochemical processes, respectively (Beard 2019). LIBs' operating voltage, efficiency, and rate capability are all related to their impedance. It also has the potential to have an effect on the battery's useful capacity. Impedance will normally increase as a function of the aging of the cell. This is because the materials that make up the electrodes, the electrolyte, and the electrical connections all degrade as the cell ages (Beard 2019). EIS is a strong and popular noninvasive diagnostic technique that may be utilized for characterizing LIBs (Dollé et al. 2001).

EIS may be used to verify state estimates, such as state of function, state of health (SOH), and state of charge, in addition to internal temperature monitoring and characterization for second-life applications over the lifespan of a battery as a prognostic or diagnostic tool. Several physicochemical techniques for the study of battery materials derived from aged cells may be used in order to get helpful insights into Li-ion cell degradation and to confirm the interpretation of EIS data. Both of these goals can be accomplished simultaneously (Waldmann et al. 2016). In order to undertake a postmortem examination of battery materials that is trustworthy, special attention should be paid to the technique for collecting materials from dismantled cells. This will ensure that reliable and accurate findings are obtained from the study. The flowchart that can be seen in Figure 8.14 reports all of the many stages that are often required to carry out a complete postmortem examination.

Li et al. (2022) presented two approaches for estimating the SOH. It is a major and tough problem to detect the SOH of LIBs quickly and efficiently as a result of the interaction of numerous degradation processes that occur throughout the aging process of LIBs. This interaction occurs during the aging process. The SOH of a battery is measured in comparison with the performance of a battery that has just been manufactured (Liu et al. 2022). The experimental study on the effects of aging on a LIB (Tröltzsch, Kanoun, and Tränkler 2006) after going through 230 discharge

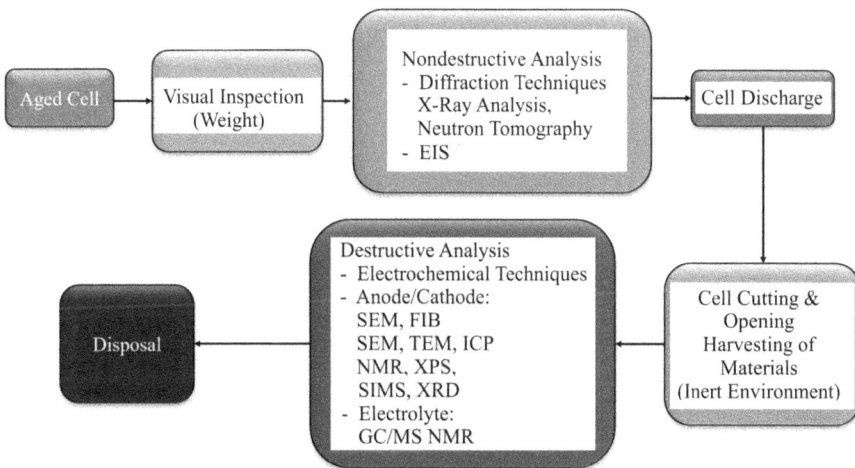

FIGURE 8.14 Flowchart illustration for Li-ion cell postmortem examination.

and charge cycles showed that the battery's capacity was found to be 14% lower. As a result of the modeling results, the values of the charge transfer resistance, series resistance, and Warburg coefficient are all altered by approximately 60%. It is necessary to consider two electrodes, electrolytes, and current collectors to model a full battery. Figure 8.15 shows the whole equivalent circuit, where the first electrode is modeled as a composite electrode represented as a $Z_{electrode}$, and the electrolyte is modeled as a series resistance R_e. An R_C combination serves as the model for the second electrode. The battery's connections and current collectors are modeled by the inductor L.

Figure 8.16 illustrates the impedance spectrum of a prismatic LIB (1.2 Ah, ICP 063465). The frequency of the measurement ranges from 3 mHz to 10,000 Hz. The average difference between what was expected of the model and what was measured was about 3.5%, and the biggest difference was 10%.

The Warburg coefficient, denoted by σ_w has an effect on the impedance spectrum at low frequencies. This influence is what causes the phenomenon known as the diffusion spike.

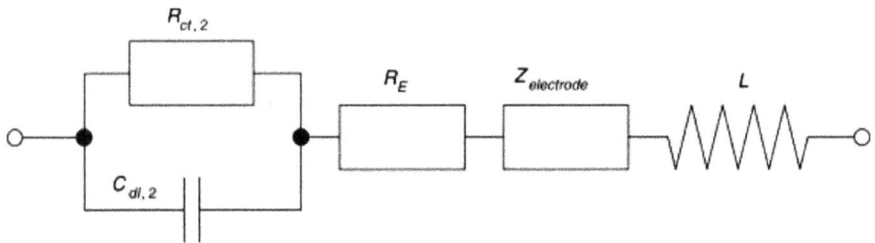

FIGURE 8.15 Battery equivalent circuit. [Reprinted with permission from Tröltzsch, Kanoun, and Tränkler (2006). Copyright (2006) Elsevier].

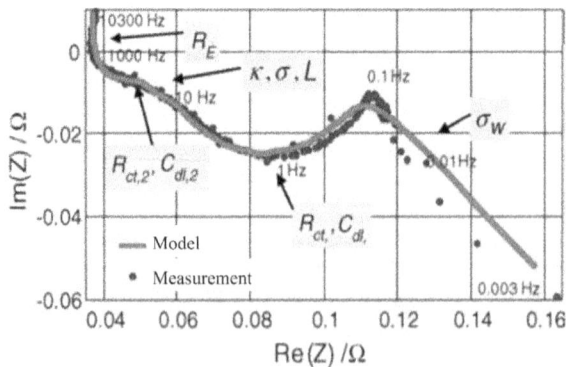

FIGURE 8.16 The influence of parameters on the impedance spectrum. [Reprinted with permission from Tröltzsch, Kanoun, and Tränkler (2006). Copyright (2006) Elsevier].

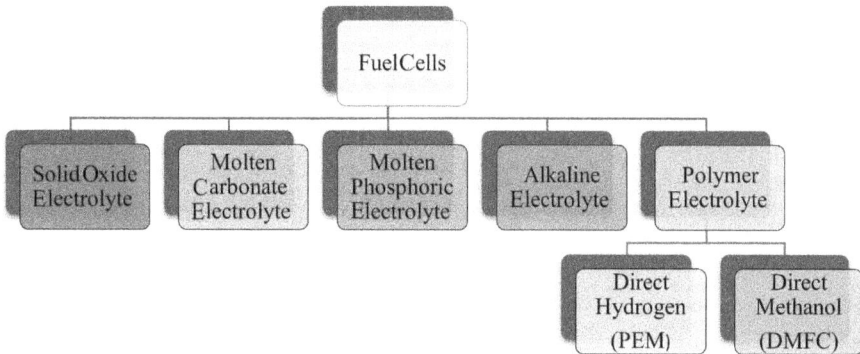

FIGURE 8.17   Major contemporary fuel-cell classification.

8.4.2 Proton-exchange membrane fuel-cell health prediction through electrochemical impedance spectroscopy

Among renewable energy sources, a fuel cell is regarded as the most promising power-producing technology. Figure 8.17 depicts a brief classification of fuel cells. The proton-exchange membrane fuel cell, sometimes known as PEMFC, is becoming an increasingly popular topic of discussion. It has found usage in many areas, including fuel-cell electric cars, as it produces no emissions, generates little noise, and has high efficiency with regard to the conversion of energy (FCEV). In comparison with other methods of diagnosis, such as the polarization curve approach, which is the most used one, EIS is the most widespread method that can be used to analyze the performance and diagnose PEMFC. EIS can be used to study the kinetics of reactants (Santarelli and Torchio 2007), the characterization of transport diffusion losses, the evaluation of ohmic resistance, and the properties of the electrode in a proton-exchange membrane. Chevalier et al. (2014) studied the cell states of health in the PEMFC stack by employing EIS. In order to determine an accurate SOH, it is necessary to examine both the static and the harmonic information. The proposed model solves the static as well as harmonic problems (Zhao et al. 2022) that are based on a thorough comparison and study done on four standard equivalent circuit models for fitting the EIS of the fuel cell. Experiments are carried out under a variety of operating circumstances, such as variations in intake pressure, stoichiometry, and air humidity, with the goals of determining the accuracy of the fitting, the changing trend of the model parameters, and the degree to which the model parameters are a good match.

8.5 IMPEDIMETRIC CHARACTERIZATION OF ELECTROACTIVE POLYMERS

Over the past 20 years, an increasing amount of focus has been placed on gaining a better understanding of the mechanisms behind the functioning of electromechanical actuators and mechanoelectric sensors as well as improving their overall performance. Electroactive polymers (EAPs) are useful materials that respond to an electrical stimulus. They are soft, flexible, and not very dense. The two most common kinds of EAPs are known as ionic electroactive polymers (IEAPs) and electronic electroactive polymers. As seen in Figure 8.18a, each of these classifications can be further divided into numerous branches. EAPs typically deposited on electrodes as thin films and ionic EAPs (IEAP) have been found to be more useful than other types of electroactive polymers.

When exposed to an external voltage, IEAP actuators, which are electromechanical devices, show mechanical deformation. The reason for this deformation is that after applying an external voltage, free ions of different sizes start to move and accumulate at the site of oppositely charged electrodes, and because of this, a volume imbalance takes place across the thickness of the structure (Montazami, Wang, and Heflin 2012). The backbone of IEAP is an ionomeric membrane and this membrane helps in transporting the ions through the structure's thickness. DuPont de Nemours, Inc., developed Nafion, the most widely used ion-exchange membrane. IEAP has been used to make soft robots (Zhang et al. 2019), artificial muscles (Bar-Cohen 2004)), on-chip fluid mixing (Price, Anderson, and Culbertson 2009), smart skins, sensors, and actuators, among other things. The five-layer structure of the ionic liquid gel soft actuator is seen in Figure 8.18b. Hong, Almomani, and Montazami (2014) reported the fabrication of IEAP actuators as well as the electromechanical and electrochemical characterization. An ionomeric membrane (Nafion) and an ionic liquid, 1-ethyl-3-methylimidazolium trifluoromethanesulfonate (EMI-Tf) were used to make IEAP actuators. Between the frequencies of 1.0×10^5 Hz to 0.1 Hz, studies of impedance spectroscopy were carried out at the set potential difference of 10 mV. Over 60-s intervals, the flow of current was measured in response to a 4 V step potential. The doped ionomer membrane's electrical conductivity (σ) is evaluated by the following equation:

$$\sigma = \frac{h}{RA}$$

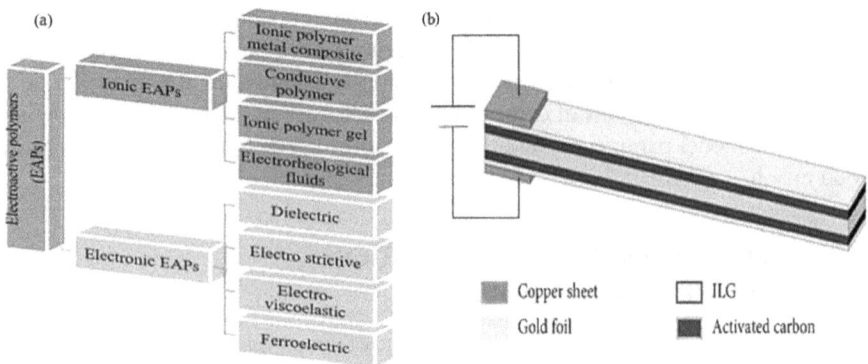

FIGURE 8.18   (a) Different types of electroactive polymers; (b) five-layer structure of the soft actuator. [Reprinted with permission from Zhang et al. (2019). Copyright (2019) Hindawi].

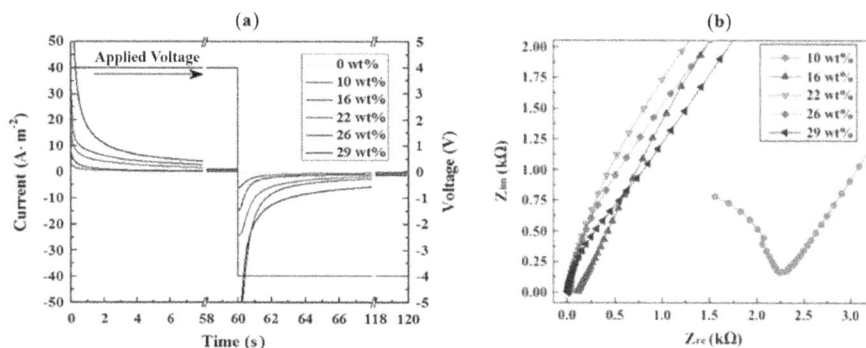

FIGURE 8.19 (a) Effect on discharging and charging current for ionic liquids having different concentrations during application of a 4-volt square wave; (b) a Nyquist plot illustrating the magnitude of the impedance of IEAP actuators with different concentrations of ionic liquid. [Reprinted with permission from Hong, Almomani, and Montazami (2014). Copyright (2014) Elsevier].

where A and h represent the area and thickness of the doped ionomer membrane, respectively, and R represents resistance. As illustrated in Figure 8.19a, the magnitude of the displaced increases when the amount of ionic liquid present in the samples goes up to a higher concentration, indicating that the current flow is caused by mobilized ions in addition to the fact that more ions are displaced in samples with a higher ionic liquid concentration. The x axis has been asymptotically reached by all curves (about zero current) after about 55 s ($60 > t > 55$ and $120 > t > 115$) in Figure 8.19a, showing that it has been completely charged. Nyquist plot for the IEAP actuators is depicted in Figure 8.19b. Figure 8.19b makes it clear that the electrochemical systems displayed almost pure resistance behavior at higher frequencies that were close to the origin of the x axis. This may be noticed by looking at the plot. While increasing the content of ionic liquids in IEAP, the electrochemical response improved.

8.6 CONCLUSION AND FUTURE PERSPECTIVES

Multifunctional sensors are becoming more popular in the sensor industry these days. These sensors can measure more than one quality parameter of an analyte at the same time, which is an exciting development. Impedimetric sensors can deliver additional information while simultaneously measuring impedance within a certain frequency spectrum. When a large amount of data is created for a single value of the measured quantity, exciting opportunities arise for mitigating confounding factors, bolstering the integrity of the data, and even achieving simultaneous measurement of many parameters. By using impedance spectroscopy in this manner, several conventional capacitive, resistive, or inductive sensors can achieve a greater level of performance. This chapter describes diverse nonbiological applications of the EIS technique and its ability to characterize electrochemical systems. It includes the basic elements of impedance measurement, equivalent circuit modeling, and transfer function modeling for the EIS technique. The characterization of energy devices and the impedance spectra of lithium batteries makes it possible to get information in a manner that is not damaging and to differentiate between various phenomena relating to the electrolyte and the electrodes. The life prediction of industrial lubricants, EOs, LIBs, etc., has been discussed in-depth. Corrosion monitoring within metals and reinforced concrete has been detailed, and the required interpretation of impedance measurement data has also been discussed. Moreover, the application of the EIS technique to characterizing EAPs is detailed in and validated by the recent literature. As polymer composites are very suitable for soft robotics applications, the EIS technique can provide very precise and accurate measurement of process variables, which finally enables researchers to design and fabricate smart and intelligent polymer robotic systems for biomedical applications.

REFERENCES

Ahmad, Zaki. 2006. *Principles of Corrosion Engineering and Corrosion Control. Principles of Corrosion Engineering and Corrosion Control.* Elsevier. https://doi.org/10.1016/B978-0-7506-5924-6.X5000-4.

Arora, Pankaj, Ralph E. White, and Marc Doyle. 1998. "Capacity Fade Mechanisms and Side Reactions in Lithium-Ion Batteries." *Journal of The Electrochemical Society* 145 (10): 3647–67. https://doi.org/10.1149/1.1838857.

Bar-Cohen, Yoseph. 2004. "EAP as Artificial Muscles: Progress and Challenges." *Smart Structures and Materials 2004: Electroactive Polymer Actuators and Devices (EAPAD)* 5385: 10–16. https://doi.org/10.1117/12.538698.

Bard Allen, J., and R. Faulkner Larry. 2001. *Electrochemical Methods: Fundamentals and Applications.* Wiley New York.

Beard, Kirby W. 2019. *Linden's Handbook of Batteries.* McGraw-Hill Education.

Bhatt, Geeta, and Shantanu Bhattacharya. 2019. "Biosensors on Chip: A Critical Review from an Aspect of Micro/Nanoscales." *Journal of Micromanufacturing* 2 (2): 198–219. https://doi.org/10.1177/2516598419847913.

Bhatt, Geeta, Swati Gupta, Gurunath Ramanathan, and Shantanu Bhattacharya. 2021. "Integrated DEP Assisted Detection of PCR Products with Metallic Nanoparticle Labels through Impedance Spectroscopy." *IEEE Transactions on Nanobioscience* 21 (4): 502–10.

Bhatt, Geeta, Keerti Mishra, Gurunath Ramanathan, and Shantanu Bhattacharya. 2019. "Dielectrophoresis Assisted Impedance Spectroscopy for Detection of Gold-Conjugated Amplified DNA Samples." *Sensors and Actuators, B: Chemical* 288: 442–53. https://doi.org/10.1016/j.snb.2019.02.081.

Bhattacharya, Shantanu, Avinash Kumar Agarwal, Om Prakash, Shailendra Singh, Mohit Pandey, and Rishi Kant. 2019. "Introduction to Sensors for Aerospace and Automotive Applications." In *Sensors for Automotive and Aerospace Applications,* edited by Shantanu Bhattacharya, Avinash Kumar Agarwal, Om Prakash, and Shailendra Singh, 1–6. Singapore: Springer Singapore. https://doi.org/10.1007/978-981-13-3290-6_1.

Bock, C., and V. I. Birss. 1999. "Irreversible Decrease of Ir Oxide Film Redox Kinetics." *Journal of The Electrochemical Society* 146 (5): 1766–72. https://doi.org/10.1149/1.1391840.

Chauhan, Pankaj Singh, Mohit Pandey, and Shantanu Bhattacharya. 2019. "Paper Based Sensors for Environmental Monitoring." In *Paper Microfluidics: Theory and Applications,* edited by Shantanu Bhattacharya, Sanjay Kumar, and Avinash K. Agarwal, 165–81. Singapore: Springer Singapore. https://doi.org/10.1007/978-981-15-0489-1_10.

Chevalier, S., B. Auvity, J. C. Olivier, C. Josset, D. Trichet, and M. Machmoum. 2014. "Detection of Cells State-of-Health in PEM Fuel Cell Stack Using EIS Measurements Coupled with Multiphysics Modeling." *Fuel Cells* 14 (3): 416–29. https://doi.org/10.1002/fuce.201300209.

Choudhury, Sagnik Sarma, Mohit Pandey, and Shantanu Bhattacharya. 2021. "Recent Developments in Surface Modification of PEEK Polymer for Industrial Applications: A Critical Review." *Reviews of Adhesion and Adhesives* 9 (3): 401–33. https://doi.org/10.47750/RAA/9.3.03.

Citti, Isabelle, Omar Aaboubi, Jean Paul Chopart, Claude Gabrielli, Alain Olivier, and Bernard Tribollet. 1997. "Impedance of Laminar Free Convection and Thermal Convection at a Vertical Electrode." *Journal of the Electrochemical Society* 144 (7): 2263.

Ciucci, Francesco. 2019. "Modeling Electrochemical Impedance Spectroscopy." *Current Opinion in Electrochemistry* 13: 132–39. https://doi.org/10.1016/j.coelec.2018.12.003.

Dollé, Mickaël, François Orsini, Antoni S. Gozdz, and Jean-Marie Tarascon. 2001. "Development of Reliable Three-Electrode Impedance Measurements in Plastic Li-Ion Batteries." *Journal of The Electrochemical Society* 148 (8): A851. https://doi.org/10.1149/1.1381071.

Encinas-Sánchez, V., M. T. de Miguel, M. I. Lasanta, G. García-Martín, and F. J. Pérez. 2019. "Electrochemical Impedance Spectroscopy (EIS): An Efficient Technique for Monitoring Corrosion Processes in Molten Salt Environments in CSP Applications." *Solar Energy Materials and Solar Cells* 191 (August 2018): 157–63. https://doi.org/10.1016/j.solmat.2018.11.007.

Farrington, Amiel M., and Jonathan M. Slater. 1997. "Monitoring of Engine Oil Degradation by Voltammetric Methods Utilizing Disposable Solid Wire Microelectrodes." *The Analyst* 122 (6): 593–96. https://doi.org/10.1039/a608022g.

Fernandez, A. G., and A. Mallco. 2018. "Corrosion Monitoring by Electrochemical Impedance Spectroscopy Test of Low-Cr Alloy Steel T 22 and High-Ni Alloy HR 224 in Nitrate Molten Salt." *Insights Anal Electrochem* 4 (1): 7. https://doi.org/10.21767/2470-9867.100028.

Fog, Agner, and Richard P. Buck. 1984. "Electronic Semiconducting Oxides as PH Sensors." *Sensors and Actuators* 5 (2): 137–46. https://doi.org/10.1016/0250-6874(84)80004-9.

Franceschetti, Donald R., J. Ross Macdonald, and Richard P. Buck. 1991. "Interpretation of Finite-Length-Warburg-Type Impedances in Supported and Unsupported Electrochemical Cells with Kinetically Reversible Electrodes." *Journal of the Electrochemical Society* 138 (5): 1368.

Gabrielli, C., J. J. Garcia-Jareno, and H. Perrot. 2001. "Charge Compensation Process in Polypyrrole Studied by Ac Electrogravimetry." *Electrochimica Acta* 46 (26–27): 4095–103.

Gerischer, Heinz. 1951. "Wechselstrompolarisation von Elektroden Mit Einem Potentialbestimmenden Schritt Beim Gleichgewichtspotential I." *Zeitschrift Für Physikalische Chemie* 198 (1): 286–313.

Hong, Wangyujue, Abdallah Almomani, and Reza Montazami. 2014. "Influence of Ionic Liquid Concentration on the Electromechanical Performance of Ionic Electroactive Polymer Actuators." *Organic Electronics* 15 (11): 2982–87. https://doi.org/10.1016/j.orgel.2014.08.036.

Kant, Rishi, Pankaj Singh Chauhan, Geeta Bhatt, and Shantanu Bhattacharya. 2019. "Corrosion Monitoring and Control in Aircraft: A Review." In *Sensors for Automotive and Aerospace Applications, Springer*: 39–53. https://doi.org/10.1007/978-981-13-3290-6_3.

Kant, Rishi, Himanshu Singh, Monalisha Nayak, and Shantanu Bhattacharya. 2013. "Optimization of Design and Characterization of a Novel Micro-Pumping System with Peristaltic Motion." *Microsystem Technologies* 19 (4): 563–75. https://doi.org/10.1007/s00542-012-1658-y.

Kim, Je-Kyoung, Seong-Hoon Kee, Cybelle M Futalan, and Jurng-Jae Yee. 2019. "Corrosion Monitoring of Reinforced Steel Embedded in Cement Mortar under Wet-and-Dry Cycles by Electrochemical Impedance Spectroscopy." *Sensors* 20 (1): 199.

Kumar, Sanjay, Pulak Bhushan, Mohit Pandey, and Shantanu Bhattacharya. 2019. "Additive Manufacturing as an Emerging Technology for Fabrication of Microelectromechanical Systems (MEMS)." *Journal of Micromanufacturing* 2 (2): 175–97. https://doi.org/10.1177/2516598419843688.

Lassali, T. A. F., J. F.C. Boodts, and L. O. S. Bulhões. 2000. "Faradaic Impedance Investigation of the Deactivation Mechanism of Ir-Based Ceramic Oxides Containing TiO2 and SnO2." *Journal of Applied Electrochemistry* 30 (5): 625–34. https://doi.org/10.1023/A:1003901520705.

Li, Dezhi, Dongfang Yang, Liwei Li, Licheng Wang, and Kai Wang. 2022. "Electrochemical Impedance Spectroscopy Based on the State of Health Estimation for Lithium-Ion Batteries." *Energies* 15 (18): 6665. https://doi.org/10.3390/en15186665.

Liu, Chunli, Dezhi Li, Licheng Wang, Liwei Li, and Kai Wang. 2022. "Strong Robustness and High Accuracy in Predicting Remaining Useful Life of Supercapacitors." *APL Materials* 10 (6): 061106. https://doi.org/10.1063/5.0092074.

Liu, Yi Shao, Padmapriya P. Banada, Shantanu Bhattacharya, Arun K. Bhunia, and Rashid Bashir. 2008. "Electrical Characterization of DNA Molecules in Solution Using Impedance Measurements." *Applied Physics Letters* 92 (14): 143902. https://doi.org/10.1063/1.2908203.

Lvovich, Vadim F., and Matthew F. Smiechowski. 2006. "Impedance Characterization of Industrial Lubricants." *Electrochimica Acta* 51: 1487–96. https://doi.org/10.1016/j.electacta.2005.02.135.

Mansfeld, F. 1988. "Don't Be Afraid of Electrochemical Techniques - but Use Them with Care!" *Corrosion* 44 (12): 856–68. https://doi.org/10.5006/1.3584957.

Marcus, Philippe, ed. 2002. *Corrosion Mechanisms in Theory and Practice*. CRC Press. https://doi.org/10.1201/9780203909188.

Montazami, Reza, Dong Wang, and James R Heflin. 2012. "Influence of Conductive Network Composite Structure on the Electromechanical Performance of Ionic Electroactive Polymer Actuators." *International Journal of Smart and Nano Materials* 3 (3): 204–13.

Morcillo, M., D. De la Fuente, I. Díaz, and H. Cano. 2011. "Atmospheric Corrosion of Mild Steel." *Revista de Metalurgia* 47 (5): 426–44. https://doi.org/10.3989/revmetalm.1125.

Orazem, Mark E., and Bernard Tribollet. 2008. *Electrochemical Impedance Spectroscopy*. Wiley, New Jersey, 383–89.

Pandey, Mohit, Mohammed Rashiku, and Shantanu Bhattacharya. 2021. "Chapter 10- Recent Progress in the Development of Printed Electronic Devices." In *Chemical Solution Synthesis for Materials Design and Thin Film Device Applications*, edited by Soumen Das and Sandip Dhara, 349–68. Elsevier. https://doi.org/https://doi.org/10.1016/B978-0-12-819718-9.00008-X.

Pandey, Mohit, Krutika Shahare, Mahima Srivastava, and Shantanu Bhattacharya. 2019. "Paper-Based Devices for Wearable Diagnostic Applications." In *Paper Microfluidics: Theory and Applications*, edited by Shantanu Bhattacharya, Sanjay Kumar, and Avinash K. Agarwal, 193–208. Singapore: Springer Singapore. https://doi.org/10.1007/978-981-15-0489-1_12.

Pandey, Mohit, Mahima Srivastava, Krutika Shahare, and Shantanu Bhattacharya. 2019. "Paper Microfluidic-Based Devices for Infectious Disease Diagnostics." In *Paper Microfluidics: Theory and Applications*, edited by Shantanu Bhattacharya, Sanjay Kumar, and Avinash K. Agarwal, 209–25. Singapore: Springer Singapore. https://doi.org/10.1007/978-981-15-0489-1_13.

Pandey, Mohit, Poonam Sundriyal, Shreyansh Tatiya, and Shantanu Bhattacharya. 2022. "Polymer-Based Electrolytes for Solid-State Batteries: Current Status and Future Challenges in Emerging Applications." *Trends in Applications of Polymers and Polymer Composites, AIP*: 5–22. https://doi.org/10.1063/9780735424555_005.

Pandey, Mohit, Shreyansh Tatiya, Shantanu Bhattacharya, and Shailendra Singh. 2019a. "Sensors in Assembly Shop in Automobile Manufacturing." In *Sensors for Automotive and Aerospace Applications*, edited by Shantanu Bhattacharya, Avinash Kumar Agarwal, Om Prakash, and Shailendra Singh, 193–207. Singapore: Springer Singapore. https://doi.org/10.1007/978-981-13-3290-6_10.

———. 2019b. "Sensors in the Joining and Welding Process in Automobile Manufacturing." In *Sensors for Automotive and Aerospace Applications*, edited by Shantanu Bhattacharya, Avinash Kumar Agarwal, Om Prakash, and Shailendra Singh, 241–56. Singapore: Springer Singapore. https://doi.org/10.1007/978-981-13-3290-6_13.

Pandey, Mohit, Shreyansh Tatiya, and Shantanu Bhattacharya. 2021. "Design and Development of MEMS-Based Sensors for Wearable Diagnostic Applications." *MEMS Applications in Biology and Healthcare, Springer*: 10–34. https://doi.org/10.1063/9780735423954_010.

Price, Alexander K., Kristen M. Anderson, and Christopher T. Culbertson. 2009. "Demonstration of an Integrated Electroactive Polymer Actuator on a Microfluidic Electrophoresis Device." *Lab on a Chip* 9 (14): 2076. https://doi.org/10.1039/b823465e.

Price, Richard J., and Lionel J. Clarke. 1991. "Chemical Sensing of Amine Antioxidants in Turbine Lubricants." *The Analyst* 116 (11): 1121. https://doi.org/10.1039/an9911601121.

Ribeiro, D. V., and J. C.C. Abrantes. 2016. "Application of Electrochemical Impedance Spectroscopy (EIS) to Monitor the Corrosion of Reinforced Concrete: A New Approach." *Construction and Building Materials* 111: 98–104. https://doi.org/10.1016/j.conbuildmat.2016.02.047.

Santarelli, M. G., and M. F. Torchio. 2007. "Experimental Analysis of the Effects of the Operating Variables on the Performance of a Single PEMFC." *Energy Conversion and Management* 48 (1): 40–51.

Scrosati, Walter A. van Schalkwijk Bruno. 2002. *Advances in Lithium-Ion Batteries*. Springer Science \& Business Media.

Shih, Hong, and Florian Mansfeld. 1989. "A Fitting Procedure for Impedance Data of Systems with Very Low Corrosion Rates." *Corrosion Science* 29 (10): 1235–40. https://doi.org/10.1016/0010-938X(89)90070-X.

Simon Araya, Samuel, Fan Zhou, Simon Lennart Sahlin, Sobi Thomas, Christian Jeppesen, and Søren Knudsen Kær. 2019. "Fault Characterization of a Proton Exchange Membrane Fuel Cell Stack." *Energies* 12 (1): 152. https://doi.org/10.3390/en12010152.

Smiechowski, Matthew F., and Vadim F. Lvovich. 2002. "Electrochemical Monitoring of Water–Surfactant Interactions in Industrial Lubricants." *Journal of Electroanalytical Chemistry* 534 (2): 171–80. https://doi.org/10.1016/S0022-0728(02)01106-3.

———. 2003. "Iridium Oxide Sensors for Acidity and Basicity Detection in Industrial Lubricants." *Sensors and Actuators, B: Chemical* 96 (1–2): 261–67. https://doi.org/10.1016/S0925-4005(03)00542-2.

———. 2005. "Characterization of Non-Aqueous Dispersions of Carbon Black Nanoparticles by Electrochemical Impedance Spectroscopy." *Journal of Electroanalytical Chemistry* 577 (1): 67–78. https://doi.org/10.1016/j.jelechem.2004.11.015.

Stehouwer, D. M., and R. D. Hudgens. 1987. "Coolant Contamination of Diesel Engine Oils." *SAE transactions*: 118–37 https://doi.org/10.4271/870645.

Sundriyal, Poonam, Mohit Pandey, and Shantanu Bhattacharya. 2020. "Plasma-Assisted Surface Alteration of Industrial Polymers for Improved Adhesive Bonding." *International Journal of Adhesion and Adhesives* 101: 102626. https://doi.org/10.1016/j.ijadhadh.2020.102626.

Tatiya, Shreyansh, Mohit Pandey, Shantanu Bhattacharya, and Shailendra Singh. 2019. "Sensors Used in Automotive Paint Shops." In *Sensors for Automotive and Aerospace Applications*, edited by Shantanu Bhattacharya, Avinash Kumar Agarwal, Om Prakash, and Shailendra Singh, 257–64. Singapore: Springer Singapore. https://doi.org/10.1007/978-981-13-3290-6_14.

Tribollet, Bernard, and John Newman. 1983. "The Modulated Flow at a Rotating Disk Electrode." *Journal of the Electrochemical Society* 130 (10): 2016.

Tröltzsch, Uwe, Olfa Kanoun, and Hans Rolf Tränkler. 2006. "Characterizing Aging Effects of Lithium Ion Batteries by Impedance Spectroscopy." *Electrochimica Acta* 51: 1664–72. https://doi.org/10.1016/j.electacta.2005.02.148.

Ulrich, Christian, Henrik Petersson, Hans Sundgren, Fredrik Björefors, and Christina Krantz-Rülcker. 2007. "Simultaneous Estimation of Soot and Diesel Contamination in Engine Oil Using Electrochemical Impedance Spectroscopy." *Sensors and Actuators, B: Chemical* 127 (2): 613–18. https://doi.org/10.1016/j.snb.2007.05.014.

Waldmann, Thomas, Amaia Iturrondobeitia, Michael Kasper, Niloofar Ghanbari, Frédéric Aguesse, Emilie Bekaert, Lise Daniel et al. 2016. "Review—Post-Mortem Analysis of Aged Lithium-Ion Batteries: Disassembly Methodology and Physico-Chemical Analysis Techniques." *Journal of The Electrochemical Society* 163 (10): A2149–64. https://doi.org/10.1149/2.1211609jes.

Wang, Simon S. 2001. "Road Tests of Oil Condition Sensor and Sensing Technique." *Sensors and Actuators B: Chemical* 73 (2–3): 106–11. https://doi.org/10.1016/S0925-4005(00)00660-2.

Wang, Simon S., and Han-S. Lee. 1994. "The Development of in Situ Electrochemical Oil-Condition Sensors." *Sensors and Actuators B: Chemical* 17 (3): 179–85. https://doi.org/10.1016/0925-4005(93)00867-X.

Wang, Simon S., and Han Sheng Lee. 1997. "The Application of a.c. Impedance Technique for Detecting Glycol Contamination in Engine Oil." *Sensors and Actuators, B: Chemical* 40 (2–3): 193–97. https://doi.org/10.1016/S0925-4005(97)80261-4.

Yao, Sheng, Min Wang, and Marc Madou. 2001. "A PH Electrode Based on Melt-Oxidized Iridium Oxide." *Journal of The Electrochemical Society* 148 (4): H29. https://doi.org/10.1149/1.1353582.

Zhang, Chenghong, Bin He, Zhipeng Wang, Yanmin Zhou, and Aiguo Ming. 2019. "Application and Analysis of an Ionic Liquid Gel in a Soft Robot." *Advances in Materials Science and Engineering* 2019 (May): 1–14. https://doi.org/10.1155/2019/2857282.

Zhao, Lei, Haifeng Dai, Fenglai Pei, Pingwen Ming, Xuezhe Wei, and Jiangdong Zhou. 2022. "A Comparative Study of Equivalent Circuit Models for Electro-Chemical Impedance Spectroscopy Analysis of Proton Exchange Membrane Fuel Cells." *Energies* 15 (1): 386. https://doi.org/10.3390/en15010386.

Part III

Technological integrations and developments

9 CMOS-based electrochemical impedance sensors

Manoj Bhatt, Mayank Punetha, Mitesh Upreti,
Manoj Singh Adhikari, and Sanjay Mathur

CONTENTS

9.1 INTRODUCTION

The metal oxide semiconductor (MOS) logic-circuit family, particularly complementary MOS (CMOS), is the most widely used technology when it comes to the creation of large-scale integrated circuits or digital circuits. Compared with other families of logic-circuitry, such as transistor–transistor logic (TTL) and emitter-coupled logic, it captures the majority of the IC market. Because of its tiny size, ease of manufacture, low power dissipation, high input resistance, and other factors, it is becoming more and more popular and performing better every day. The MOS-based logic-circuit family is the greatest choice for digital circuit design because of these properties, which also enable extremely high levels of integration for both logic and memory circuits.

MOS transistors are the fundamental building blocks of CMOS logic circuits (especially enhancement-type MOS transistors). The n-channel and p-channel MOS transistors are voltage-controlled resistance devices with three terminals (gate, drain, and source). On an n-type substrate, a PMOS transistor is created with heavily doped p regions for the source and drain. In contrast to NMOS transistors, which are constructed on p-type substrates with highly doped n regions for the source and drain, PMOS transistors have p-type channels with holes as the charge carrier. An n-type channel and electrons serve as the charge carriers in NMOS transistors. Although both MOS transistors function on the same principle, the p-channel MOS has a negative threshold voltage while the n-channel MOS has a positive threshold voltage. Despite functioning under opposing conditions, the two MOS transistors have the same basic operating principle. Figure 9.1 displays the circuit symbols for the p-channel and n-channel MOS transistors. Due to the insulating material, the input impedance of MOS transistors is extremely high and is represented by a gap between the gate and the other two MOS terminals (i.e., the drain-source terminals), as illustrated in Figure 9.1. This gap also shows that, regardless of the value of V_{gs}, there is no current flow from gate to source or from gate to drain.

A MOS transistor functions as a switch in digital logic applications, switching between the "on" and "off" states. When it is "on," the resistance is very low, providing no obstruction to the current as it flows from source to drain and giving the appearance of an "on" switch. When a MOS is in the "off" state, it has an extremely high resistance that creates a very high barrier to current. As a result, no current flow from source to drain is visible, making the MOS appear to be an "off" switch. The voltage between the gate and source (V_{gs}) terminals of the MOS transistor regulates the high and low resistance characteristics of this semiconductor. V_{gs} in an NMOS transistor is typically either zero or positive. If V_{gs} is zero, MOS's proposed resistance from drain to source, or R_{ds}, is quite high. As V_{gs} rises from zero to a positive value, R_{ds} falls to a very low value. The same observation is made using V_{gs} for PMOS, although V_{gs} for PMOS only works with negative voltage values. In other words, we can claim that PMOS and NMOS exhibit identical behavior for V_{gs} polarities that are either complementary or contrary to one another. There are three unique

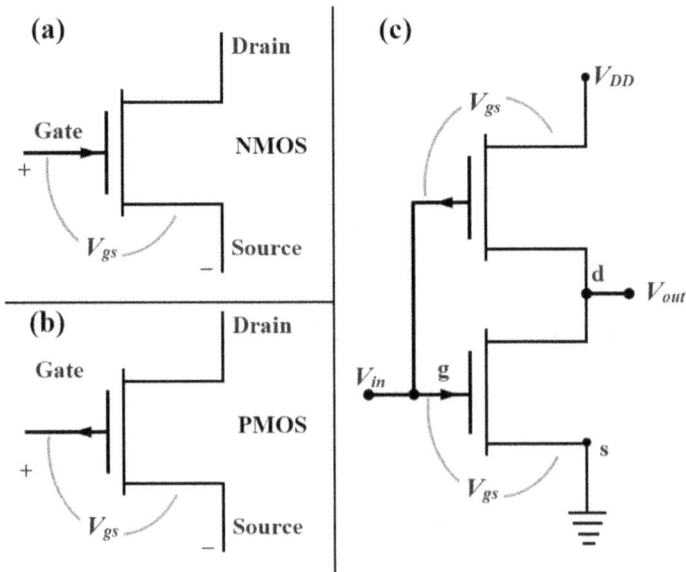

FIGURE 9.1   Circuit symbols for MOS transistor (a) p-channel MOS (PMOS); (b) n-channel MOS (PMOS); (c) CMOS inverter circuit symbol.

operating areas on the I_d–V_{ds} MOS transistor output characteristic curve (explains the IV CMOS feature as well). These operating regions are the cutoff region, the triode region, and the saturation region. In MOS, the saturation region is used for amplification purposes or when MOS is used as an amplifier. For operation as a digital element or as a switch, it is operated in cutoff and triode regions.

Any logic circuit used in digital operation handles electric signals as levels of voltage to produce digital or binary output, such as 0 (low) or 1 (high). Any voltage value between 0 and 1.5 V is interpreted by a CMOS-based logic circuit as logic 0 or low, and any voltage value between 3.5 and 5 V is interpreted as logic 1 or high. CMOS-based logic circuits are typically powered by a 5 V power source. The undefined voltage level is the voltage range between these two defined levels, and it typically happens during signal transitions. Depending on how the circuit is interpreted, the output corresponding to the undefined voltage level is either 0 or 1. The simplest CMOS-based digital logic circuit is the CMOS inverter circuit, as shown in Figure 9.1c, which only needs one NMOS and PMOS transistor. From Figure 9.1c, it is clear that the input to the transistor is applied at the junction point where the gates of the two MOS transistors are joined, and the output of the inverter is computed at this point. From the same figure, it can be deduced that the inverter's output is about V_{DD} or 1 (high) when NMOS is "off" and PMOS is "on," as opposed to 0 or low when NMOS is "on" and PMOS is "off." It can also be seen in Table 9.1's truth table for CMOS inverter circuits. The CMOS inverter circuit's operation is typically explained by the following two scenarios because it only has one input:

(1) When $V_{in} = 0$ V, the NMOS is "off", because the voltage between the gate-to-source terminal is 0 or in other words, $|V_{gs} < V_t|$ and the PMOS is "on" because $|V_{gs}| > |V_t|$ where $V_{gs} = -V_{DD}$. Because under "on" condition, the MOS circuit offers very low resistance, and, due to this, the output terminal is directly connected to V_{DD}, and, therefore, the output voltage is equal to V_{DD} or high.

(2) When $V_{in} = 5$ V, the NMOS is "on", because the voltage between the gate-to-source terminal is equal to V_{in} or +5 V, that is, $|V_{gs} > V_t|$, and, on the other hand, PMOS is "off" because $V_{gs} = 0$, that is, $|V_{gs}| < |V_t|$. Because NMOS is under "on" condition here, it offers very low resistance and, as a result of this, output terminal is directly connected to the ground terminal, and the output voltage is equal to 0 or low.

It is evident from the foregoing justification and the truth Table 9.1 that the logic circuit behaves like an inverter. Because a low input voltage of zero generates a high voltage, such as 5 V, and vice versa.

TABLE 9.1
Truth table for CMOS inverter circuit

V_{in}	NMOS	PMOS	V_{out}
0 or low	• off	• On	• 5v(1) or high
• 5v (1) or high	• on	• Off	0 or low

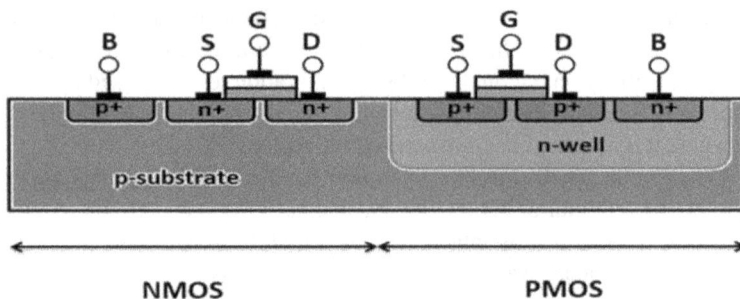

FIGURE 9.2   Cross-section of a CMOS chip.

As the name suggests, complementary MOS makes use of the capabilities of both NMOS and PMOS on adjacent regions of the chip. The NMOS transistor is directly implemented in the p-type substrate as presented in Figure 9.2, whereas the PMOS transistor is imposed in the specifically designed n-region, also known as the n-well. Two MOS are isolated through a thick portion of the oxide layer that serves as an insulator in this illustration (in the figure, these MOS are not shown as isolated so they cannot be used here).

The usage of CMOS technology in sensors is rapidly growing due to reduced power loss, cost, and area and enhanced capability. Keeping the high-utility response of CMOS sensors in mind, this chapter in detail discusses the integration aspect of CMOS with biosensors, their development methods, and their applicability domain.

9.2   CMOS-BASED IMPEDANCE SENSING

9.2.1   Electrochemical biosensors

As modern biosensors are reliable and very efficient for the quick and accurate detection of contaminated bioagents in the surroundings, they have a vital role in the field of the healthcare sector, agriculture, biotechnology, military, and bioterrorism detection. The global study done on worldwide trends (Grand View Research 2019) mentioned that biosensors' worldwide cost requirement would have approached US$20 billion by 2020, which has extensively exceeded the estimate. Due to their cost-effective nature, less complexity, better performance with regard to response time and ability to miniaturization, electrochemical techniques/biosensors (ECBs) are one of the most widely used techniques among all the existing techniques used for biosensing or to record biorecognition events and converts the obtained response into electronic form. Electrochemical-based biosensors are mainly utilized for the identification of hybridized DNA (Chiti, Marrazza, and Mascini 2001; Cai et al. 2004Gu and Hasebe 2012), enzyme reactions (Hrapovic and Luong 2003; Kotanen et al. 2012), glucose concentration, bacteria (Setterington and Alocilja 2011), and DNA-binding drugs. The principal/main idea behind the operation of electrochemical biosensors is that several chemical reactions generate or ingest electrons or ions, which produce some deviation in the electrical properties of the solution that can be sensed and used as measuring parameters.

FIGURE 9.3 Fundamental components related to general integrated electrochemical biosensors. [Reprinted with permission from Li et al. (2017). Copyright (2017) MDPI].

The four fundamental components in electrochemical biosensors that can be used to describe them are analyte, transducer, bio-interface, and instrumentation, as shown in Figure 9.3. The target molecule intended for measurement, such as bacteria/cell, DNA, or protein, is referred to as the analyte and the bio-interface, also known as the biorecognition element, enables the selective recognition of the target analyte. The electrical signal is generated by the transducer, which interprets the interactions between the analyte and bio-interface. In electrochemical biosensors, an electrode serves as the typical transducer, converting the ion/electron flow created by a biorecognition event (interaction of the analyte with biorecognition element) into a measurable electrical signal in the form of voltage or current. The instrumentation, which is the fourth component, is commonly made up of electronic circuitry to capture, boost, and record biorecognition signals generated by the transducer. At last, the electrolyte solution (represented as the liquid environment in Figure 9.3) provides a base for the analyte to perform a biorecognition event and as a medium for the analytes to reach the transducer. A substrate or support is also a part of the electrochemical biosensor, which provides physical support for the other components, and also can have an important role in the sensitivity and selectivity of the biosensor.

We can classify electrochemical biosensors as amperometric, conductimetric, impedimetric, and potentiometric on the basis of the electrical parameters that are being measured. From this chapter, we are likely to gain an in-depth insight into electrochemical impedimetric biosensors.

9.2.2 Impedimetric Biosensors and Their Importance

At equilibrium, impedimetric/impedance biosensors quantify the dynamic electrical impedance response of the electrode–electrolyte interface (Barsoukov and Macdonald 2005). Impedimetric biosensors use changes in the resistance and/or capacitance at the bio-interface during biorecognition events to detect changes in the biological sample. The measurement of impedance and its sensing has found inventive applications in biomedical engineering fields, like point-of-care or bedside diagnostic through the detection of the cell, DNA, protein, ions, environmental monitoring, and implantable/nonimplantable microdevices.

The principle of electrochemical/electrical impedance spectroscopy (EIS) is based on Ohm's law as depicted in Figure 9.4a. Here, in the figure, the two-electrode impedance spectroscopy system is considered, in which the first electrode is known as the working electrode (WE) and the other one is known as the counter electrode

(CE). An electric field is developed between these two electrodes due to the applied source voltage. The value of the developed electric field can be modulated by the presence/absence and physical location of any particle such as cells between both the electrodes and this is the cause of change in the value of the measured current and vice versa. It works on the principle of Ohm's law, $Z = V/I$, where Z is impedance, V is voltage, and I is current. These quantities are measured through the analyte. Figure 9.4b illustrates the theory behind the impedance measurements of biomolecules.

Electrodes are submerged in the solution and an alternating current (AC) voltage source is added to link the electrodes, where electrodes and voltage source form a closed loop. The conductivity of biomolecules is different from the medium and electrodes, and because of this, one or several biomolecules are close to or stay on the electrodes. These molecules change the current flowing through this loop. In several cases, biomolecules, being capacitive in nature, have a blocking effect on the current. An alternative configuration also uses three electrodes; however, this configuration is used in EIS, which typically relies on a reduction/oxidation or redox reaction taking place.

The impedance-sensing technique, also known as EIS or electric cell-substrate impedance sensing (Giaever and Keese 2012; J. Gu and McFarlane 2012; Xu et al. 2016; Yu et al. 2016) is a method to study and diagnose living analytics/cells exists in in vitro and in vivo environments. However, in EIS, there exists a three-electrode system as compared with two other impedance-sensing techniques, which is used for performing a chemical reaction that is crucial for the measurement, but in the case of whole cell impedance measurements, only two-electrode EIS is used.

Since the impedance-sensing technique is label-free and noninvasive and of low cost, it is a highly popular technique among existing standard biochemical assays and has many other advantages as compared with other biochemical assays, for exanple, EIS prevents situations like cell death and is proficient to monitor analytics/cell behavior in real-time. EIS sensors are less vulnerable to interference due to nonspecific binding in comparison with label-free optical biosensors. Real-time EIS measurements of a single cell or multiple populations of cells can increase the useful information extracted from an experiment. The ability to monitor a single cell is dependent on the size of the electrode. Many researchers have EIS in various applications, for example, measurement of concentrations of cells (Yang et al. 2004;

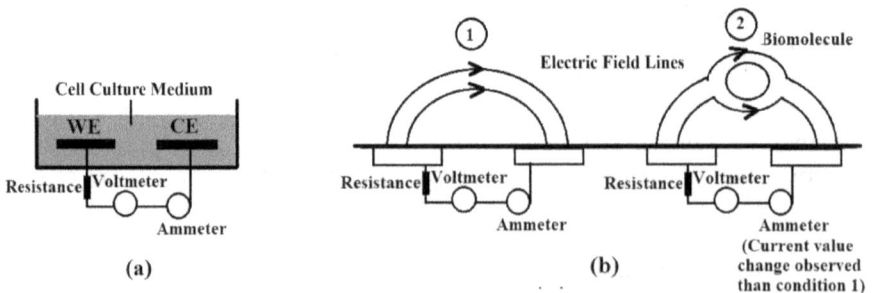

FIGURE 9.4 (a) Diagram of a two-electrode electrochemical biosensor; (b) working principle of cell impedance sensing.

Mishra et al. 2005), identifying small molecules of biological relevance (Bontidean et al. 1998; Yin 2004) and observing bilayer lipid membrane. The main advantage of the EIS technique as compared with potentiometric and amperometric biosensors is that it uses a small amount of sinusoidal (generally the range of the stimulus signal amplitude is between 5 and 10 mv) and because of which most biorecognition layers are not damaged or disturbed.

Electrochemical sensor configurations are mainly developed by two types of electrode systems, a two- or three-electrode system, as stated. Figure 9.5 shows the equivalent circuit of both electrode systems. In a two-electrode system, a stimulus voltage, V_{st} is the applied voltage between the reference electrode (RE) and the WE, and V_{cell} is the resulting potential across the electrochemical cell's electrode–electrolyte interface, and it should be the same during the electrochemical reaction. Z_{bio} is the impedance offered by the electrode–electrolyte interface. R_s is the resistance inherently possessed by the electrochemical cells that provides the ion flow path to RE through the electrolyte solution. Due to R_s, a voltage drop exists, and as a result, V_{cell} is always less than V_{st}. This creates an undesirable environmental condition for the operation. Since the resistance of the electrochemical cell cannot be eliminated, so a CE can be added to a three-electrode system to eliminate the negative effects caused by the voltage drop across R_s. In a three-electrode system, CE supplies the sensing current, I_{sn}, which ensures that there is no current flow through the RE in the steady bias state and eliminates any voltage drop across R_s.

In the case of impedimetric sensors, AC is small in amplitude, because of which the voltage drop due across the solution resistance is very small and can be ignored. So the two-electrode system is also very useful as equivalent to the three-electrode

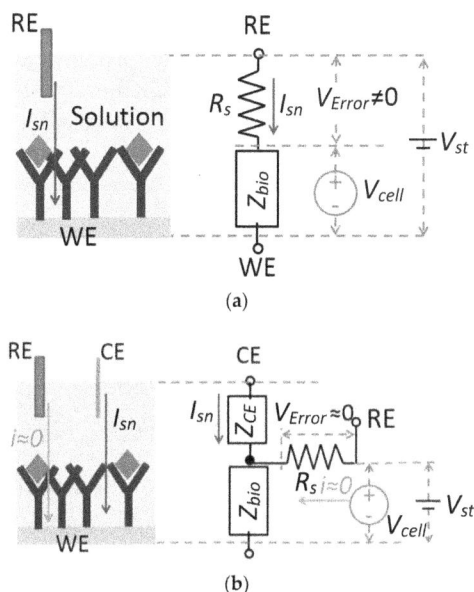

(a)

(b)

FIGURE 9.5 Equivalent circuit models of the (a) two-electrode system; (b) three-electrode system. [Reprinted with permission from (Li et al. 2017). Copyright (2017) MDPI].

system in the case of impedimetric biosensors, but in the case of amperometric techniques, the three-electrode system is often preferred because here large response currents are generally desired.

Over the previous few decades, the field of microelectronics has encountered tremendous growth and technological advancements, which improved CMOS integrated circuit technology, and it has become a robust and cost-effective electronics platform that is very well matched for the fabrication of EIS instrumentations. CMOS-based micro-electrodes have observed a lot of advancements primarily by scaling down the biosensor's size as compared to silicon-based microchips. As compared with conventional tabletop instrumentation and sensors, CMOS chips-based impedance sensors are more sensitive, consume lesser power, and provide a very fast response (Li et al. 2011). Extremely sensitive applications like DNA testing and drug screening can also be accomplished with the help of high-density CMOS–biosensor arrays (Yang et al. 2009).

9.2.3 CMOS INSTRUMENTATION FOR EIS BIOSENSORS

In biosensors, electrodes convert the biorecognition process into a quantifiable electrical signal as voltage or current, after which electronic instrumentation can be used to measure the signals and provide the necessary biasing signals to guide electrochemical reactions. Due to low cost and better performance, mainly in low-frequency-based applications, for example, biosensing, CMOS circuits completely dominate the industry related to modern microelectronics. As a result, these days, the development of maximum electrochemical instrumentation is being done through CMOS technology.

The functional blocks of CMOS-based electrochemical instrumentation for impedimetric biosensors have four main components, as depicted in Figure 9.6: potentiostat, signal generator, readout circuitry, and signal-processing unit. The desired signal shapes like triangle, constant, saw, and pulse waveforms are produced by the

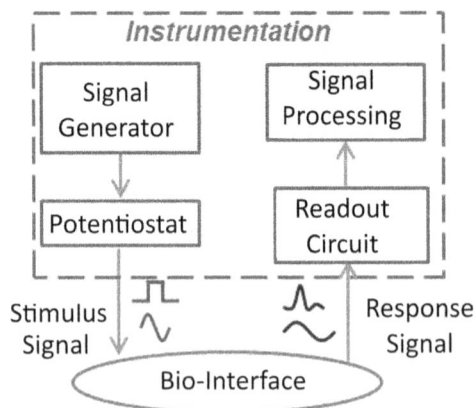

FIGURE 9.6 Basic functional block diagram of instrumentation for electrochemical biosensors. [Reprinted with permission from Li et al. (2017). Copyright (2017) MDPI].

signal generator and the biasing potential of the electrodes is controlled through the potentiostat. The response signal of the sensor is amplified by the readout circuit, and it can offer front-end signal conditioning.

The signal-processing block improves the data from the readout circuit by processes such as by filtering it. The signal-processing unit and signal generator are not connected directly to the electrode–electrolyte interface and, hence, can be applied with common circuits. Therefore, potentiostat and readout circuits are in general vital functional units responsible for the sensor performance. To start a biochemical reaction in electrochemical sensors, V_{cell} should be reached to analyte's redox potential, so there must be circuitries to precisely set V_{cell} during measurement of sensor response. The potentiostat circuit that establishes V_{cell} at the electrolyte–electrode interface is a vital functional unit for the sensor performance.

The first CMOS-based potentiostat was developed in 1987 (Turner, Harrison, and Baltes 1987) with the help of a single operational amplifier (OPA) for a two-electrode amperometric chemical sensor. This development started the use of microelectronics/CMOS in electrochemical instrumentation and started a new domain of miniaturization for point-of-care diagnostics, implantable device application, and high-output screening. The basic potentiostat configurations in the three-electrode systems are determined by the instrumentation circuit's ground RE (also said as the analog ground) (Ahmadi and Jullien 2009). The easiest form of the grounded-WE potentiostat can be realized using a single OPA, as illustrated in Figure 9.7.

Another vital unit responsible for the sensor performance is a readout circuit that measures the output signal produced by electrochemical biosensors. There are two types of current readout circuitry, viz. DC or AC readout circuitry, depending upon the type of electrochemical method employed. As defined earlier, amperometric biosensors produce DC response current, because of which they need DC readout circuits and impedimetric sensors that generate an AC response that must be quantified by AC readout circuitry. Here, we focus on impedimetric-based biosensors and AC readout circuitry.

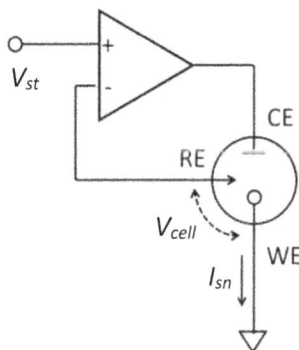

FIGURE 9.7   Grounded-WE three-electrode potentiostat using single OPA. [Reprinted with permission from Li et al. (2017). Copyright (2017) MDPI].

9.2.4   AC CURRENT READOUT CIRCUITRY

A complex impedance with capacitive and resistive components is used to model impedimetric sensors that can fluctuate with biorecognition events. Randles model circuit is commonly used to represent a typical electrochemical impedimetric sensor interface (Figure 9.8; Bard and Faulkner 2001).

The change in the phase and/or amplitude of the output response with respect to the phase and/or amplitude of the stimulus signal typically manifests the biorecognition events. The real and imaginary components of the sensor's impedance model can be algebraically calculated from the phase-amplitude response for determining the desired sensing parameters.

9.2.5   ALGORITHMS FOR EXTRACTION OF IMPEDANCE

For the extraction of real and imaginary components of the impedimetric sensors, we can use either the digital or analog domain method. The Fast Fourier Transform (FFT) algorithm is the commonly used method in the digital domain (Barsoukov and Macdonald 2005) and with the help of the Frequency Response Analyzer (FRA) method, we can calculate the sensor's components in the analog domain. The FRA method, due to its ability to be implemented with basic analog circuitry, is well-suited in sensor array microsystems (Rairigh, Mason, and Yang 2006), unlike the FFT that may require more complex digital components. Therefore, the FRA method is widely used in CMOS impedimetric current readout circuits (Min and Parve 2007; Gozzini, Ferrari, and Sampietro 2009; Manickam et al. 2010). The functional block diagram of the FRA method for calculating transducer admittance/impedance and a signal's phasor and time domain expression has been depicted in Figure 9.9.

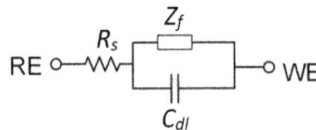

FIGURE 9.8 Randles model circuit for impedimetric sensor model where R_s represents solution resistance C_{dl} indicates double-layer capacitance, and Z_f denotes faradic components. [Reprinted with permission from Li et al. (2017). Copyright (2017) MDPI].

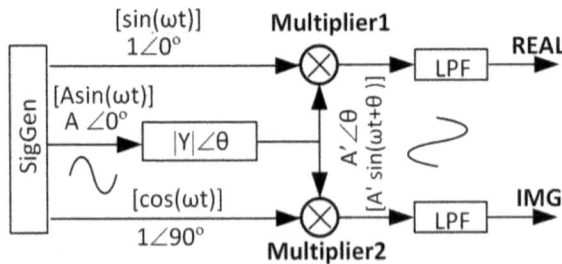

FIGURE 9.9 Functional block diagram of FRA method. [Reprinted with permission from Li et al. (2017). Copyright (2017) MDPI].

Here, $|Y|\angle\theta$ is the transducer's admittance in phasor form and $A\sin\omega t$ is the applied voltage to the electrochemical transducer, where ω is the applied signal's frequency. The output of the transducer is given by $A'\sin(\omega t + \theta)$ and the sine and cosine functions, represented by $sin(\omega t)$ and $cos(\omega t)$, respectively, are used as reference signals to separate and obtain the real and imaginary components of a signal. LPF denotes the low pass filter and removes the AC components at the output of the FRA method. REAL and IMG represent two orthogonal components of the output that represent the real and imaginary parts of impedance or admittance, respectively, and these terms are given by the following equations:

$$\text{REAL} = (A'/2)cos(\theta) \tag{9.1}$$

$$\text{IMG} = (A'/2)sin(\theta) \tag{9.2}$$

$|Y|\angle\theta$ can be evaluated through the given equations:

$$|Y| = \frac{A'}{A} = \sqrt{\text{REAL}^2 + \text{IMG}^2} \tag{9.3}$$

$$\theta = \tan^{-1}\frac{\text{REAL}}{\text{IMG}} \tag{9.4}$$

9.2.6 IMPEDANCE EXTRACTION ALGORITHM IMPLEMENTATION

Based on the lock-in technique, many EIS readout circuits have been designed. A new structure that combines the lock-in multiplier and sigma–delta technique has been developed (Yang et al. 2009) for extracting and digitizing impedance. The circuit, which can be seen in Figure 9.10, uses shared resources for impedance extraction and digitization to increase hardware efficiency, allowing for more than 100

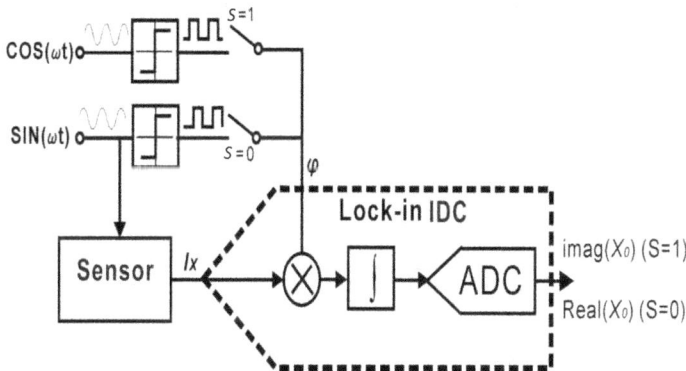

FIGURE 9.10 Diagram of lock-in impedance extraction circuit. [Reprinted with permission from Li et al. (2017). Copyright (2017) MDPI].

readout channels in a small 3 mm × 3 mm die. The real and imaginary parts of the impedance are extracted using digital multipliers integrated into the analog-to-digital converter's integrator, and high-frequency noise is eliminated by the integrator.

9.3   INTEGRATING CMOS AND BIOSENSORS

Combining biosensor interfaces and CMOS detection circuitry has numerous advantages, such as downsizing, automated detection, improved performance, and lower reagent expenses. To take advantage of these benefits, a recent inclination in CMOS biosensors is the integration of electrochemical biosensors and readout circuits to create on-chip measurement systems. To reduce the environmental noises and reduce wiring complexity, the high-density electrical metal routings of interior CMOS are generally used to attach instrumental circuits with electrode arrays. Despite the many benefits, integrating CMOS technology with biosensors can be challenging because the materials and methods used to create CMOS electronics are not always compatible with those used in biosensors. This creates a contrast in requirements and compatibility that makes it difficult to combine the two in a microsystem. Based on various integration efforts made in the past (Frey et al. 2010; Huang and Mason 2013; Datta-Chaudhuri, Abshire, and Smela 2014; L. Li and Mason 2014), the primary CMOS–biosensor integration components are depicted in Figure 9.11. The chip of CMOS behaves both as a sensor substrate and an electrochemical instrumentation platform. This means that the CMOS chip is used both to measure the electrical signals produced by the biosensor and to support the physical structure of the biosensor. The CMOS chip has sensor electrodes and probes (biological probes) attached to its surface, which are used to detect specific biomolecules or biological signals. The electrodes of the sensor are connected to the CMOS chip's electronic elements via thin film planar metal routings at pre-design electrical contact openings. It is also mentioned that biosensors require a liquid environment, so the integrated sensor needs to be connected to a fluid handling interface, in either reservoir mode (Yin 2004) or microfluidic channel mode (Huang and Mason 2013; Datta-Chaudhuri, Abshire, and Smela 2014). Furthermore, the packaging is essential to guard the active electronics of the system from corrosive biosensing liquids. In particular, it should protect the integrated sensor from potential hazards such as chemical corrosion of the reactive aluminum CMOS contact pads and electrical shorts caused by conductive electrolyte fluids. At last, the entire assembly needs both fluidic and electrical connections with the outside world, meaning that the packaging must also provide a means for the electrical signals from the sensor to be transmitted to the outside world, and for the fluid to be delivered to and from the sensor. This is crucial for the biosensor to be able to function properly. In the field of CMOS–biosensor microsystems, two major challenges are the design and integration of on-chip biosensor electrodes and the development of chip-scale packaging that can withstand liquid environments. On-chip biosensor electrodes refer to the incorporation of biological sensing elements, such as enzymes or antibodies, directly onto the surface of a CMOS chip. This is a challenge because the materials and processes used to make traditional CMOS devices are not compatible with many biological materials, and there is often a mismatch between the electrical properties of the two. While chip-scale packaging refers to protecting

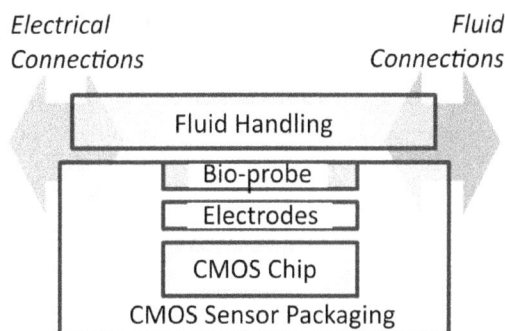

FIGURE 9.11 Key components of the CMOS–biosensor microsystem. [Reprinted with permission from Li et al. (2017). Copyright (2017) MDPI].

from the harsh conditions found in liquid environments, such as humidity, temperature changes, and chemical interference, and enclosing the biosensor chip in a liquid environment, solving these challenges is crucial for the development of effective and practical biosensor microsystems.

9.4 CONCLUSION AND FUTURE PERSPECTIVES

Researchers have been working extensively to make more versatile and smaller biochips to deliver enhanced efficacies in terms of sensitivity and selectivity. CMOS biosensors in this respect are part of an emerging field that is expanding in all dimensions, including fabrication specialization to downsize the microchip, its efficiency enhancement, and cost reduction. The chapter in detail discusses the integration of CMOS in electrochemical sensors, primarily impedimetric sensors, in terms of device assembly, optimization, and result extraction. It has been observed that the CMOS biosensors have observed applications in several biological detection and environmental monitoring domains and have been performing extremely well. The CMOS integration is primarily responsible for circuitry design aspects and also works as a substrate for microchip design.

As CMOS integration is a very prominent step in sensor technology, it is expected that in the near future, CMOS technology will be effectively utilized to develop highly versatile microchips that would observe applications in several biological/nonbiological domains.

REFERENCES

Ahmadi, Mohammad Mahdi, and Graham A. Jullien. 2009. "Current-Mirror-Based Potentiostats for Three-Electrode Amperometric Electrochemical Sensors." *IEEE Transactions on Circuits and Systems I: Regular Papers* 56 (7): 1339–48. https://doi.org/10.1109/TCSI.2008.2005927.

Bard, Allen J., and Larry R. Faulkner. 2001. *Electrochemical Methods: Fundamentals and Applications, 2nd Edition*. Wiley. https://www.wiley.com/en-us/Electrochemical+Methods%3A+Fundamentals+and+Applications%2C+2nd+Edition-p-9780471043720.

Barsoukov, Evgenij, and J. Ross Macdonald. 2005. *Impedance Spectroscopy: Theory, Experiment, and Applications.* Wiley.

Bontidean, Ibolya, Christine Berggren, Gillis Johansson, Elisabeth Csöregi, Bo Mattiasson, Jonathan R. Lloyd, Kenneth J. Jakeman, and Nigel L. Brown. 1998. "Detection of Heavy Metal Ions at Femtomolar Levels Using Protein-Based Biosensors." *Analytical Chemistry* 70 (19): 4162–69. https://doi.org/10.1021/ac9803636.

Cai, Wei, John R. Peck, Daniel W. Van Der Weide, and Robert J. Hamers. 2004. "Direct Electrical Detection of Hybridization at DNA-Modified Silicon Surfaces." *Biosensors and Bioelectronics* 19 (9): 1013–19. https://doi.org/10.1016/j.bios.2003.09.009.

Chiti, Giacomo, Giovanna Marrazza, and Marco Mascini. 2001. "Electrochemical DNA Biosensor for Environmental Monitoring." *Analytica Chimica Acta* 427: 155–64. https://doi.org/10.1016/S0003-2670(00)00985-5.

Datta-Chaudhuri, Timir, Pamela Abshire, and Elisabeth Smela. 2014. "Packaging Commercial CMOS Chips for Lab on a Chip Integration." *Lab on a Chip* 14 (10): 1753–66. https://doi.org/10.1039/c4lc00135d.

Frey, Urs, Jan Sedivy, Flavio Heer, Rene Pedron, Marco Ballini, Jan Mueller, Douglas Bakkum et al. 2010. "Switch-Matrix-Based High-Density Microelectrode Array in CMOS Technology." *IEEE Journal of Solid-State Circuits* 45 (2): 467–82. https://doi.org/10.1109/JSSC.2009.2035196.

Giaever, Ivar, and Charles R. Keese. 2012. "Electric Cell-Substrate Impedance Sensing Concept to Commercialization." *Cancer Metastasis - Biology and Treatment* 17 (1): 1–19. https://doi.org/10.1007/978-94-007-4927-6_1.

Gozzini, Fabio, Giorgio Ferrari, and Marco Sampietro. 2009. "An Instrument-on-Chip for Impedance Measurements on Nanobiosensors with Attofarad Resolution." In *Digest of Technical Papers - IEEE International Solid-State Circuits Conference*, 346–48. https://doi.org/10.1109/ISSCC.2009.4977450.

Grand View Research. 2019. "Biosensors Market Size & ShareIndustry Analysis Report, 2019–2026." Market Research Report, 2019.

Gu, Jinlong, and Nicole McFarlane. 2012. "Low Power Current Mode Ramp ADC for Multi-Frequency Cell Impedance Measurement." *Midwest Symposium on Circuits and Systems*, 1016–19. IEEE. https://doi.org/10.1109/MWSCAS.2012.6292195.

Gu, Tingting, and Yasushi Hasebe. 2012. "Novel Amperometric Assay for Drug-DNA Interaction Based on an Inhibitory Effect on an Electrocatalytic Activity of DNA-Cu(II) Complex." *Biosensors and Bioelectronics* 33 (1): 222–27. https://doi.org/10.1016/j.bios.2012.01.005.

Hrapovic, Sabahudin, and John H. T. Luong. 2003. "Picoamperometric Detection of Glucose at Ultrasmall Platinum-Based Biosensors: Preparation and Characterization." *Analytical Chemistry* 75 (14): 3308–15. https://doi.org/10.1021/ac026438u.

Huang, Yue, and Andrew J. Mason. 2013. "Lab-on-CMOS Integration of Microfluidics and Electrochemical Sensors." *Lab on a Chip* 13 (19): 3929–34. https://doi.org/10.1039/c3lc50437a.

Kotanen, Christian N., Francis Gabriel Moussy, Sandro Carrara, and Anthony Guiseppi-Elie. 2012. "Implantable Enzyme Amperometric Biosensors." *Biosensors and Bioelectronics* 35 (1): 14–26. https://doi.org/10.1016/j.bios.2012.03.016.

Li, Haitao, Xiaowen Liu, Lin Li, Xiaoyi Mu, Roman Genov, and Andrew J. Mason. 2017. "CMOS Electrochemical Instrumentation for Biosensor Microsystems: A Review." *Sensors, Switzerland* 17 (1): 74. https://doi.org/10.3390/s17010074.

Li, Lin, Xiaowen Liu, Waqar A. Qureshi, and Andrew J. Mason. 2011. "CMOS Amperometric Instrumentation and Packaging for Biosensor Array Applications." In *IEEE Transactions on Biomedical Circuits and Systems* 5: 439–48. https://doi.org/10.1109/TBCAS.2011.2171339.

Li, Lin, and Andrew Mason. 2014. "Development of an Integrated CMOS-Microfluidic Instrumentation Array for High Throughput Membrane Protein Studies." In *Proceedings - IEEE International Symposium on Circuits and Systems*: 638–41. https://doi.org/10.1109/ISCAS.2014.6865216.

Manickam, Arun, Aaron Chevalier, Mark McDermott, Andrew D. Ellington, and Arjang Hassibi. 2010. "A CMOS Electrochemical Impedance Spectroscopy (EIS) Biosensor Array." In *IEEE Transactions on Biomedical Circuits and Systems* 4: 379–90. https://doi.org/10.1109/TBCAS.2010.2081669.

Min, Mart, and Toomas Parve. 2007. "Improvement of Lock-in Electrical Bio-Impedance Analyzer for Implantable Medical Devices." *IEEE Transactions on Instrumentation and Measurement* 56 (3): 968–74. https://doi.org/10.1109/TIM.2007.894172.

Mishra, Nirankar N., Scott Retterer, Thomas J. Zieziulewicz, Michael Isaacson, Donald Szarowski, Donald E. Mousseau, David A. Lawrence, and James N. Turner. 2005. "On-Chip Micro-Biosensor for the Detection of Human CD4+ Cells Based on AC Impedance and Optical Analysis." *Biosensors and Bioelectronics* 21 (5): 696–704. https://doi.org/10.1016/j.bios.2005.01.011.

Rairigh, Daniel, Andrew Mason, and Chao Yang. 2006. "Analysis of On-Chip Impedance Spectroscopy Methodologies for Sensor Arrays." *Sensor Letters* 4 (4): 398–402. https://doi.org/10.1166/sl.2006.054.

Setterington, Emma B., and Evangelyn C. Alocilja. 2011. "Rapid Electrochemical Detection of Polyaniline-Labeled Escherichia Coli O157:H7." *Biosensors and Bioelectronics* 26 (5): 2208–14. https://doi.org/10.1016/j.bios.2010.09.036.

Turner, Robin F. B., D. Jed Harrison, and Henry P. Baltes. 1987. "A CMOS Potentiostat for Amperometric Chemical Sensors." *IEEE Journal of Solid-State Circuits* 22 (3): 473–78. https://doi.org/10.1109/JSSC.1987.1052753.

Xu, Youchun, Xinwu Xie, Yong Duan, Lei Wang, Zhen Cheng, and Jing Cheng. 2016. "A Review of Impedance Measurements of Whole Cells." *Biosensors and Bioelectronics* 77: 824–36. https://doi.org/10.1016/j.bios.2015.10.027.

Yang, Chao, Sachin R. Jadhav, R. Mark Worden, and Andrew J. Mason. 2009. "Compact Low-Power Impedance-to-Digital Converter for Sensor Array Microsystems." *IEEE Journal of Solid-State Circuits* 44 (10): 2844–55. https://doi.org/10.1109/JSSC.2009.2028054.

Yang, Liju, Yanbin Li, Carl L. Griffis, and Michael G. Johnson. 2004. "Interdigitated Microelectrode (IME) Impedance Sensor for the Detection of Viable Salmonella Typhimurium." *Biosensors and Bioelectronics* 19 (10): 1139–47. https://doi.org/10.1016/j.bios.2003.10.009.

Yin, Fan. 2004. "A Novel Capacitive Sensor Based on Human Serum Albumin-Chelant Complex as Heavy Metal Ions Chelating Proteins." *Analytical Letters* 37 (7): 1269–84. https://doi.org/10.1081/AL-120035897.

Yu, Yongchao, Khandaker Abdullah Al Mamun, Aysha Siddique Shanta, Syed Kamrul Islam, and Nicole McFarlane. 2016. "Vertically Aligned Carbon Nanofibers as a Cell Impedance Sensor." *IEEE Transactions on Nanotechnology* 15 (6): 856–61. https://doi.org/10.1109/TNANO.2016.2558102.

10 Application of polymers in impedance sensors

Jasdeep Bhinder and Manoj Bhatt

CONTENTS

10.1 INTRODUCTION

The primary function of a sensor is to provide information on the physical, chemical, and biological environment. To date, semiconductors, semiconducting metal oxides, solid electrolytes, insulators, ionic membranes, organic semiconductor metals, and catalytic materials have been used for making sensors. But, with advancements in the field of polymer science, polymers have also marked their presence in the field of sensors. Physical and chemical properties of polymers can be altered according to a particular need, thereby increasing the importance of polymers in the construction of sensor devices. Physicochemical properties of polymers make them extensively suitable for a wide range of applications. Composites have also gained interest as they can reversibly or irreversibly modify their characteristics through externally acting parameters (e.g., change in pH, temperature, magnetic/electric fields, and light radiations attributable to the presence of specific ions/molecules). Insulating properties of polymers are also being exploited by researchers to use them as sensors in the

DOI: 10.1201/9781003358091-13

electronic industry. Conducting polymers are primarily utilized as either a coating material or an encapsulating material for the electrode/substrate surface; whereas nonconducting polymers are commonly used for facilitating specific receptor immobilization in various sensor devices.

Polymers have gained enormous attention in the field of artificial sensors to mimic natural sense organs. Advancement in the field of nanotechnology has further revolutionized the advancement of polymer nanocomposites as sensors. Polymer nanocomposites have the advantage of better selectivity and rapid measurements over classical sensor materials. Polymers play their part in sensors directly either by participating in the sensing procedure or they assist in immobilizing the various sensing-associated components. This chapter extensively discusses various polymers that are used in sensors and initially discusses their field of applications in general usage devices, such as temperature sensors, pH sensors, gas sensors, ion-selective sensors, stress sensors, and bio-inspired sensors. The chapter explores the usage of various functional polymers and further discusses about the application of polymers in impedance sensors in detail. Polymer-based impedance sensors observe extensive applications in areas of wearable or implantable sensors, biosensors, and environmental monitoring, which are discussed and reviewed in detail in this chapter.

10.2 CLASSIFICATION OF POLYMERS FOR SENSOR APPLICATIONS

Polymers can exist in various forms, such as solutions, gels, films, solids, or self-assembled nanoparticles.

10.2.1 General polymer usage as sensors

Numerous applications of polymers in sensors can be studied, but the few notable applications are those of temperature sensors (Uchiyama et al. 2003), pH sensors (Munkholm et al. 1986), gas sensors (Chauhan and Bhattacharya 2018), ion-selective sensors, stress sensors (Basu et al. 2019), and bio-inspired sensors (Bhatt et al. 2017). These sensors are easy to fabricate, and they possess a high level of sensitivity and reproducibility, which makes them adaptable to discrete situations. A much-detailed description of these sensors is provided in the following paragraphs.

10.2.1.1 Temperature sensors

Temperature sensors, generally termed thermometers, as the name implies, are sensors responsible for recording any change in temperature of the analyte/substrate/environment. In terms of polymer application, polymer-based fluorescent temperature sensors were first fabricated in 2003 (Uchiyama et al. 2003), and since then, several modifications have evolved in this field. Various structures of polymers in pure or doped form can be used in thermometers. Figure 10.1 shows an example of a thermosensitive, doped fluorescent polymer [P(DMAPAM-co-NTBAM-co-DBDAE)] used in thermometers.

Polydimethylsiloxane (PDMS) and its modifications are extensively utilized polymers in sensor devices (Bhattacharya et al. 2007). PDMS doped with active interferometer elements-based optical microfibers is also used in temperature sensors

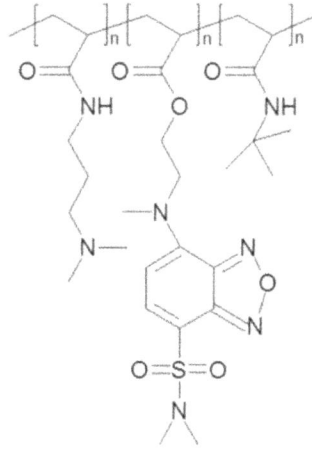

FIGURE 10.1   Structure of P(DMAPAM-co-NTBAM-co-DBDAE) thermosensitive copolymer. [Reprinted with permission from Cichosz, Masek, and Zaborski (2018). Copyright (2018) Elsevier].

(Hernandez-Romano et al. 2015). The role of this polymer is to act as a protective material and assist in preventing pollution and aging. This device can measure temperatures from 20°C to 48°C (sensitivity: 3101.8 pm °C^{-1}).

10.2.1.2   pH sensors

Monitoring pH is extremely important in biochemistry/chemistry as determining the actual pH of the environment of chemical reactions enormously helps in ascertaining the product's final form. The most suited example of polymers for pH sensing is of photochemical polymerizable copolymers that are fabricated through a combination of methylene-bis-acrylamide and acrylamide (Munkholm et al. 1986). This polymer contains amino fluorescence, which is covalently attached to the surface of the fiber and acts as a pH sensor.

The pH sensor is also used in blood analysis (Huber et al. 2004), where it forms a coating sensitive to environmental changes. This coating is attained by reacting 1-hydroxypyrrole-3,6,8-trisulphochlorane and amino ethyl cellulose fibers. The surface obtained is further deposited with a polyester film and a polyurethane (PU)-based ion-permeable hydrogel. Conductivity of the fabricated assembly is changed to a quantifiable extent as hydrogen ions come in contact, thereby facilitating pH quantification (Adhikari and Majumdar 2004). Poly(p-phenylenediamine) (PPV) is a conjugated polymer and has a conduction ability attributable to the presence of both delocalized π bonds on the aromatic rings and double bonds in the carbon chain, because of which PPV is also a very favorable candidate for electronic applications (Pistor et al. 2007).

10.2.1.3   Gas sensors

Conductive polymers along with their variants/composites like PMMA and PVC find high utility in applications like gas sensors that can measure potential electrode

TABLE 10.1

Various kinds of moisture sensors based on polymers

Polymer	Application	Properties	Ref
Tetraethyl orthosilicate modified 2-acrylamide-2-methylpropane	Electrical property change measurement	2% linearity with low hysteresis	Cichosz, Masek, and Zaborski (2018)
Iron oxide doped polypyrrole nanoparticles	Electrical property change measurement	Increased sensitivity reported with increasing polypyrrole concentration	Suri et al. (2002)

changes (Chauhan et al. 2021). The prerequisite condition for designing gas sensors is to have an active functional group in their structures (Adhikari and Majumdar 2004) that further facilitates the sensing procedure. Gas sensors have the add-on advantage of being used as moisture sensors that can determine relative humidity values in an environment. These sensors have found usage in a wide range of fields, viz. industry, medicine, and households. Any change in the electrical conductivity of these materials or sensors as they come in contact with water vapors is used as the underlying principle for quantification. Table 10.1 summarizes some examples of moisture sensors based on polymeric materials.

Applications of certain polymers attributable to the presence of hydrophilic moieties [−COOH, −SO$_3$H, and −N+ (R)$_3$Cl] are restricted at higher humidity in the air because of which these polymers are soluble in water. In order to solve this issue, a suitable blend of hydrophobic polymers is made. Moreover, combining two polymers to obtain a new composite material possessing entirely different properties is also a possible solution, and one such example is sodium polystyrene sulfonate (PSSNa) (Zhao et al. 2022).

10.2.1.4 Ion-selective sensors

Ion sensors comprise an ion exchange or interaction instance with the measuring electrodes/system as the conductive polymer comes in contact with the analyte. This instance is quantified as an electrical signal and the electrode that facilitates such an event is known as an ion-selective electrode. The primary objective of such sensors is to detect the presence of specific ions in a solution (Adhikari and Majumdar 2004). The most suitable example of such sensors is a PU/PVC-based copolymer that, when incorporated with silicone rubber, can be used to detect Na$^+$ ions in body fluids (Goldberg et al. 1994). Calcium ions can be detected using polyaniline-based (PANI) electrodes. The membrane of these electrodes is created using a conductive polymer matrix with organic phosphate derivatives, for example, bis[4-(1,1,3,3-tetramethyl butyl)] phenyl phosphoric acid (DTMBP-PO4H), as a lipophilic additive (Lindfors and Ivaska 2001). PVC in the form of electrodes or membranes is also used extensively to devise ion-selective sensors (Armstrong and Horvai 1990). Table 10.2 further lists some more polymers used as ion-selective sensors.

TABLE 10.2

Applications of various polymers in ion-selective sensors

Polymer	Detected ion	Ref
Modified poly-vinyl chloride	Nickel (II) phosphates	Arsalan et al. (2019)
Lipophilic acrylic resin	Calcium, magnesium	Numata et al. (2001)
Poly(3,4-ethylenedioxythiophene) doped with poly(4-styrenesulfonate) ions (PEDOT–PSS, Baytron P) (disposable plastic)	Calcium, potassium	Michalska and Maksymiuk (2004)

10.2.1.5 Stress sensors

Stress sensors relatively belong to the recent compilation of intelligent materials and are based on the photoluminescence phenomenon. The nature of forces (shearing or stretching) applied onto the sensor results in a photoluminescence effect, change in color, or fluorescence emissions (De Baere, Van Paepegem, and Degrieck 2009; Sagara and Kato 2009), which is further quantized to record the applied forces/

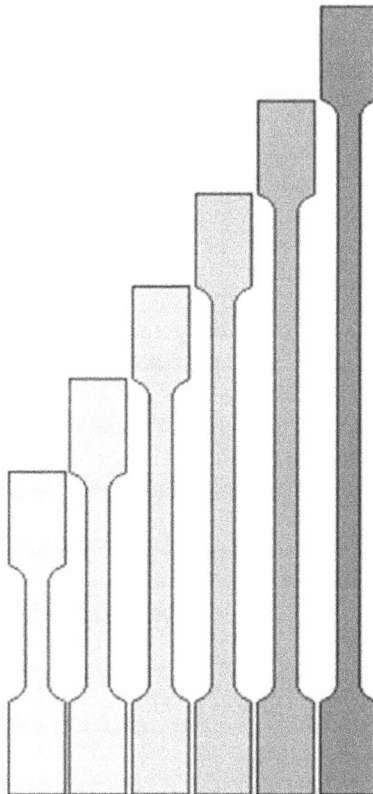

FIGURE 10.2 Color change observed in the extended sample. [Reprinted with permission from Cichosz, Masek, and Zaborski (2018). Copyright (2018) Elsevier].

stresses. A scheme for color change in a stress sensor is shown in Figure 10.2, which shows that an increase in length (tensile stress) promotes color darkening.

These sensors can finely record variation in stress and, hence, can be efficiently used for cardiac recording by noticing blood pressure changes (Shu et al. 2015). The easy-to-wear flexible pressure sensors in collaboration with computerized systems can also help in determining the heart rate, which is further recorded through the measured strain changes.

10.2.1.6 Biosensors

Biosensors collectively comprise a sensor, a bioreceptor, and a biological component and are programmed to convert biological reactions into electrical signals (Bhatt and Bhattacharya 2018; Bhatt et al. 2019). Similar to stress sensors, biosensors also find extended applications in environmental pollution monitoring and bedside/medical diagnostics (Adhikari and Majumdar 2004; Bhatt et al. 2021). The operating principle of the biosensor is shown in Figure 10.3, which shows that the interaction of an analyte sample with a bioreceptor molecule generates a particular kind of response, which is amplified and recorded through the electronic assembly. Hence, the impulse captured by a suitable detector is the result of a specific enzymatic reaction.

Biosensors can be used for quantifying the amount of glucose, hemoglobin, urea, lipids, etc. in the analyte sample. The most commonly used polymeric material in most polymer biosensors is polyaniline (PANI). The most important aspect of biosensor design is to build a stable polymer matrix comprising sufficient specific functional groups packed in an appropriate structure.

10.2.2 FUNCTIONAL POLYMERS

Functional polymeric materials possess biological activity, pharmacological properties, chemical reactivity, biocompatibility, photosensitivity, catalytical properties, and electrical conductivity that make them an integral part of the polymer family. The main focus in today's research is on the preparation methodologies with a great deal of emphasis on the development of new polymerization. Theoretically, polymerization can be classified into two categories, viz. chain growth and step growth. Chain growth polymerization requires the initiation of monomers to start chain growth, involving the addition of monomers to growing free radicals and cationic or

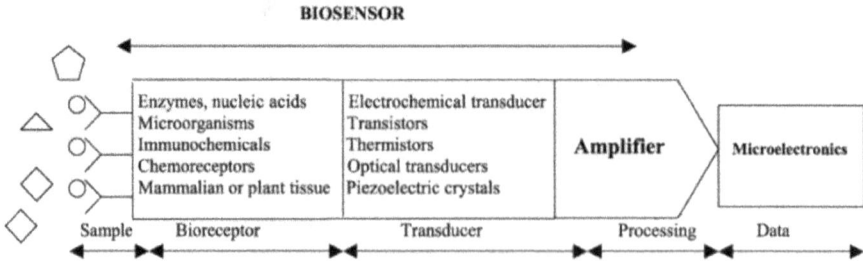

FIGURE 10.3 Operating principle of a biosensor. [Reprinted with permission from Adhikari and Majumdar (2004). Copyright (2004) Elsevier].

anionic chains. Step-growth polymerization is generally exhibited by condensation polymerization and alkyne-based click polymerization.

Over the past two decades, polymeric materials are physically or chemically modified by introducing appropriate functional groups, which has been of great interest to the research community. Moreover, the recent progress in a more controlled synthesis of these polymeric materials has allowed for the fabrication of new chemical and biological sensor systems. These modified polymers are very often used for manufacturing biosensor tools or lab-on-chip devices. These sensors based on polymers have more structural stability, ease of processing and integration with detection devices, dispersion in water, and biocompatibility (Hu and Liu 2010). The most commonly used compounds for the development of sensors are poly (3,4-ethylene dioxythiophene) polystyrene sulfonate (PEDOT: PSS), polypyrrole (PPy), and polyaniline (PANI). Due to their conductive properties, these polymers are widely used in electrochemical biosensor devices. Figure 10.4 represents the idea of the detection mechanism of a biosensor built using a functional polymer material. It shows that the primary detection principle, that is, the bidirectional electron flow facilitated between various components of the sensor and polymeric film plays a significant role in accomplishing efficient electron transfer.

To develop biosensors, it is very important to select the most suited immobilization method, ensuring the optimum performance of biosensors (storage stability, high sensitivity, short response time). Hence, the selection of an appropriate functional monomer/polymer is an extremely important step for the development of biosensors. The functional monomer creates a direct bond between the monomers and the target analyte molecules of the matrix (Yoshikawa, Tharpa, and Dima 2016). The matrix interacts through a covalent bond or noncovalent bond with a monomer that has a very stable binding and immensely increases the specificity of the sensor. The detection method for circulating tumor cells (CTCs) is described by An et al., which can pave a way for the early detection of cancer (An et al. 2018). In this work, they used a benzo boric acid-modified gold-plated substrate (Manczak et al. 2016) as shown in Figure 10.5.

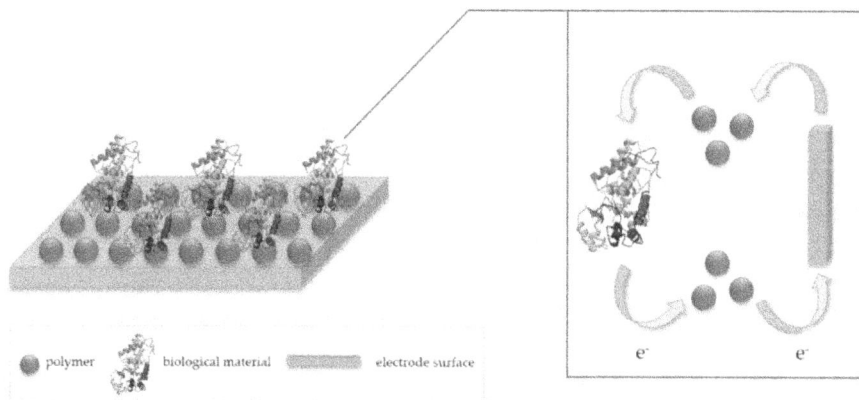

polymer   biological material   electrode surface   e⁻   e⁻

FIGURE 10.4 Detection mechanism of polymeric materials. [Reprinted with permission from Li and Liu (2012). Copyright (2012) MDPI].

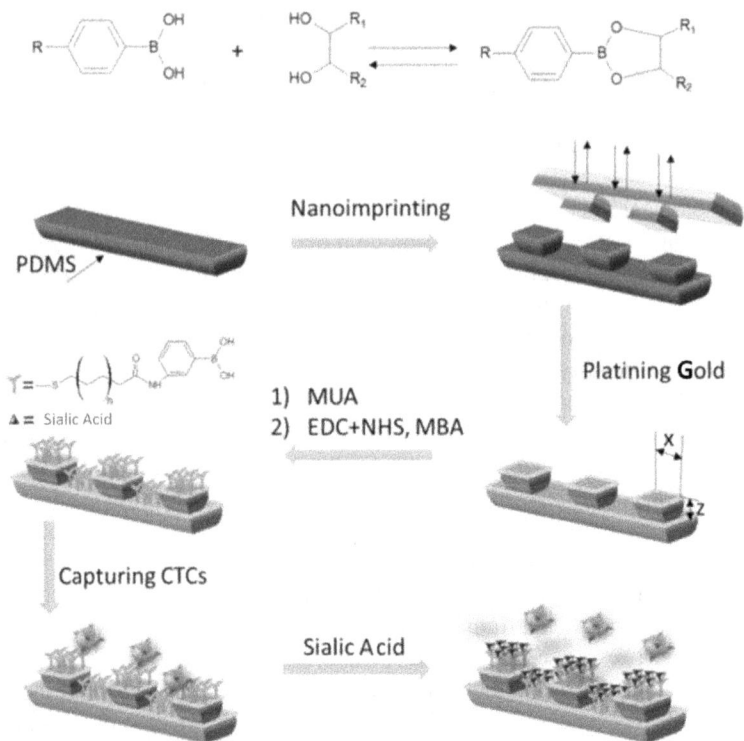

FIGURE 10.5 Modification of gold-plated substrate, determination, and circulating tumor cells (CTCs) release. [Reprinted with permission from Li and Liu (2012). Copyright (2012) MDPI].

The polymeric substrate was covered with a 3D microarray using a nanoimprinting technique. The magnetron was used to deposit the Au layer on the surface of the substrate. The higher capture efficiency is shown by the 3D surface in comparison with the smooth substrate. The presented solution reports a method of early detection of cancer. Hence, because of their high-utility applications, functional polymers linked with biosensor applications are becoming an important area of research.

10.2.2.1 Electroactive polymers

Electroactive polymers (EAPs) change shape and dimension when an electric field is applied to them (Bar-Cohen 2004; Carpi and Smela 2009). EAPs belong to the family of electroresponsive polymers that show electrically coupled reactions (Bharti et al. 2006). Broadly, EAPs are classified as ionic EAPs and electronic EAPs (Bar-Cohen 2004). Ionic EAPs require a low voltage of 5V for actuation (Jean-Mistral, Basrour, and Chaillout 2010) and are mainly synthesized by using polymers, such as polypyrrole (PPy) and poly(3,4-ethyl-lenedioxythiophene) (PEDOT), which are biocompatible, thereby making ionic EAPs most suited for applications related to biological environments (Carpi and Smela 2009). Ionic EAPs operate with an ion reservoir to transport ions or molecules within them and because of the presence of ions, the stress/strain causes an ion migration within them.

10.2.2.2 Conducting polymers

Most polymers used for sensor applications are conductive in nature. The electrical properties of these sensors can be tailored to generate a measurable signal that may be registered. Conductive polymers offer a great design variety and flexibility and can form selective layers (as in gas sensors) in which the primary charge of the physical parameter in the transduction mechanism is generated between the analyte gas and conducting matrix (Tyler McQuade, Pullen, and Swager 2000). The disadvantage associated with conducting polymers is that they are generally porous and are easily penetrated by gases, thereby changing their electrical properties (Janata and Josowicz 2003).

Conducting polymers either mainly comprise an intrinsically conducting polymer or are formulated by mixing the polymer with a conductive filler (Lange, Roznyatovskaya, and Mirsky 2008). Conducting polymers are generally made from monomers and dopants that tend to polymerize during the oxidation process. The oxidation process is started by using oxidizing agents or applying an anodic current to polymer monomers in electrochemical cells. The add-on advantage of conducting polymers is that they also serve as a suitable substrate for the immobilization of biosensing enzymes, covalent binding, and cross-linking (Matadi Boumbimba et al. 2012). The most promising application of conductive polymers is for sensing gases. Conducting polymers doped with other polymers (PMMA, PVC) or with polymers having active functional groups can also be used efficiently for detecting gases.

10.3 APPLICATIONS OF POLYMERS IN IMPEDANCE SENSORS

Polymers are extensively being used in sensing (Bhatt et al. 2022), primarily in wearable or implantable sensors, in vitro biosensors, food monitoring, and environmental monitoring because of the enormous capabilities of applications of polymers in the sensors domain, such as flexibility, manufacturability, low cost, and high diversity. These fields utilize polymers with different capabilities as they are considered apt for the required detection need. The following sections elaborate on the utilization of polymers in the stated domains of impedance sensing.

10.3.1 WEARABLE OR IMPLANTABLE SENSORS

This category comprises two kinds of sensors: one is the wearable sensor that is directly in contact with the human body externally to record the biological data associated with body condition, whereas the other is the implantable sensor that fits within the human skin for signal recording.

While designing a wearable sensor, various factors like comfort, skin compatibility, and high efficiency are the primary considerations. Although conventional metal electrode-based sensors are durable and convenient to fabricate, these electrodes can extensively cause skin irritation and discomfort along with reduced efficacy under dry conditions. This has generated the need to explore more skin-compatible, flexible, breathable, durable yet efficient solutions to the wearable sensor domain. Conducting polymer-based wearable devices are very popularly used in this domain, in both the modalities, intrinsically conducting polymer and polymer introduced with conducting filler material like chitosan or graphene. The principle of charge transfer in intrinsically conducting polymers includes interactions of π orbitals while preserving the chain structure via σ bonds.

Figure 10.6 shows the brief classification of polymer-based sensors and their applicability in the wearable sensors domain. It can be observed that polymers can behave as active material or substrate material when used in wearable sensors. Active materials, like polymer gels (Hu et al. 2019) can simultaneously respond depending upon the absorption/desorption (swelling) characteristics, while in substrate material capability, the polymer further promotes immobilization of some biological species, which helps in detection. Various designs like contact lens sensors, tattoo-based sensors, patched sensors, and textile sensors are the few primary applications of polymers in wearable sensors, where tattoo, patched, and textile sensors efficiently use electrochemical impedance sensing.

Polymers like liquid crystalline polymers (LCP) (Wang et al. 2016), polymer gels (Hu et al. 2019), conducting polymers (Harito et al. 2020), polymer composites (Harito et al. 2019), and piezoelectric polymers are the common ones that are used as the active material application while thermoplastic/thermosetting polymers (Massy 2017) and LCPs (Barón et al. 2002) are commonly used as substrate materials. In electrochemical wearable sensors, instead of analyzing interstitial fluid in invasive sensors, primarily sweat, saliva, and tears are studied. Sweat is the primary body fluid that is released on the skin as a result of the physical activities of the body. It

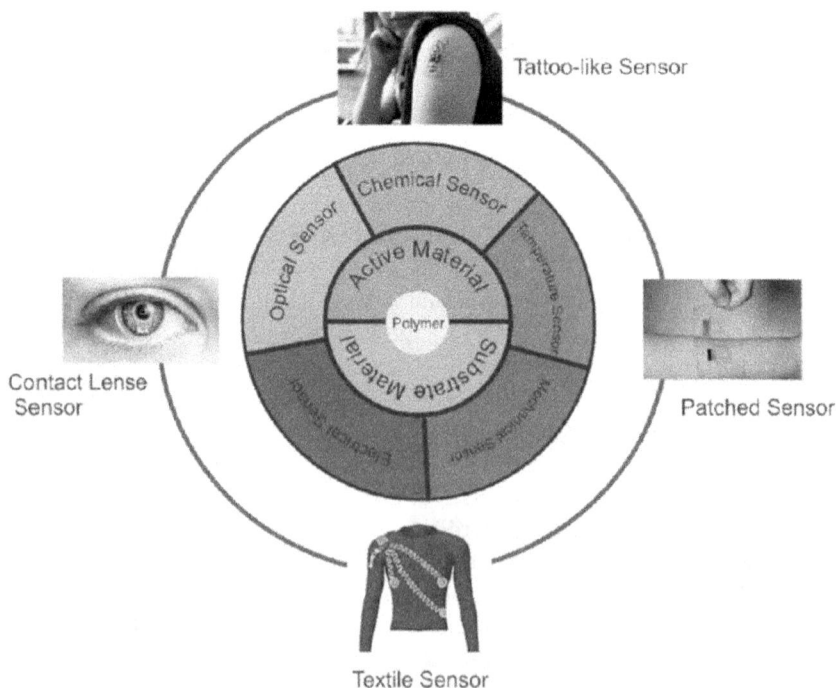

FIGURE 10.6 Utilization of polymers for wearable sensors. [Reprinted with permission from Harito et al. (2020). Copyright (2020) IOP science].

comprises much information regarding body metabolism, like various electrolyte components (potassium and sodium ions with a specific pH) and metabolities like lactic acid and glucose. The pH of the sweat indicates the hydration/dehydration level, body odor, and skin diseases of the body as a function of the sodium ion level. Among the conducting polymer-based tattoo-based wearable sensors, polypyrrole and polyaniline electrodes observe the greatest pH sensitivity (Korostynska et al. 2007). Tattoo-based wearable sensor wisely integrates electrode, signal processing unit, transceiver, and power supply unit to get the desired result. These sensors were initially fabricated using clean-room techniques, but these sensors are extensively fabricated using robust techniques like screen printing, direct writing/inkjet printing, and cut-and-paste methods (Harito et al. 2020). Various biological/derived components, such as sweat-like glucose oxidase, lactate oxidase, and alcohol oxidase enzyme, can be efficiently detected through these wearable sensors. Temperature sensing as a pre-indicator to avoid heat stroke is also being integrated into these sensors.

Patch-type sensors are the next category of the electrochemical wearable sensor, where polymers are commonly used as electrolyte gels in wet electrodes. As gel may lead to skin irritation, patch-type sensors in dry electrode form or microneedle patterned contact form are widely used (O'Mahony et al. 2012). Figure 10.7 shows the configuration of a patch-type sensor under wet conditions and a microneedle-based dry electrode along with the electrical equivalent circuit in both cases. The various significant components in an electrical circuit are the half-cell potential of the electrode (E_{el}), capacitance and resistance of the electrode–electrolyte interface (C_{el} and R_{el}), electrolyte gel resistance (R_{gel}), half-cell potential of the electrolyte–skin interface (E_{sc}), capacitance and resistance of skin (C_{sc} and R_{sc}), and the resistance of subcutaneous tissue and dermis (R_{sg}). It can be observed from the figure that in the case of dry electrodes, the equivalent circuit shows the inclusion of impedance from various significant sources as compared to microneedle-based dry electrode condition, which enhances the interface impedance overall.

Under wet conditions, the electric current signal is distorted at the electrode–gel interface as well as at the gel–skin interface, which weakens the signal strength and hence a lower signal-to-noise ratio is observed in this case. But, in the case of dry

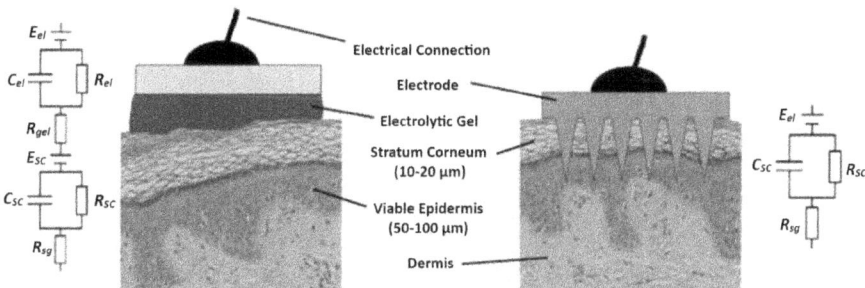

FIGURE 10.7 Electrical model with the corresponding sketch for wet (left) and microneedle-based dry electrodes (right). [Reprinted with permission from O'Mahony et al. (2012). Copyright (2012) Elsevier].

electrodes, the direct contact between the electrode and skin removes the interfacial impedance value and, hence, presents an enhanced signal-to-noise ratio. Polymer electrodes instead of metal electrodes are again recommended in these cases because of their highly flexible nature, which makes them take shape with respect to the skin. The conducting polymer microneedle-type electrodes can be fabricated using photolithography, soft lithography, replica molding, injection molding, 3D printing, and embossing. These sensors can be effectively used for monitoring various body fluids like lactate, glucose, and drugs like theophylline (Sharma et al. 2017) and can also be used for drug delivery applications (Goud et al. 2019).

Textile-type electrochemical wearable sensors primarily used conductive electrodes sewed, printed, or coated over the textile where body fluid (sweat) acts as an electrolyte to complete the sensing device. As active material sensors, these sensors facilitate reversible bonding/releasing reactions, while in passive stimulus response sensors, direct change in resistance or capacitance can be measured as a function of the ions/body fluid component concentration during bending or stretching (Seyedin et al. 2019). The electrical conductivity of the textile fiber can be significantly modified through spinning, knitting, or coating of conductive materials like silver, carbon-based materials like graphene or carbon fiber/nanoparticle/nanotube, or conducting polymer. Conducting polymers are generally coated over the fiber with an interface elastomer-like PDMS to avoid cracking due to the brittle nature of the conducting polymer. Washable textile-impedance sensors with efficient durability are also fabricated these days with screen printing (Jose et al. 2021). Figure 10.8 shows the fabricated interdigitated electrode assembled with a heating platform on the textile. The corresponding results show that the sensor is capable of efficiently measuring the impedance variation for 0.5 μL body fluid sample, over a range of temperatures.

Recently, wearable impedance sensors are also observing their applications in several diverse and extremely important domains like the diagnosis of congestive heart failures (Lee et al. 2015) and gesture recognition with machine learning (Yao et al. 2020).

10.3.2 Biosensors

Although the wearable sensor belongs to a category of biosensors/biological impedance sensors, biosensing is not limited to wearable sensors. There can be a variety of sensors that are utilized for screening biological samples, viz. pathogens, cells, DNA, RNA, proteins, drugs, and antibodies/antigens (Bhatt et al. 2016; Bhatt and Bhattacharya 2019) in primarily in vitro mode, where the samples are carried to a fabricated microdevice that analyzes the sample to produce a detection signal. These detection signals are extensively utilized to study the behavior and characteristics of the analyte. Biosensors are even capable of carrying out single-molecule detection, which indicates the highly efficient nature of these sensors. Polymer-based impedance sensors have a similar kind of motivation to achieve, along with the inherent advantages they offer like cost-effectiveness, easy fabrication, flexibility, and the possibility of easily manipulating substrate characteristics as per analyte requirements.

Polymer-based impedance biosensor applications can again be observed as active or substrate material, whereas as an active material, the polymer layer reversibly/

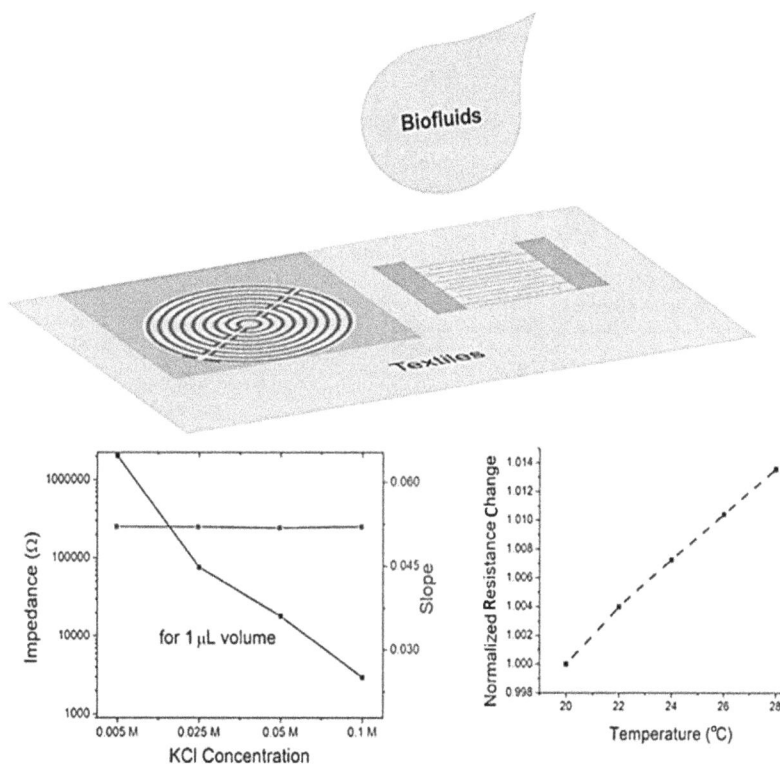

FIGURE 10.8 Screen-printed interdigitated electrode set with the heating platform and corresponding impedance change with concentration and over a temperature range. [Reprinted with permission from Jose et al. (2021). Copyright (2021) American Chemical Society].

irreversibly participates in the detection event while as a substrate material, it further supports the subsequent detection event through the intermediate layer by facilitating a functionalization or hybridization event. Figure 10.9 shows a schematic of polymer application in biosensing (Alberti et al. 2021). It can be observed that both natural and synthetic polymers can be efficiently utilized for devising biosensors for food monitoring, healthcare, and environmental monitoring application. The first two applications have been explained in this section, while the environmental monitoring application has been detailed in the next section.

Along the line, Wei et al. (2018) proposed a hydrogel film-based temperature-sensitive sensor for the impedimetric detection of bovine serum albumin (protein). Reversible structure change of the molecularly imprinted polymer layer after applying a temperature stimulus is used to sense the protein using a ferrocene redox label. Polymeric impedance sensors are also commonly used for humidity sensing, which observes applications in food, agriculture, industrial, and health monitoring. These eco-friendly sensors are developed by printing interdigitated electrodes through silver ink over a flexible polyethylene terephthalate substrate and one prominent health application of these sensors is human breathing monitoring (Soomro et al.

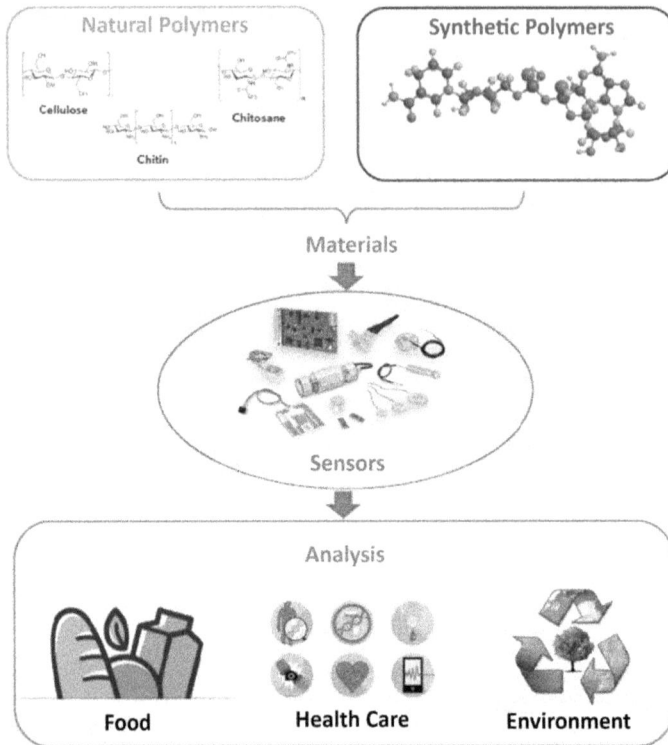

FIGURE 10.9   Applications of polymers (natural/synthetic) in biosensors to observe various food/healthcare/environmental monitoring applications. [Reprinted with permission from Alberti et al. (2021). Copyright (2021) MDPI].

2019). These sensors observed a significant reduction in impedance as a result of an increase in the humidity ratio.

These sensors are often integrated with various nanoparticles for enhancing their detection capability. Polyaniline-gold nanoparticle-based composite deposited over glassy carbon electrodes is employed for simultaneously determining the presence of uric acid and epinephrine markers (Zou et al. 2013).

10.3.3   Environmental Monitoring

Environmental monitoring is one of the important applications of sensors and their integration with polymers is an extensively studied field as the limit of toxic gases is extensively monitored in most of the applications because they can severely damage the process conditions. The presence of various toxic gases can be efficiently monitored by observing a change in the electrical parameter (voltage/current/impedance) at the electrode–gas interface. Conducting polymers have observed promising results with respect to sensing gases showing oxidizing or acid–base characteristics. These gases are also sensed through conducting polymer composites with conventional polymers like PMMA or PVC that have active functional groups and solid polymer electrolytes. Polyacetylene (III) is generally observed as the first known conducting polymer, which is primarily resistive. The doping of this polymer with iodine helps in enhancing the conductivity of the polymer to the extent of metallic behavior. Research further led to the development of heterocyclic polymers for conductivity enhancement (Diaz, Kanazawa, and Gardini 1979). Polydimethylsiloxane (PDMS) being a permeable polymer helps in selecting the gases to detect as compared to other hydrophobic species.

Diniz et al. (2019) fabricated a hybrid polyaniline, poly(o-methoxyaniline)-vanadium pentoxide film for sensing ammonia. An increased sensitivity level with up to 800% in real components and 3000% in imaginary ones of impedance was recorded as compared to conventional polyaniline polymer, expressing the utility of hybrid polymers in detection. Biopolymers like cellulose nanofibers are also being utilized as ethanol gas sensors because of their effective, flexible, and stretchable yet biodegradable nature, which also makes them apt for being used as wearable sensor devices (Rakhmania et al. 2022). These sensors are fabricated using drop-casting multiwalled carbon nanotubes (MWCNT) on cellulose nanofiber sheets for conductivity enhancement, which is further dropped with enzyme solutions for ethanol vapor detection.

Along with the polymer electrode configuration, polymers are also used commonly as an ionic gel in between the metal electrodes for gas-sensing applications. In this direction, Ishizu et al. (2013) proposed a platinum metal electrode-based CO_2 gas sensor using ionic gel, which is a mixture of ionic liquid and polymer. Willa et al. (2017) fabricated a poly(ionic liquid)/alumina composite-based room temperature CO_2 sensor over interdigitated platinum electrodes on a flexible polyimide substrate. The rare metal-free composite layer over the electrodes acts as an active layer for the sensing procedure. Figure 10.10 shows the device schematic and the corresponding results related to the sensor device. It can be observed from the SEM cross-section (Figures 10.10b,c) that the composite layer deposited through the spray-coating method is highly uniform as compared with the spin-coated composite layer, which shows the varied thickness of the composite at various places. The Nyquist and Warburg impedance plots (Figure 10.10d,e) corresponding to the sensor signify that the lower frequency impedance is governed through the diffusion of the ionic species (primarily protons) and the overall charge transfer resistance increases with the increase in CO_2 concentration. Similar behavior has also been observed with Warburg impedance.

Along with the polymer film, paper-based impedance sensors are also being explored extensively, which efficiently reduces the cost of detection microchips. In paper-based sensors, conducting electrodes are normally printed or painted using various techniques for carrying out the detection, where the efficiency enhancement is further done by integrating several additional chemistries. Carbon electrodes are the most common type of electrodes that are employed in paper-based impedance sensors. In this line, hydrogel-modified paper-based carbon interdigitated electrodes have been utilized for detecting monosaccharides like fructose and glucose in solution through impedance spectroscopy (Daikuzono et al. 2017). The hydrogel layer over the screen-printed electrodes swells in the presence of fructose and glucose further modifying the impedance value. Polyaniline/graphene electrodes have been

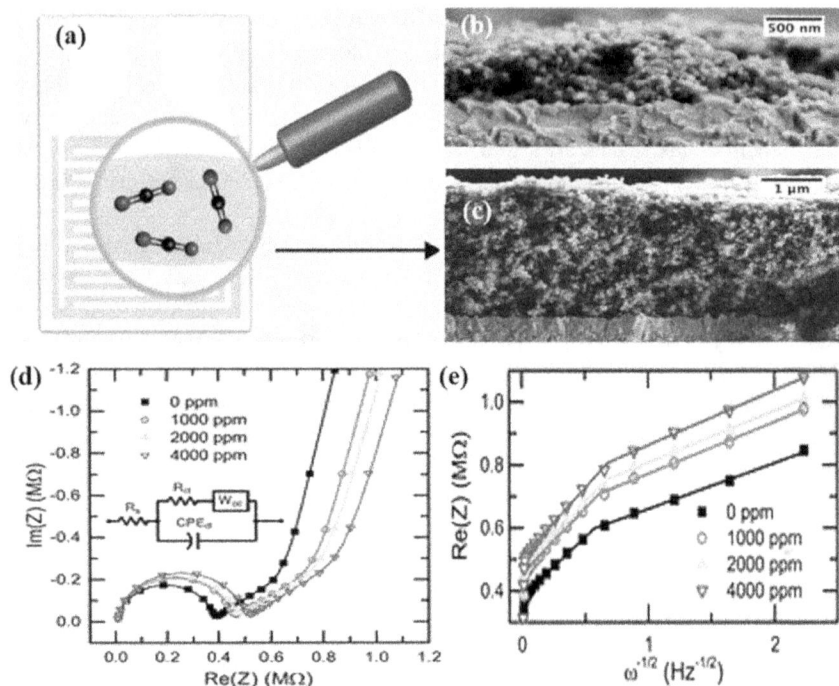

FIGURE 10.10 (a) Schematic of the sensor device with composite coating; cross-section of SEM image of P[VBTMA][PF$_6$]/Al$_2$O$_3$ composite film coated through (b) spin-coating; (c) spray-coating methods; (d) Nyquist plot and equivalent circuit diagram with composite-coated sensing layer for different concentrations of CO$_2$; (e) corresponding Warburg plot related to the impedance data. [Reprinted with permission from Willa et al. (2017). Copyright (2017) American Chemical Society].

utilized for detecting human interferon-gamma (Ruecha et al. 2019). The screen-printed graphene electrodes modified with polyaniline demonstrated enhanced efficiency as compared with only polyaniline electrodes. Li et al. synthesized ZnO nanowires on paper-based carbon electrodes using a hydrothermal approach to demonstrate the testing of COVID-19 IgG antibody testing via impedance measurements and observed an efficacy level of up to 10 ng mL^{-1} within a very short duration (Li et al. 2021). Simple folding paper-based impedance sensors have also been utilized for assessing cardiovascular risk by analyzing C-reactive protein presence (Boonyasit, Chailapakul, and Laiwattanapaisal 2019). A label-free detection using carbon electrodes has attained an efficacy level up to the standard methods.

10.4 CONCLUSION AND FUTURE PERSPECTIVES

The future of polymer-based impedance sensing is promising as impedance spectroscopy is a fast-growing research domain, thanks to its rapid and cost-effective yet integrable nature. Polymer application is a very adaptive innovation because of its flexible, durable, and eco-friendly qualities; the possibilities of attaining required characteristic features and ease of fabrication are its very attractive features. Polymer materials have the desired physical and chemical properties that enable fabrication through various techniques. Their adaptability to several diverse detection needs in active material or substrate material mode is being extensively studied. Small-scale wireless device formats, like ImpediSense (Dheman et al. 2021), have been rigorously integrating the various sensing, analysis, and communication modules while consuming fairly less power. Even self-powered devices and small-sized rechargeable battery-based devices are extensively being used in invasive and noninvasive modes.

REFERENCES

Adhikari, Basudam, and Sarmishtha Majumdar. 2004. *Polymers in Sensor Applications*. Progress in Polymer Science, Oxford. https://doi.org/10.1016/j.progpolymsci.2004.03.002.

Alberti, Giancarla, Camilla Zanoni, Vittorio Losi, Lisa Rita Magnaghi, and Raffaela Biesuz. 2021. "Current Trends in Polymer Based Sensors." *Chemosensors* 9 (5): 108. https://doi.org/10.3390/chemosensors9050108.

An, Li, Guangtong Wang, Yu Han, Tianchan Li, Peng Jin, and Shaoqin Liu. 2018. "Electrochemical Biosensor for Cancer Cell Detection Based on a Surface 3D Micro-Array." *Lab on a Chip* 18 (2): 335–42. https://doi.org/10.1039/c7lc01117b.

Armstrong, R. D., and G. Horvai. 1990. "Properties of PVC Based Membranes Used in Ion-Selective Electrodes." *Electrochimica Acta* 35 (1): 1–7. https://doi.org/10.1016/0013-4686(90)85028-L.

Arsalan, Mohd, Aiman Zehra, Mohammad Mujahid Ali Khan, and Rafiuddin. 2019. "Preparation and Characterization of Polyvinyl Chloride Based Nickel Phosphate Ion Selective Membrane and Its Application for Removal of Ions through Water Bodies." *Groundwater for Sustainable Development* 8: 41–48. https://doi.org/10.1016/j.gsd.2018.06.008.

Baere, I. De, W. Van Paepegem, and J. Degrieck. 2009. "On the Nonlinear Evolution of the Poisson's Ratio under Quasi-Static Loading for a Carbon Fabric-Reinforced Thermoplastic. Part I: Influence of the Transverse Strain Sensor." *Polymer Testing* 28 (2): 196–203. https://doi.org/10.1016/j.polymertesting.2008.12.002.

Bar-Cohen, Yoseph. 2004. *Electroactive Polymer (EAP) Actuators as Artificial Muscles: Reality, Potential, and Challenges,* Second Edition. SPIE Digital Library. https://doi.org/10.1117/3.547465.

Barón, M., R. F. T. Stepto, C. Noël, V. P. Shibaev, M. Hess, A. D. Jenkins, I. Il Jin et al. 2002. "Definitions of Basic Terms Relating to Polymer Liquid Crystals (IUPAC Recommendations 2001)." *Pure and Applied Chemistry* 74: 493–509. https://doi.org/10.1351/pac200274030493.

Basu, Aviru Kumar, Amar Nath Sah, Asima Pradhan, and Shantanu Bhattacharya. 2019. "Poly-L-Lysine Functionalised MWCNT-RGO Nanosheets Based 3-d Hybrid Structure for Femtomolar Level Cholesterol Detection Using Cantilever Based Sensing Platform." *Scientific Reports* 9 (1): 3686. https://doi.org/10.1038/s41598-019-40259-5.

Bharti, Vivek, Yoseph Bar-Cohen, Zhong-Yang Cheng, Qiming Zhang, and John Madden. 2006. "Electroresponsive Polymers and Their Applications." *Materials Research Society Symposium Proceedings* 889: 105–192.

Bhatt, Geeta, and Geeta Bhattacharya. 2018. "DNA Based Sensors." In *Environmental, Chemical, and Medical Sensors*, edited by Shantanu Bhattacharya, Avinash Agarwal, Nripen Chanda, Ashok Pandey, and Ashis Sen, 343–70. Springer Nature Singapore Pte Ltd. https://doi.org/10.1007/978-981-10-7751-7_15.

Bhatt, Geeta, and Shantanu Bhattacharya. 2019. "Biosensors on Chip: A Critical Review from an Aspect of Micro/Nanoscales." *Journal of Micromanufacturing* 2 (2): 198–219. https://doi.org/10.1177/2516598419847913.

Bhatt, Geeta, Swati Gupta, Gurunath Ramanathan, and Shantanu Bhattacharya. 2021. "Integrated DEP Assisted Detection of PCR Products with Metallic Nanoparticle Labels through Impedance Spectroscopy." *IEEE Transactions on Nanobioscience* 21 (4): 502–10.

Bhatt, Geeta, Rishi Kant, Keerti Mishra, Kuldeep Yadav, Deepak Singh, Ramanathan Gurunath, and Shantanu Bhattacharya. 2017. "Impact of Surface Roughness on Dielectrophoretically Assisted Concentration of Microorganisms over PCB Based Platforms." *Biomedical Microdevices* 19 (2): 28. https://doi.org/10.1007/s10544-017-0172-5.

Bhatt, Geeta, Sanjay Kumar, Poonam Sundriyal, Pulak Bhushan, Aviru Basu, Jitendra Singh, and Shantanu Bhattacharya. 2016. "Microfluidics Overview." In *Microfluidics for Biologists: Fundamentals and Applications*, edited by C. K. Dixit, and A. Kaushik, 33–83. Springer International Publishing Switzerland. https://doi.org/10.1007/978-3-319-40036-5_2.

Bhatt, Geeta, Vinay Kumar Patel, Rishi Kant, and Shantanu Bhattacharya. 2022. "Polymer Microfabrication for Biomedical Applications." *Trends in Fabrication of Polymers and Polymer Composites*. AIP Publishing: 5.1–5.23. https://doi.org/10.1063/9780735423916_005.

Bhatt, Geeta, Keerti Mishra, Gurunath Ramanathan, and Shantanu Bhattacharya. 2019. "Dielectrophoresis Assisted Impedance Spectroscopy for Detection of Gold-Conjugated Amplified DNA Samples." *Sensors and Actuators, B: Chemical* 288 (June): 442–53. https://doi.org/10.1016/j.snb.2019.02.081.

Bhattacharya, Shantanu, Y. Gao, V. Korampally, M. T. Othman, S. A. Grant, S. B. Kleiboeker, K. Gangopadhyay, and S. Gangopadhyay. 2007. "Optimization of Design and Fabrication Processes for Realization of a Optimization of Design and Fabrication Processes for Realization of a PDMS-SOG-Silicon." *Journal of Microelectromechanical Systems* 16 (2) (April): 404–10.

Boonyasit, Yuwadee, Orawon Chailapakul, and Wanida Laiwattanapaisal. 2019. "A Folding Affinity Paper-Based Electrochemical Impedance Device for Cardiovascular Risk Assessment." *Biosensors and Bioelectronics* 130: 389–96. https://doi.org/10.1016/j.bios.2018.09.031.

Carpi, Federico, and Elisabeth Smela. 2009. *Biomedical Applications of Electroactive Polymer Actuators*. Wiley. https://doi.org/10.1002/9780470744697.

Chauhan, Pankaj Singh, and Shantanu Bhattacharya. 2018. "Highly Sensitive $V_2O_5 \cdot 1.6H_2O$ Nanostructures for Sensing of Helium Gas at Room Temperature." *Materials Letters* 217: 83–87. https://doi.org/10.1016/j.matlet.2018.01.056.

Chauhan, Pankaj Singh, Aniket Mishra, Geeta Bhatt, and Shantanu Bhattacharya. 2021. "Enhanced He Gas Detection by V_2O_5-Noble Metal (Au, Ag, and Pd) Nanocomposite with Temperature Dependent n- to p-Type Transition." *Materials Science in Semiconductor Processing* 123: 105528. https://doi.org/10.1016/j.mssp.2020.105528.

Cichosz, Stefan, Anna Masek, and Marian Zaborski. 2018. "Polymer-Based Sensors: A Review." *Polymer Testing* 67 (May): 342–48. https://doi.org/10.1016/j.polymertesting.2018.03.024.

Daikuzono, C. M., C. Delaney, H. Tesfay, L. Florea, O. N. Oliveira, A. Morrin, and D. Diamond. 2017. "Impedance Spectroscopy for Monosaccharides Detection Using Responsive Hydrogel Modified Paper-Based Electrodes." *Analyst* 142 (7): 1133–39. https://doi.org/10.1039/c6an02571d.

Dheman, Kanika, Philipp Mayer, Manuel Eggimann, Simone Schuerle, and Michele Magno. 2021. "ImpediSense: A Long Lasting Wireless Wearable Bio-Impedance Sensor Node." *Sustainable Computing: Informatics and Systems* 30: 100556. https://doi.org/10.1016/j.suscom.2021.100556.

Diaz, A. F., K. Keiji Kanazawa, and Gian Piero Gardini. 1979. "Electrochemical Polymerization of Pyrrole." *Journal of the Chemical Society, Chemical Communications* 14: 635–36. https://doi.org/10.1039/C39790000635.

Diniz, M. O., A. F. Golin, M. C. Santos, R. F. Bianchi, and E. M. Guerra. 2019. "Improving Performance of Polymer-Based Ammonia Gas Sensor Using POMA/V2O5 Hybrid Films." *Organic Electronics* 67: 215–21. https://doi.org/10.1016/j.orgel.2019.01.039.

Goldberg, Howard D., Richard B. Brown, Dong P. Liu, and Mark E. Meyerhoff. 1994. "Screen Printing: A Technology for the Batch Fabrication of Integrated Chemical-Sensor Arrays." *Sensors and Actuators: B. Chemical* 21 (3): 171–83. https://doi.org/10.1016/0925-4005(94)01249-0.

Goud, K. Yugender, Chochanon Moonla, Rupesh K. Mishra, Chunmei Yu, Roger Narayan, Irene Litvan, and Joseph Wang. 2019. "Wearable Electrochemical Microneedle Sensor for Continuous Monitoring of Levodopa: Toward Parkinson Management." *ACS Sensors* 4 (8): 2196–2204. https://doi.org/10.1021/acssensors.9b01127.

Harito, Christian, Dmitry V. Bavykin, Brian Yuliarto, Hermawan K. Dipojono, and Frank C. Walsh. 2019. "Polymer Nanocomposites Having a High Filler Content: Synthesis, Structures, Properties, and Applications." *Nanoscale* 11 (11): 4653–82. https://doi.org/10.1039/c9nr00117d.

Harito, Christian, Listya Utari, Budi Riza Putra, Brian Yuliarto, Setyo Purwanto, Syed Z. J. Zaidi, Dmitry V. Bavykin, Frank Marken, and Frank C. Walsh. 2020. "Review—The Development of Wearable Polymer-Based Sensors: Perspectives." *Journal of The Electrochemical Society* 167 (3): 037566. https://doi.org/10.1149/1945-7111/ab697c.

Hernandez-Romano, Ivan, David Monzon-Hernandez, Carlos Moreno-Hernandez, David Moreno-Hernandez, and Joel Villatoro. 2015. "Highly Sensitive Temperature Sensor Based on a Polymer-Coated Microfiber Interferometer." *IEEE Photonics Technology Letters* 27 (24): 2591–94. https://doi.org/10.1109/LPT.2015.2478790.

Hu, Jinming, and Shiyong Liu. 2010. "Responsive Polymers for Detection and Sensing Applications: Current Status and Future Developments." *Macromolecules* 43 (20): 8315–30. https://doi.org/10.1021/ma1005815.

Hu, Liang, Qiang Zhang, Xue Li, and Michael J. Serpe. 2019. "Stimuli-Responsive Polymers for Sensing and Actuation." *Materials Horizons* 6 (9): 1774–93. https://doi.org/10.1039/c9mh00490d.

Huber, Christian, Tobias Werner, Otto Wolfbeis, Bell Douglas, and Susannah Young. 2004. Optical-Chemical sensor, *US Patent*: US6835351B2, issued 2004.

Ishizu, K., Y. Takei, M. Honda, K. Noda, A. Inaba, T. Itoh, R. Maeda, K. Matsumoto, and I. Shimoyama. 2013. "Carbon Dioxide Gas Sensor with Ionic Gel." In *2013 Transducers and Eurosensors XXVII: The 17th International Conference on Solid-State Sensors, Actuators and Microsystems, TRANSDUCERS and EUROSENSORS 2013*. https://doi.org/10.1109/Transducers.2013.6627097.

Janata, Jiri, and Mira Josowicz. 2003. "Conducting Polymers in Electronic Chemical Sensors." *Nature Materials* 2 (1): 19–24. https://doi.org/10.1038/nmat768.

Jean-Mistral, C., S. Basrour, and J. J. Chaillout. 2010. "Comparison of Electroactive Polymers for Energy Scavenging Applications." *Smart Materials and Structures* 19 (8): 085012. https://doi.org/10.1088/0964-1726/19/8/085012.

Jose, Manoj, Gilles Oudebrouckx, Seppe Bormans, Paula Veske, Ronald Thoelen, and Wim Deferme. 2021. "Monitoring Body Fluids in Textiles: Combining Impedance and Thermal Principles in a Printed, Wearable, and Washable Sensor." *ACS Sensors* 6 (3): 896–907. https://doi.org/10.1021/acssensors.0c02037.

Korostynska, Olga, Khalil Arshak, Edric Gill, and Arousian Arshak. 2007. "Review on State-of-the-Art in Polymer Based PH Sensors." *Sensors* 7 (12): 3027–42. https://doi.org/10.3390/s7123027.

Lange, Ulrich, Nataliya V. Roznyatovskaya, and Vladimir M. Mirsky. 2008. "Conducting Polymers in Chemical Sensors and Arrays." *Analytica Chimica Acta* 614 (1): 1–26. https://doi.org/10.1016/j.aca.2008.02.068.

Lee, Seulki, Gabriel Squillace, Christophe Smeets, Marianne Vandecasteele, Lars Grieten, Ruben De Francisco, and Chris Van Hoof. 2015. "Congestive Heart Failure Patient Monitoring Using Wearable Bio-Impedance Sensor Technology." In *Proceedings of the Annual International Conference of the IEEE Engineering in Medicine and Biology Society, EMBS*, 2015–November: 438–41. https://doi.org/10.1109/EMBC.2015.7318393.

Li, Changhua, and Shiyong Liu. 2012. "Polymeric Assemblies and Nanoparticles with Stimuli-Responsive Fluorescence Emission Characteristics." *Chemical Communications* 48 (27): 3262–78. https://doi.org/10.1039/c2cc17695e.

Li, Xiao, Zhen Qin, Hao Fu, Ted Li, Ran Peng, Zhijie Li, James M. Rini, and Xinyu Liu. 2021. "Enhancing the Performance of Paper-Based Electrochemical Impedance Spectroscopy Nanobiosensors: An Experimental Approach." *Biosensors and Bioelectronics* 177: 112672. https://doi.org/10.1016/j.bios.2020.112672.

Lindfors, Tom, and Ari Ivaska. 2001. "Calcium-Selective Electrode Based on Polyaniline Functionalized with Bis[4-(1,1,3,3-Tetramethylbutyl)Phenyl]Phosphate." *Analytica Chimica Acta* 437 (2): 171–82. https://doi.org/10.1016/S0003-2670(01)00996-5.

Manczak, Rémi, Marc Fouet, Rémi Courson, Paul Louis Prof Fabre, Armelle Montrose, Jan Sudor, Anne Marie Gué, and Karine Reybier. 2016. "Improved On-Chip Impedimetric Immuno-Detection of Subpopulations of Cells toward Single-Cell Resolution." *Sensors and Actuators, B: Chemical* 230: 825–31. https://doi.org/10.1016/j.snb.2016.02.070.

Massy, Jim. 2017. "Thermoplastic and Thermosetting Polymers." *A Little Book About BIG Chemistry*: 19–26. https://doi.org/10.1007/978-3-319-54831-9_5.

Matadi Boumbimba, R., M. Bouquey, R. Muller, L. Jourdainne, B. Triki, P. Hébraud, and P. Pfeiffer. 2012. "Dispersion and Morphology of Polypropylene Nanocomposites: Characterization Based on a Compact and Flexible Optical Sensor." *Polymer Testing* 31 (6): 800–09. https://doi.org/10.1016/j.polymertesting.2012.05.002.

Michalska, Agata, and Krzysztof Maksymiuk. 2004. "All-Plastic, Disposable, Low Detection Limit Ion-Selective Electrodes." *Analytica Chimica Acta* 523 (1): 97–105. https://doi.org/10.1016/j.aca.2004.07.020.

Munkholm, Christiane, David R. Walt, Fred P. Milanovich, and Stanley M. Klainer. 1986. "Polymer Modification of Fiber Optic Chemical Sensors as a Method of Enhancing Fluorescence Signal for PH Measurement." *Analytical Chemistry* 58 (7): 1427–30. https://doi.org/10.1021/ac00298a034.

Numata, Manami, Keiko Baba, Akihide Hemmi, Hiromitsu Hachiya, Satoshi Ito, Takashi Masadome, Yasukazu Asano et al. 2001. "Determination of Hardness in Tapwater and Upland Soil Extracts Using a Long-Term Stable Divalent Cation Selective Electrode Based on a Lipophilic Acrylate Resin as a Membrane Matrix." *Talanta* 55 (3): 449–57. https://doi.org/10.1016/S0039-9140(01)00421-0.

O'Mahony, Conor, Francesco Pini, Alan Blake, Carlo Webster, Joe O'Brien, and Kevin G. McCarthy. 2012. "Microneedle-Based Electrodes with Integrated through-Silicon via for Biopotential Recording." *Sensors and Actuators, A: Physical* 186: 130–36. https://doi.org/10.1016/j.sna.2012.04.037.

Pistor, P., V. Chu, D. M. F. Prazeres, and J. P. Conde. 2007. "PH Sensitive Photoconductor Based on Poly(Para-Phenylene-Vinylene)." *Sensors and Actuators, B: Chemical* 123 (1): 153–57. https://doi.org/10.1016/j.snb.2006.08.005.

Rakhmania, Citra Dewi, Shaimah Rinda Sari, Yosyi Izuddin Azhar, Airi Sugita, and Masato Tominaga. 2022. "Cellulose Nanofiber Platform for Electrochemical Sensor Device: Impedance Measurement Characterization and Its Application for Ethanol Gas Sensor." *Teknomekanik* 5 (1): 57–62. https://doi.org/10.24036/teknomekanik.v5i1.12872.

Ruecha, Nipapan, Kwanwoo Shin, Orawon Chailapakul, and Nadnudda Rodthongkum. 2019. "Label-Free Paper-Based Electrochemical Impedance Immunosensor for Human Interferon Gamma Detection." *Sensors and Actuators, B: Chemical* 279: 298–304. https://doi.org/10.1016/j.snb.2018.10.024.

Sagara, Yoshimitsu, and Takashi Kato. 2009. "Mechanically Induced Luminescence Changes in Molecular Assemblies." *Nature Chemistry* 1 (8): 605–10. https://doi.org/10.1038/nchem.411.

Seyedin, Shayan, Peng Zhang, Maryam Naebe, Si Qin, Jun Chen, Xungai Wang, and Joselito M. Razal. 2019. "Textile Strain Sensors: A Review of the Fabrication Technologies, Performance Evaluation and Applications." *Materials Horizons* 6 (2): 219–49. https://doi.org/10.1039/c8mh01062e.

Sharma, Sanjiv, Anwer Saeed, Christopher Johnson, Nikolaj Gadegaard, and Anthony EG Cass. 2017. "Rapid, Low Cost Prototyping of Transdermal Devices for Personal Healthcare Monitoring." *Sensing and Bio-Sensing Research* 13: 104–08. https://doi.org/10.1016/j.sbsr.2016.10.004.

Shu, Yi, Cheng Li, Zhe Wang, Wentian Mi, Yuxing Li, and Tian Ling Ren. 2015. "A Pressure Sensing System for Heart Rate Monitoring with Polymer-Based Pressure Sensors and an Anti-Interference Post Processing Circuit." *Sensors (Switzerland)* 15 (2): 3224–35. https://doi.org/10.3390/s150203224.

Soomro, Afaque Manzoor, Faiza Jabbar, Muhsin Ali, Jae Wook Lee, Seong Woo Mun, and Kyung Hyun Choi. 2019. "All-Range Flexible and Biocompatible Humidity Sensor Based on Poly Lactic Glycolic Acid (PLGA) and Its Application in Human Breathing for Wearable Health Monitoring." *Journal of Materials Science: Materials in Electronics* 30 (10): 9455–65. https://doi.org/10.1007/s10854-019-01277-1.

Suri, Komilla, S. Annapoorni, A. K. Sarkar, and R. P. Tandon. 2002. "Gas and Humidity Sensors Based on Iron Oxide-Polypyrrole Nanocomposites." *Sensors and Actuators, B: Chemical* 81 (2–3): 277–82. https://doi.org/10.1016/S0925-4005(01)00966-2.

Tyler McQuade, D., Anthony E. Pullen, and Timothy M. Swager. 2000. "Conjugated Polymer-Based Chemical Sensors." *Chemical Reviews* 100 (7): 2537–74. https://doi.org/10.1021/cr9801014.

Uchiyama, Seiichi, Yuriko Matsumura, A. Prasanna De Silva, and Kaoru Iwai. 2003. "Fluorescent Molecular Thermometers Based on Polymers Showing Temperature-Induced Phase Transitions and Labeled with Polarity-Responsive Benzofurazans." *Analytical Chemistry* 75 (21): 5926–35. https://doi.org/10.1021/ac0346914.

Wang, Tiesheng, Meisam Farajollahi, Yeon Sik Choi, I. Ting Lin, Jean E. Marshall, Noel M. Thompson, Sohini Kar-Narayan, John D.W. Madden, and Stoyan K. Smoukov. 2016. "Electroactive Polymers for Sensing." *Interface Focus* 6 (4): 20160026. https://doi.org/10.1098/rsfs.2016.0026.

Wei, Yubo, Qiang Zeng, Qiong Hu, Min Wang, Jia Tao, and Lishi Wang. 2018. "Self-Cleaned Electrochemical Protein Imprinting Biosensor Basing on a Thermo-Responsive Memory Hydrogel." *Biosensors and Bioelectronics* 99: 136–41. https://doi.org/10.1016/j.bios.2017.07.049.

Willa, Christoph, Alexander Schmid, Danick Briand, Jiayin Yuan, and Dorota Koziej. 2017. "Lightweight, Room-Temperature CO_2 Gas Sensor Based on Rare-Earth Metal-Free Composites - An Impedance Study." *ACS Applied Materials and Interfaces* 9 (30): 25553–58. https://doi.org/10.1021/acsami.7b07379.

Yao, Jiafeng, Huaijin Chen, Zifei Xu, Jingshi Huang, Jianping Li, Jiabin Jia, and Hongtao Wu. 2020. "Development of a Wearable Electrical Impedance Tomographic Sensor for Gesture Recognition with Machine Learning." *IEEE Journal of Biomedical and Health Informatics* 24 (6): 1550–56. https://doi.org/10.1109/JBHI.2019.2945593.

Yoshikawa, Masakazu, Kalsang Tharpa, and Ştefan Ovidiu Dima. 2016. "Molecularly Imprinted Membranes: Past, Present, and Future." *Chemical Reviews* 116 (19): 11500–28. https://doi.org/10.1021/acs.chemrev.6b00098.

Zhao, Huijie, Lijie Hong, Kaiyue Han, Mujie Yang, and Yang Li. 2022. "In Situ Prepared Composite of Polypyrrole and Multi-Walled Carbon Nanotubes Grafted with Sodium Polystyrenesulfonate as Ammonia Gas Sensor with Wide Detection Range." *Journal of Polymer Engineering* (November 10) 43 (1): 53–65. https://doi.org/https://doi.org/10.1515/polyeng-2022-0106.

Zou, Lina, Yinfeng Li, Shaokui Cao, and Baoxian Ye. 2013. "Gold Nanoparticles/Polyaniline Langmuir-Blodgett Film Modified Glassy Carbon Electrode as Voltammetric Sensor for Detection of Epinephrine and Uric Acid." *Talanta* 117: 333–37. https://doi.org/10.1016/j.talanta.2013.09.035.

11 Recent progress in aptamer-based smart detection techniques for agriculture

Vibhas Chugh, Adreeja Basu, Ramesh N. Pudake,
Rajkumar Saha, and Aviru Kumar Basu

CONTENTS

11.1 INTRODUCTION

Andy Ellington coined the term "aptamer" in 1990 (Shoeib et al. 2020) from the Latin word "aptus," which means "to fit" and "meros," which means "part." Aptamers are single-stranded oligonucleotide RNA or DNA structures that may attach to other molecules with specificity and a strong affinity. They are generally 100 m long. Aptamers

DOI: 10.1201/9781003358091-14

can bind to proteins, peptides, carbohydrates, tiny compounds, poisons, and even live cells, and because of their proclivity to form helices and single-stranded loops, they can take on a variety of geometries (Wolter and Mayer 2017). They are incredibly adaptable and have a high level of specificity and selectivity when it comes to binding onto specific targets (Bai et al. 2020). The binding of the aptamer is governed by its tertiary structure rather than its basic sequence. Three-dimensional, hydrophobic interactions, intercalation, base-stacking, and shape-dependent interactions, are all involved in target recognition and binding (Elshafey, Siaj, and Zourob 2015).

A tiny single-stranded RNA (ssRNA) or DNA (ssDNA) oligonucleotide aptamer can self-fold into a specified 3D-spatial configuration. Aptamers can only attach to their targets by electrostatic interactions, hydrogen bonding, π–π stacking, hydrophobic effects, or van der Waals forces with these particular conformations. As a result, aptamers are sometimes known as 'chemical antibodies,' and they have certain benefits over antibodies. In a nutshell, oligonucleotide aptamers provide batch uniformity, resilience at room temperature, lower immunogenicity, and functionalization options to improve transport characteristics (Zhou, Jimmy Huang et al. 2014). Small molecules, such as metal ions, amino acids, antibiotics and peptides, chemical dyes, entire cells, proteins of varied sizes and functions, viruses and virus-affected cells, and bacteria, have all been targeted with aptamers (Khosravi et al. 2020). A variety of applications have taken advantage of the selective interactions between aptamers and their many targets fields, including medicine, horticulture, biotechnology, microbiology, cell biology, and chemistry (Famulok and Mayer 2014). Aptamers have, for example, been successfully used in therapeutic applications, such as targeted therapies, detection, and diagnostics. Properties such as longer shelf-life and cyclical or regulated conformational changes and renaturation have increased the versatility of aptamers in diverse methodological approaches using biosensors (Cho et al. 2013) (Paul et al. 2010). Because of their high sensitivity, cost-effectiveness, compliance with innovative nanofabrication methods, and intrinsic downsizing, aptasensors (aptamer-based sensor) are appealing devices. A variety of biosensors can be utilized for this purpose based on different transduction mechanisms (Singh et al. 2023) (Elshafey, Siaj, and Zourob 2014). As a result, a variety of electrochemical aptasensors have been developed by employing a variety of methodologies, such as potentiometry, electrochemical impedance spectroscopy (EIS), and differential pulse voltammetry.

Population growth raises worldwide concerns of food security in terms of both amount and reliability. However, the effects of climatic change have resulted in the appearance of new strains of viruses and insects, as well as a state of famine in some regions and high rainfall in others. The growth of microorganisms disrupts the environmental integrity and infects food crops with numerous types of pathogens (Yadav, Parashar, and Aggarwal 2019). The growing use of chemicals to curb the effects of these microorganisms has resulted in the presence of pesticides, microbes, antibiotics, toxins, and other harmful chemical entities, such as heavy metals in food or associated ingredients that pose a grave threat to human health. Safety-monitoring from the above pollutants threshold in food from manufacturing to ingestion calls for the creation of new technology. Nuclear magnetic resonance, infrared (IR), immunological techniques, high-performance liquid chromatography (HPLC), conventional biosensors, gas chromatography (GC), and other conventional methods, are available for testing. Even if there are screening procedures using traditional techniques that are highly specific and selective, there is still a demand for simpler, quicker, and more affordable techniques (Handy et al. 2013). A viable solution in this situation is aptamer-based technology. The new technique for the identification and evaluation of contaminations, such as pesticide quantity, antibiotics, agricultural products, food nutrition status, and other food safety-related issues, is certainly an aptamer-based solution. Chemically created aptamers provide a number of advantages when it comes to reagent effectiveness. They can be adjusted through multiple selection rounds or certain selection rounds in order to fulfill demanding application requirements. Without using any biologics or animals, a chemical reaction is used to create aptamers. They are not constrained by lot-to-lot differences because of their chemical production. Once the sequencing is characterized and the assay is designed and validated, aptamer production costs are fairly cheap. The use of smart technologies based on artificial intelligence (AI), machine learning (ML), and Internet of Things has been adding to the advancements in aptamer-based detection.

11.2   METHOD FOR SELECTION OF APTAMERS – SELEX

SELEX – Sequential Evolution of Ligands by Exponential Enrichment – is a technique that selects aptamers with an affinity for a specific target from a huge oligonucleotide pool. For aptamer selection, various proprietary oligonucleotide pools are maintained (Gong et al. 2013). The capacity of these tiny oligonucleotides to fold into distinct 3D structures that can engage with a given target with great specificity and affinity controls the SELEX process (Figure 11.1).

Aptamers can be compared to antibodies because of their capacity to recognize targets, selective binding, and affinity. On the other hand, aptamers offer more freedom in their development and application because of their unique properties (Chen et al. 2013). The SELEX technique, in particular, takes a relatively short amount of time to generate aptamers. Furthermore, aptamers are chemically produced, allowing for biochemical manipulation to include different functional groups and particular

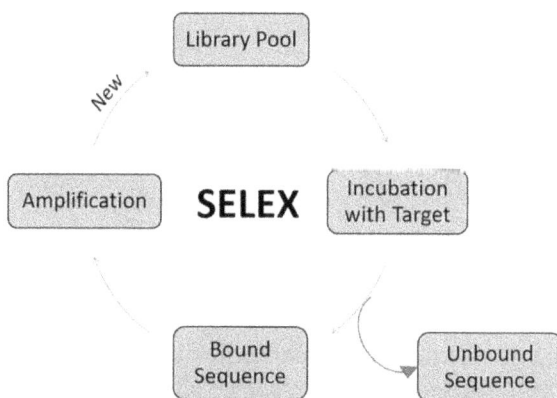

FIGURE 11.1   Basic processes involved in SELEX.

components, such as biotin, amino, carboxyl, and thiol, the majority of which have no effect on the aptamer's identification of the target (Bae et al. 2013). This has made it possible to conjugate aptamers with medicinal molecules and nanomaterials and modify nuclease-resistant bases. In the SELEX process, nonbinding aptamers are removed, whereas in the iterative process, aptamers that bind to the proposed target are extended. Positive selection rounds are occasionally followed by negative selection rounds. This improves the selectivity of the resultant aptamer candidates. To enrich the oligonucleotide pool, many rounds of SELEX are done with increasing stringency (Zhuo et al. 2017).

(1) Aptamers are discovered using an in vitro method. In a nutshell, this entails incubating the DNA pool library with the target molecule. The binding complex (target and nucleic acid sequences) is separated from the unbound sequences after incubation. The sequences that bind to the target are then eluted and incubated with the control, while the remaining segments are multiplied using the polymerase chain reaction (PCR) process, which is continued until the pool contains sequences that recognize the target specifically. Individual sequences are obtained by cloning the enriched pool and sequencing positive clones. Copies of these sequences are pharmacologically produced, tagged with a fluorescent dye, and tested against the target to discover possible aptamer candidates (present in an enhanced pool) (Bayat et al. 2018).

All of the assays utilized in the identification of possible aptamer candidates and the mode of selection and methodology employed to manufacture these aptamers are heavily influenced by the target molecule. As a result, a variety of selection modes have arisen, each of which is strategically designed to meet the needs of a certain purpose. The use of a cell-based selection of aptamer, also known as cell-SELEX, is one such method (Figure 11.2). Cell-SELEX refers to the process of selecting aptamers for target identification using living cells. As a result, cell-SELEX is a potential selection approach for a variety of applications, including cancer diagnostics and treatment (Ray and White 2017). Specifically, by using cell-SELEX, relevant probes to distinguish tumor cells from normal cells and to distinguish between two types of malignancies can be produced. This approach has so far utilized live cancer cells, and, as a result, a large number of aptamers have been created for the majority of malignancies investigated (Zhong et al. 2020).

Essentially, oligonucleotides bind to molecules on the extracellular surface in the cell-SELEX method, which is significant for the following reasons:

(2) Aptamers bind to target compounds in their natural state. Aptamers are designed for molecules on the cell surface that are in their original conformation and, hence, represent their natural packing structures and dispersion; as a result, all post-translational alterations of proteins are left unchanged, and the aptamer will attach to the true folded conformation. Given that the cell surface has a net negative charge, repulsion between the

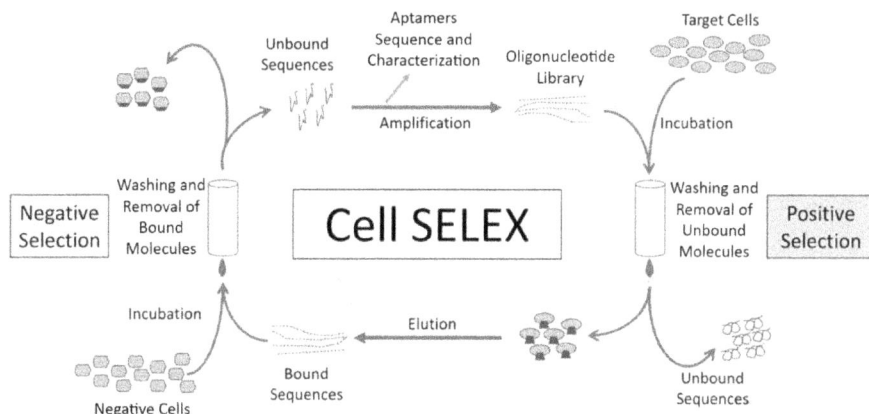

FIGURE 11.2 Schematic of cell-SELEX showing positive and negative selection.

DNA polyanion and the cell should be expected; consequently, building oligonucleotide aptamers for cells will be problematic. However, it can be expected that the aptamer-target structure-related binding will be powerful enough to resist the repulsive force. Furthermore, the involvement of divalent cations may aid in protecting DNA-negative ions and limiting their activities.

(3) It is not necessary to have any prior information about the target. In cell-SELEX, it is not necessary to know the molecular composition of the surface of the cell. Instead, the diverse cell types employed in the selection process of aptamers are a significant parameter, as the chosen aptamers will be useful for specialized cell recognition.

(4) Aptamers can be used to discover biomarkers. Because cell-SELEX has the ability to develop aptamers for unknown compounds, these probes can be utilized for the purification of the target for aptamer recognition and then for disease biomarker discovery via selective purification.

(5) Aptamers can be made for a variety of targets. The cell membrane surface is a complicated system with a plethora of chemicals, particularly proteins. Each of these compounds is a possible target in cell-SELEX. Aptamers are created for a variety of targets after a successful selection. This trait is critical because any of these compounds could have a function in the cell's development or the sickness they produce.

The phosphodiester framework of DNA can engage nonspecifically with lysine and arginine and other side-chain compounds, potentially lowering the selectivity of the aptamers formed. Nevertheless, because SELEX is an ongoing adaptation, the particular binding of the 3D structure of DNA sequences takes precedence over the arbitrary nonspecific binding. Furthermore, as the selection process advances and the procedure's rigor improves, the impact of conformationally changed binding sequences, if any, is reduced to the bare minimum (Bakhtiari et al. 2021) (Pleiko et al. 2019) (Ohuchi 2012) (Lyu et al. 2016).

The experimental design for the fabrication of aptamers is listed below:

(1) Choosing the cell lines – positive or negative: the identification of cell lines is determined by the goal and objective of the selection.
(2) Maintaining the cell culture: because cell-SELEX utilizes live cells as its threshold, adequate cell culture upkeep is critical. Cell mortality is increased in overgrown cell lines, and cell shape and protein expression are likely to change, which is not desirable.
(3) Binding sequences elution: the sequences that attach to target cells are recovered by heating the cell-DNA combination to 95°C. Any DNase produced following cell breakdown at the elevated temperature is inactivated at 95°C and therefore cannot degrade DNA.
(4) PCR: PCR is an important aspect of this protocol, and its effectiveness can assist the selection process with success by optimizing annealing temperature and PCR reagents concentration.
(5) Single-stranded DNA preparation: for the preparatory PCR, a similar PCR yields double-stranded DNA, which must be segregated and the tagged DNA sequence must be retrieved in order to keep the selection process going.
(6) Negative selection: following the second phase of selection, the chosen DNA pool could be utilized to perform negative selection, removing sequences that may bind to the target on the exterior of both the molecules and control cell cultures. Negative selection makes it nearly impossible to eliminate all of these sequences from the pool, and the remaining sequences can grow in quantity following PCR. Negative selection, however, on the other hand, can delete the majority of these sequences, preventing them from populating the selected pool. Negative selection is performed using all selected DNA target pools obtained at the completion of each cycle of negative selection.
(7) Monitoring the progression of selection: following the second phase of selection, the chosen DNA target pool could be utilized to perform negative selection, removing sequences that may bind to the target on the exterior of both the molecules and control cell cultures. Negative selection makes it nearly impossible to eliminate all of these sequences from the pool, and the remaining sequences can grow in quantity following PCR. Negative selection, however, on the other hand, can delete the majority of these sequences, preventing them from populating the selected pool. Negative selection is performed using all of the selected DNA target pools obtained at the completion of each cycle of negative selection.
(8) Concluding the selection process: when there is considerable variation in the fluorescence signal amplitude of the control background conditions and the particular DNA pools, and no significant signal difference can be observed between two or three sequential selected pools, the selection process is complete. After the selected pools have been supplemented with prospective DNA aptamer strands, the successful copycats are analyzed to identify particular aptamer targets.

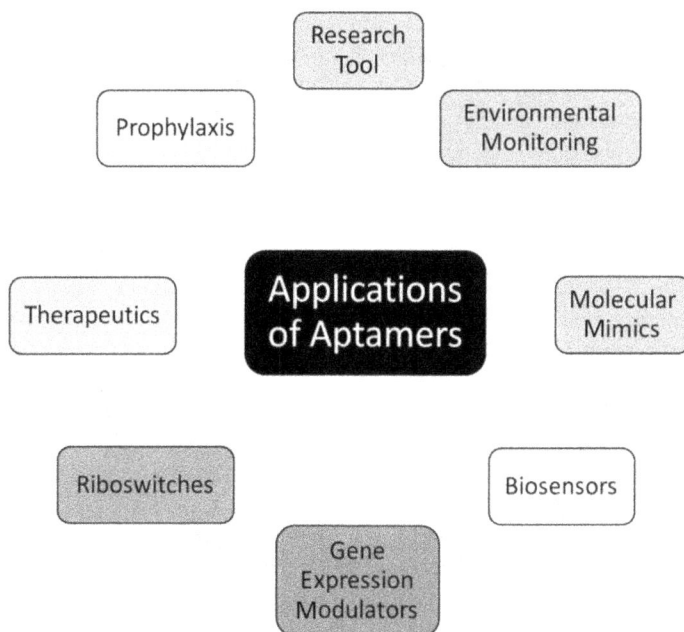

FIGURE 11.3   Applications of aptamers.

Aptamers have been advancing in different fields because of their structure-dependent chemical and physical properties, biocompatibility, and potential to exhibit biological signaling and propagation pathways. Various applications of aptamers are listed in Figure 11.3.

11.3   APTAMER-BASED DETECTION TECHNIQUES

11.3.1   ELECTROCHEMICAL IMPEDANCE SPECTROSCOPY

Physics, chemistry, medicine, and electronics are all integrated into electrochemical aptasensors. These fields complement and influence one another, which has aided in the electrochemical aptasensor's progressive growth. Fast analysis, a straightforward way of operation, strong selectivity, and high sensitivity are all benefits of electrochemical aptamer sensors. It is a molecular recognition compound called an aptamer that is combined with electrochemical sensing to create an analytical detection device (Ebrahimi et al. 2012). As a result, electrochemical aptamer sensors have slowly gained popularity in research. An effective method for analyzing the interfacial characteristics of biorecognition processes, including aptamer-target deformation that takes place at changed surfaces, is Electrochemical impedance spectroscopy (EIS) (Li et al. 2008). A redox probe couple is used in biosensing using EIS to measure changes in the electron-transfer barrier. Along with ongoing technological advancements, the growing number of biomedical applications for electronic bio-impedance assessments has prompted several scientists to start working on mobile and even wearable electrical bio-impedance monitoring systems (Rodriguez, Kawde, and Wang 2005). Researchers would be inspired to create and use light-weight and wearable electrical bio-impedance devices for plant healthcare screening applications, opening up a brand-new and sizable market for medicinal technological advancement (Tabrizi and Acedo 2022).

11.3.2 ARTIFICIAL INTELLIGENCE/MACHINE LEARNING-BASED SELECTION AND DETECTION

To choose aptamer alternatives with specificity and affinities, various computational methods have been inspired by artificial intelligence (AI), including machine learning (ML) algorithms. It has been demonstrated that several ML techniques surpass a variety of traditional methods for predicting binding affinities, including molecular docking and high-throughput screening technologies (Heredia, Roche-Lima, and Parés-Matos 2021). The SELEX procedure can be influenced by a wide range of variables. As a result, AI has a lot of potential in this field. The ability of ML techniques to incorporate domain data to drastically limit the remedy space of optimizing search issues, such as addressing the biomolecular inverse concerns, has been proved in recent AI research. Therefore, these in silico techniques have the ability to offer a financially viable option to increase the dependability and efficiency of the aptamer design process. An AI-assisted aptamer design capability is being developed using a variety of methodologies in an effort to speed up the discovery of high-performing aptamers for the detection of novel biological markers (Schoukroun-Barnes, Wagan, and White 2014). The creation of sensitive, specific, structure-switching aptamer-based electrochemical sensors that can identify a broad range of target analytes is being pursued using a number of different strategies. The binding capabilities of an electrode-based DNA or RNA aptamer biorecognition element enable the specificity. The structure and flexibility of the aptamer probes are changed by the target, which results in alterations in electron transport efficacy in this category of sensors. Electrochemical monitoring makes it simple to track these changes. Due to this signaling process, there are several strategies for enhancing the analytical properties of the aptamer sensor, such as its limit of detection, selectivity, specificity, and observed attachment affinity (Chen et al. 2021).

11.3.3 ENZYME-LINKED OLIGONUCLEOTIDE ASSAY

Following the broad evolution of ELISA devices, an enzyme-linked oligonucleotide assay (ELONA) mimic was created to detect analyte molecules. The fundamental distinction between ELISA and ELONA is that ELONA makes use of only aptamers rather than antibodies as the principal biorecognition element (Sypabekova et al. 2017). Any ELISA-based identification assay could conceivably be carried out with ELONA and the required aptamers. In ELONA, the utilization of aptamers rather than antibodies has various advantages. Biotin or chemically synthesized dyes, for example, can be used to mark oligonucleotide aptamers without affecting their specificity or the binding ability for target moieties. As a result, ELONA can be made with just one labeled aptamer. This setup is cost-effective as compared to direct ELISA and more straightforward than indirect ELISA. In direct ELONA, sandwich ELONA can be utilized to reduce backdrop noise caused by target immobility (Wang et al. 2015).

Sandwich ELONA (Torrini et al. 2019) can be used as a favorable option in cases when analytes of interest do not have two recognition antibodies. The immobilization of the principal aptamer on the surface of substrate support is a crucial and difficult phase for both competitive (Guan et al. 2015) and sandwich ELONA. Primary aptamers must be suitably immobilized for specific targets, with high binding selectivity and specificity (Fu et al. 2011). Chemical and physical reactions, such as physical adsorption, self-assembly, covalent binding, avidin–biotin immobilization, and hybridization, can all be used to immobilize aptamers. ELONA helps well in detecting analytes where antibody-based detection fails due to a suitable immobilization approach.

11.3.4 APTAMER-BASED LATERAL FLOW ASSAY

Lateral flow assay (LFA), a paper-based system for detecting targets in liquid samples, makes use of aptamers. The capture of the target molecule at a prominent line by an immobilized sensor is based on affinity interactions similar to those of ELISA (Liu et al. 2020) (Phung et al. 2020). LFA can be classified as a direct and competitive assay. Target-consisting samples enter the conjugate patch, where the recognition components conjugated with colored or fluorescent particles (CPs) bind to targets in direct assays. As the bound targets comprise an immobile sensor, CPs stay within the test line. The arrested CP-bound molecules collect and keep the remaining CPs (Dalirirad, Han, and Steckl 2020), and, as a result, both the control and test lines appear to be colored. If the sample does not contain the target, just the control line appears to be colored (Huang et al. 2021). With a similar principle, aptamer-based lateral flow assays (ALFAs) have better scalability and can detect a broader spectrum of targets. Sandwich ALFA, for instance, should be evaluated for tiny compounds with fewer antigenic determinants that cannot be detected by antibodies (Kaiser et al. 2018). Furthermore, aptamers, split aptamers, antibodies, biotin or streptavidin, and complementary ssDNA probes can be utilized as both capture agents and the membrane-immobilized sensors at the test line in ALFA. More promising and referential ALFA designs can be comprehensively introduced into the lateral flow assays, based on aptamer designs (Jauset-Rubio et al. 2017).

11.3.5 APTAMER-BASED FLUOROPHOTOMETRY

A flow cytometer, a fluorometer, or a fluorescence microscope can be used to identify targets of concern utilizing fluorophore-labeled aptamers, aptamer-conjugated quantum dots, or excitonic nanoparticles as probes. Labeling cancer cells with particular fluorophore-labeled aptamers produced by cell-SELEX is a standard approach for cancer cell identification. Aptamers can also be used tofluorophore label other tiny entities such as bacteria, poisons, and other macromolecules (Rogers et al. 2015). It can be seen how aptamers can be used to identify macromolecules and microbes (1 m), such as viruses, mycoplasma, and chlamydia, which are difficult to be identified

using conventional centrifugation methods, aptamer-added magnetic beads, or aga-
rose gel beads. The amount of target present is reflected in the intensity of the ensu-
ing fluorescent signal (Dolgosheina et al. 2014). Though aptamer-based fluorescence
probe assays are cost-effective, easy, and quick, they do have drawbacks, such as
aptamer immobilization, reliance on experimental equipment precision, the need for
logical control setup, and reduced specificity (Arora, Sunbul, and Jäschke 2015).

Some advances have been made to structure-switching aptamers depending on
fluorescence resonance energy transfer (FRET) (Pehlivan et al. 2019). For target
detection, researchers have divided a DNA aptamer into two portions: one is biotin-
labeled, whereas the other is fluorophore-labeled. The integrated components of the
target can be caught by streptavidin-labeled beads, resulting in fluorescence (Jepsen
et al. 2018). If a sample does not contain the target, the streptavidin-labeled beads
will only catch the biotin-labeled portion of the sample, leaving the beads nonfluo-
rescent. The fluorescence of the tagged aptamer has been quenched using graphene
oxide (GO), an energy absorber (Yue et al. 2021). After the aptamer attaches to its
substrate, a transformation process in the aptamer causes aptamer-conjugated fluoro-
phores to be removed from GO, enabling fluorescence recovery. Based on the fluorescent components employed, other quenchers such as TAMRA dye, cytochrome C
protein, and gold nanoparticles (GNPs) can be explored (Xu et al. 2019).

11.4   BIOSENSORS AND APTASENSORS

Conventional techniques used for the detection of a variety of biomolecules, microorganisms, and contaminants have various loopholes with respect to their application,
functioning, and reliability. Modern-day aptamer-based biosensors can be used to
overcome these shortcomings. These biosensors are not only reliable but also rapid,
cost-effective, and easy to operate (Bhalla et al. 2016). They outperform conventional
techniques and assays in the detection of materials. These aptamer-based devices
produce efficient results that can be validated by comparing with those from conventional techniques (Challier et al. 2012). Biosensors are biological analyte detection
devices that include a biorecognition interface element, a signal transducer, and a
signal-processing unit as shown in Figure 11.4. Immobilized proteins, antibodies,
enzymes, nucleic acid, or protein aptamers, or even organisms can all be used as
a specific biorecognition-sensing component on which a biosensor is built (Basu,
Basu, and Bhattacharya 2020) (Di Pietrantonio, Cannatà, and Benetti 2019) (Das
Bhowmik, Basu, and Sahoo 2016). Biochemical reactions are converted into read-
able signals that may be monitored and studied immediately using a combination of

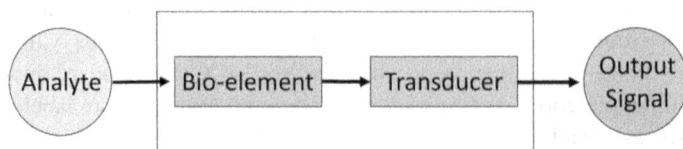

FIGURE 11.4   Biosensor – basic design.

chemical and physical principles and computer technology. Aptasensors are biosensors that use aptamers as recognition elements.

Modern-day aptasensors possess various transduction mechanisms based on the properties utilized by nucleic acids, such as fluorescence, electrochemistry, luminescence, colorimetry, amplification, and mass sensitivity (Hianik 2018).

According to the modifications caused by aptamers, optical aptasensors are classified as labeled or unlabeled. Aptamers are labeled with optically active compounds like fluorescent markers or colorimetric or luminous materials. Labeled optical entities are triggered or quenched, and clustered or scattered depending on the conformational nature of the aptamer following target attachment, which affects their optical characteristics (Lan hua Liu, Zhou, and Shi 2015). Biorecognition causes a change in the component surface that an optical system detects in an annotated aptamer model. Surface plasmon resonance, subwavelength fluorescence, and optical transmission diffraction gratings are some of the clinical findings used in this concept (Costantini et al. 2019) (Basu, Basu, and Bhattacharya 2021). The fundamentals of electrochemical aptasensors utilize the concepts of basic electrochemistry. A signal transduction unit, to put it simply, converts microvoltage or microcurrent changes into visible signals. Biorecognition causes the aptamer to change conformation or brings covalently coupled enzymes to the surface of the electrode. Such changes in redox molecules cause a charge flow, which is subsequently monitored and quantified. Amperometric (voltammetric) and impedimetric sensors are by far the most often used electrochemical sensors (Basu, Basu, and Bhattacharya 2019). There are many ways to magnify the signal to enhance the sensitivity of electrochemical aptasensors.

Piezoelectric effect-based quartz crystal microbalance (QCM) is the most used common mass-sensitive or mechanical aptasensor. QCM identifies gravitational perturbations on the crystals by frequency monitoring in the resonator (Basu, Basak, and Bhattacharya 2020; Basu et al. 2019). Pathogen-detection QCM aptasensors have also been produced (Brockman et al. 2013). Conformational modification of the structure of aptamers is a common technique that aids in the binding of the ligand to small aptamer molecules. Proper knowledge of the structure of aptamers is required to rationally design a structure-switching biosensor approach (Breaker 2012).

Aptamers binding to molecules is attributed to their small size; therefore, the sensor interface should be capable of binding only to the specific molecule out of the multiple recognition sites of the target, which makes the choosing of aptamers a tedious task. A solution to circumvent the issue is to select aptamers that do not consider the target or a different aptamer alone but specifically the aptamer–target complex (Maduraiveeran and Jin 2019). Detection of infections and assessment of heavy metals, allergies, and toxins are all part of the monitoring of the plant's production environment. Pathogens are the primary agents reducing agricultural productivity and causing increased economic losses. As the initial stage in managing a plant's disease, pathogen detection is critical. Aptasensors appear to have potential use for detecting infections in a sensitive, quick, and accurate manner (Hosseinzadeh and Mazloum-Ardakani 2020). Furthermore, because consumers and healthcare organizations are watchful and sensitive to hazardous heavy metals, pesticides, mycotoxins, and allergens in plants and biomedications, determining trace quantities of these pollutants in medicinal plants and biopharmaceuticals has become highly significant.

As a result, developing a reliable analytical aptasensor is important for a quick and accurate determination of biodrug contamination (Purkayastha et al. 2010) (Moreno 2019).

11.5 APPLICATIONS OF APTAMER-BASED SENSORS

11.5.1 Small toxic molecules

Aptamers can be selected for various small toxic molecules that can be classified as cyanotoxins and mycotoxins that are produced by fungus and water-grown Cyanobacteria, respectively, and other toxic molecules are obtained from dinoflagellates (Handy et al. 2013).

Mycotoxins are fungus-based toxins that can contaminate grain and products obtained from grain. They are extremely carcinogenic and can be the cause of various diseases, because of which they are classified as a risk to animal and plant health by the Food and Agricultural Organization, United Nations. There are various aptamer sensors that focus on different mycotoxins: one of the most common use of aptamers is for the recognition of ochratoxin A (Ha 2015). A plethora of techniques in combination with aptasensors is also available for the detection of aflatoxin M1 and B1 (Guo et al. 2014). Dipstick assays, quantitative PCR, and split aptazymes are used for the easy workflow management of aptasensors. Furthermore, gold nanomaterial-based sensors can be utilized in electrochemical colorimetric and chemiluminescent sensors for the detection of real or spiked samples. Very similar results can be obtained from aptasensors compared to those from conventional immunoassays. Aptasensors have also been used to detect other mycotoxins like fumonisin, T-2, and lysergamine (Chen et al. 2014). Detection of specific contaminants and the source of the toxins is of utmost importance as various fungus species can be manufactured by a single mycotoxin, and the same mycotoxin can generate various other fungus species. To reach this goal, the simultaneous detection of various toxins is a must (Pothoulakis and Ellis 2015).

Aptamer sensors that are capable of the simultaneous detection of ochratoxin A and fumonisin B1 have been developed based on their fluorescence property (Ha 2015). The efficiency of these aptamer biosensors can be validated using the usual enzyme-linked immunosorbent assay (ELISA).

Cyanotoxins show hepatotoxicity and can even result in a test of the exposed organism. An example of cyanotoxins is microcystin LR, a cyclic heptapeptide with a molecular weight of nearly 1000 dalton. Microcystin LR was successfully detected in spiked water by the use of a sandwich assay that combined an aptamer and horse radish peroxidase-labeled antibody. Microcystin LR can also be successfully detected by the use of an electrochemical aptamer-based sensor. Onsite testing can be qualified by the use of colorimetric aptasensors that require just 5 min of analysis time (Chinnappan et al. 2019). All sensors mentioned above can reach a very low limit of detection as prescribed by the World Health Organization. An electrochemical transduction mechanism is implemented in the switch structure-based biosensor. For such a sensor, aptamers can be selected for two of the smaller toxins – cylindrospermopsin, which is an alkaloid, and anatoxin A, which is a neurotoxin.

Studies have focused on selecting aptamers for electrochemical biosensors to detect neurotoxins like brevetoxin-2 and saxitosin (Eissa, Siaj, and Zourob 2015).

11.5.2 Industry-originated pollutants

Industrial pollution plays a vital role in the contamination of the environment, food, and water. Increasing demand for the production of industrial goods is the major cause of growing contamination due to industrial-origin toxic molecules. Industrial pollutants like polychlorinated biphenyls (PCBs) that are used as coolants and lubricants in the industry have lipophilic properties and higher stability that aids in their accumulation in the environment (Chen and Zhou 2016). These contaminants are carcinogenic and toxic, and their products were banned, but they still persist in the environment; therefore, detection devices are needed to specifically control them. Various varieties of PCBs are known for their chemical heterogeneity. An ideal device would detect all of them and would need a decreased aptamer specificity, which is in contrast with that of the conventional design. During various attempts to simultaneously detect PCBs, some were able to detect six different PCBs, but their interaction was weak. A similar kind of detection mechanism was also used in the binding of chloramphenicol and PCBs to aptamers (Feng et al. 2015). The number of targets that can be simultaneously detected is limited by the available applicable ions as Tracers' data used to determine the target were metal ions. Gold nanomaterial-based sensor can be designed for PCB 77, which is one of the most toxic PCBs. However, the aptamer chosen to detect PCB I was not very specific and recognized a few additional PCBs (Fu et al. 2015). The affinity of the aptamer helped achieve the micromolar range and a very low limit of detection.

Bisphenol A, once used in the production process of plastics, spreads from plastic-packaging products, specifically bottles for drinks and food, and it has been termed a contaminant in the environment. To detect this contaminant in water and milk, different aptasensors were used successfully (Chung et al. 2015). The high nitrogen content of melamine has led to its use in the food industry as an adulterant that probably increases the content of protein in the diet. Complementary sequences to the aptamer were conjugated to the DNA strand that was further bonded to invertase and magnetic beads. As soon as melamine is added, the structure of the aptamer is changed, and it dissociates the complementary DNA strands that enable the separation of invertase and magnetic beads. The sensor can also be quantified with a glucometer that efficiently converts sucrose into glucose (Chen et al. 2013). Methylenedinitroaniline (MDA) is a product of the thermal degradation of polyurethanes and is used as an industrial chemical in plastic production. It is known to be carcinogenic and is suspected to damage DNA. It is also a byproduct of producing dialysis machines, resulting from polyurethane sterilization. Work is in progress to choose a suitable aptamer that can detect the presence of MDA (Pfeiffer and Mayer 2016).

11.5.3 Specific secondary metabolites identification

Synthesis of secondary metabolites with medicinal chemistry and commercial benefit is costly and often impossible, and biomedicines are considerably more cost-effective and advantageous than synthetic ones. Hence, locating medicinal plant sources that contain biomedicines of appropriate quantity and quality is essential. Implementation of detection technologies such as HPLC, GC, and GC/MS is time-consuming and costly. Accordingly, aptasensor is highly suggested for determining the antioxidant potential of medical phytoconstituents and quantifying valuable plant compounds and secondary metabolites, including vinorelbine (Zhou, Zhou et al. 2014).

11.5.4 DRUGS

Antibiotic drugs are used in large quantities in livestock, agriculture, and farming as well as in the medical practice, and because of the large quantities used, they can pose a problem to the environment. Aptamer-based sensors have been successful against drugs like chloramphenicol (Alibolandi et al. 2015). Selection of aptamers for antibiotics like danofloxacin is underway (Han, Yu, and Lee 2014). Various aptasensors make use of malachite green, a dye that is used for its antiparasitic and antifungal properties in biological applications. Simultaneous detection of chloramphenicol and malachite green can be done with aptasensors (Feng et al. 2015). One of the most commonly used drugs is diclofenac, which is also present in the biomagnification cycle in our environment. Aptamers that detect the drug are reported by various research groups (Joeng et al. 2009; Zou et al. 2022).

11.5.5 PESTICIDES

Excessive use of herbicides and pesticides in fodder crops poses environmental contamination. Using aptamer sensors, organophosphorus-based pesticides are detected after an elaborate and tedious protocol of extraction. This focuses on the need for aptasensor optimization for the analytical detection procedures of various other chemical moieties. Fan et al. have successfully used metal nanomaterials in an aptamer-based fluorescent screening assay of acetamiprid (Guo et al. 2016).

11.5.6 EXPLOSIVES

Detection of commonly used explosives is required for the safety of the public and to keep a check on environmental contamination. An electrochemical aptamer-based sensor is used for the detection of 2,4,6-trinitrotoluene (TNT). FRET-based transduction mechanism for the optic sensors can also be used to detect TNT in the environment (Torshizi et al. 2016).

11.6   NATURAL APTAMER SENSORS – RIBOSWITCHES

Riboswitches contain aptamers and natural RNA element-based expression platforms. The binding of a particular identified molecule triggers a structural change in the expression platform that activates it (Yokobayashi 2019) and results in the transcriptional downregulation, upregulation, or translational inhibition of the mRNA-containing riboswitch (Ge and Marchisio 2021). This demonstrates that RNA can attach to tiny molecules with high affinities, a trait that applies to aptamers as well.

The only noticeable difference between the two is size, as riboswitches' aptamer domains are often longer than those of chosen aptamers, allowing for more complicated three-dimensional structures (Schneider and Suess 2016). Selecting aptamers with longer libraries would be one solution. However, given that the number of aptamers with high affinities is continually increasing even without lengthier libraries, it is more likely a matter of applying the proper selection procedure rather than a general tedious problem (Breaker 2012).

Riboswitches are increasingly recognized as a possible solution for antibiotic drugs since they are a widespread and necessary mechanism for gene control in bacteria, which has been encouraging their usage in the medication development field.

11.7 LIMITATIONS RELATED TO APTAMER SELECTION BINDING TO SMALL MOLECULES

When selecting aptamers that interact with tiny molecules, several specific issues arise, although they do not, when selecting aptamers that interact with proteins and other target molecules. To begin with, the ligand and aptamer interaction is quite limited because of their small size (Yoo, Jo, and Oh 2020). As a solution, if target selection is not done, as is usually the case, one of its functional groups will have to be used for immobilization, reducing the number of possible interactions with a future aptamer even more (Kalra et al. 2018). Furthermore, immobilization could result in the formation of a new epitope that is essential for aptamer interaction. Because of specific bond requirements, limited functional groups are available to bond (Zhou, Jimmy Huang et al. 2014). The capacity of the aptamer to discriminate between closely related compounds that differ only at the immobilization location could also be harmed. The use of nucleobase-modified nucleic acid pools to pick protein targets has recently been shown to greatly boost the success rate of selections, most likely because of the increased interaction possibilities offered by the change (McKeague and Derosa 2012). As previously indicated, when using tiny molecule targets, the restricted number of interaction possibilities is frequently an issue. Applying these methods to the selection of aptamers that bind to tiny molecules could improve the chances of selecting even the most difficult targets (Lubin Liu et al. 2021). The high target concentration requirement may not be always possible because of the target molecule's solubility, leaving binding to the nonimmobilized target to chance. Furthermore, identifying binding in solution may be difficult if the target's solubility in the largely aqueous buffers employed for the selection and binding tests is insufficient in comparison with the aptamer's affinity (Rozenblum, Pollitzer, and Radrizzani 2019).

11.8 CONCLUSION AND FUTURE PERSPECTIVES

The use of aptamers, especially as chemical antibodies, has gained a lot of interest in biomedical sectors because of their high precision, binding tenacity, freedom of construction, and relatively low cost. We highlighted recent advances and possible applications of aptamer-based technologies in the horticulture and plant business in this chapter. Aptamers have become a major participant in agricultural product detection, including sensing, tracking, drug delivery, and biological drug refinement. It should be mentioned that potential researchers and sectors specializing in biotechnologies are interested in selecting aptamers with the greatest sensitivity and selectivity for a variety of targets, particularly secondary metabolites with curative value. With the ubiquitous use of aptamer-modified systems, more researchers and scientists from industrial businesses are realizing their immense practical potential, and we are certain that aptamer-based devices would join the commercial market soon.

REFERENCES

Alibolandi, Mona, Farzin Hadizadeh, Fereshteh Vajhedin, Khalil Abnous, and Mohammad Ramezani. 2015. "Design and Fabrication of an Aptasensor for Chloramphenicol Based on Energy Transfer of CdTe Quantum Dots to Graphene Oxide Sheet." *Materials Science and Engineering C* 48: 611–19. https://doi.org/10.1016/j.msec.2014.12.052.

Arora, Ankita, Murat Sunbul, and Andres Jäschke. 2015. "Dual-Colour Imaging of RNAs Using Quencher- and Fluorophore-Binding Aptamers." *Nucleic Acids Research* 43 (21): 1–9. https://doi.org/10.1093/nar/gkv718.

Bae, Hyunjung, Shuo Ren, Jeehye Kang, Minjung Kim, Yuanyuan Jiang, Moonsoo M. Jin, Irene M. Min, and Soyoun Kim. 2013. "Sol-Gel SELEX Circumventing Chemical Conjugation of Low Molecular Weight Metabolites Discovers Aptamers Selective to Xanthine." *Nucleic Acid Therapeutics* 23 (6): 443–49. https://doi.org/10.1089/nat.2013.0437.

Bai, Jiuyuan, Yao Luo, Xin Wang, Shi Li, Mei Luo, Meng Yin, Yuanli Zuo et al. 2020. "A Protein-Independent Fluorescent RNA Aptamer Reporter System for Plant Genetic Engineering." *Nature Communications* 11 (1): 1–14. https://doi.org/10.1038/s41467-020-17497-7.

Bakhtiari, Hadi, Abbas Ali Palizban, Hossein Khanahmad, and Mohammad Reza Mofid. 2021. "Novel Approach to Overcome Defects of Cell-SELEX in Developing Aptamers Against Aspartate β-Hydroxylase." *ACS Omega* 6 (16): 11005–14. https://doi.org/10.1021/acsomega.1c00876.

Basu, Aviru Kumar, Anup Basak, and Shantanu Bhattacharya. 2020. "Geometry and Thickness Dependant Anomalous Mechanical Behavior of Fabricated SU-8 Thin Film Micro-Cantilevers." *Journal of Micromanufacturing* 3 (2): 113–20. https://doi.org/10.1177/2516598420930988.

Basu, Aviru Kumar, Adreeja Basu, and Shantanu Bhattacharya. 2019. "Study of PH Induced Conformational Change of Papain Using Polymeric Nano-Cantilever." *AIP Conference Proceedings* 2083 (1): 030001. https://doi.org/10.1063/1.5094311.

———. 2020. "Micro/Nano Fabricated Cantilever Based Biosensor Platform: A Review and Recent Progress." *Enzyme and Microbial Technology* 139: 109558. https://doi.org/10.1016/j.enzmictec.2020.109558.

———. 2021. "Recent Trends and Progress in Mems-Based Bioinspired/Biomimetic Sensors." *MEMS Applications in Biology and Healthcare* (December): 2–1. https://doi.org/10.1063/9780735423954_002.

Basu, Aviru Kumar, Amar Nath Sah, Asima Pradhan, and Shantanu Bhattacharya. 2019. "Poly-L-Lysine Functionalised MWCNT-RGO Nanosheets Based 3-d Hybrid Structure for Femtomolar Level Cholesterol Detection Using Cantilever Based Sensing Platform." *Scientific Reports* 9 (1): 3686. https://doi.org/10.1038/s41598-019-40259-5.

Bayat, Payam, Rahim Nosrati, Mona Alibolandi, Houshang Rafatpanah, Khalil Abnous, Mostafa Khedri, and Mohammad Ramezani. 2018. "SELEX Methods on the Road to Protein Targeting with Nucleic Acid Aptamers." *Biochimie* 154: 132–55. https://doi.org/10.1016/j.biochi.2018.09.001.

Bhalla, Nikhil, Pawan Jolly, Nello Formisano, and Pedro Estrela. 2016. "Introduction to Biosensors." *Essays in Biochemistry* 60 (1): 1–8. https://doi.org/10.1042/EBC20150001.

Bhowmik, Sudipta Shekhar Das, Adreeja Basu, and Lingaraj Sahoo. 2016. "Direct Shoot Organogenesis from Rhizomes of Medicinal Zingiber Alpinia Calcarata Rosc. and Evaluation of Genetic Stability by RAPD and ISSR Markers." *Journal of Crop Science and Biotechnology* 19 (2): 157–65. https://doi.org/10.1007/S12892-015-0119-4.

Breaker, Ronald R. 2012. "Riboswitches and the RNA World." *Cold Spring Harbor Perspectives in Biology* 4 (2): a003566. https://doi.org/10.1101/cshperspect.a003566.

Brockman, Luke, Ronghui Wang, Jacob Lum, and Yanbin Li. 2013. "QCM Aptasensor for Rapid and Specific Detection of Avian Influenza Virus." *Open Journal of Applied Biosensor* 02 (04): 97–103. https://doi.org/10.4236/ojab.2013.24013.

Challier, Lylian, François Mavré, Julie Moreau, Claire Fave, Bernd Schöllhorn, Damien Marchal, Eric Peyrin, Vincent Noël, and Benoit Limoges. 2012. "Simple and Highly Enantioselective Electrochemical Aptamer-Based Binding Assay for Trace Detection of Chiral Compounds." *Analytical Chemistry* 84 (12): 5415–20. https://doi.org/10.1021/ac301048c.

Chen, Junhua, and Shungui Zhou. 2016. "Label-Free DNA Y Junction for Bisphenol A Monitoring Using Exonuclease III-Based Signal Protection Strategy." *Biosensors and Bioelectronics* 77: 277–83. https://doi.org/10.1016/j.bios.2015.09.042.

Chen, Xiujuan, Yukun Huang, Nuo Duan, Shijia Wu, Xiaoyuan Ma, Yu Xia, Changqing Zhu, Yuan Jiang, and Zhouping Wang. 2013. "Selection and Identification of SsDNA Aptamers Recognizing Zearalenone." *Analytical and Bioanalytical Chemistry* 405 (20): 6573–81. https://doi.org/10.1007/s00216-013-7085-9.

Chen, Xiujuan, Yukun Huang, Nuo Duan, Shijia Wu, Yu Xia, Xiaoyuan Ma, Changqing Zhu, Yuan Jiang, and Zhouping Wang. 2014. "Screening and Identification of DNA Aptamers against T-2 Toxin Assisted by Graphene Oxide." *Journal of Agricultural and Food Chemistry* 62 (42): 10368–74. https://doi.org/10.1021/jf5032058.

Chen, Zihao, Long Hu, Bao Ting Zhang, Aiping Lu, Yaofeng Wang, Yuanyuan Yu, and Ge Zhang. 2021. "Artificial Intelligence in Aptamer–Target Binding Prediction." *International Journal of Molecular Sciences* 22 (7): 3605. https://doi.org/10.3390/IJMS22073605.

Chinnappan, Raja, Razan AlZabn, Khalid M. Abu-Salah, and Mohammed Zourob. 2019. "An Aptamer Based Fluorometric Microcystin-LR Assay Using DNA Strand-Based Competitive Displacement." *Microchimica Acta* 186 (7): 1–10. https://doi.org/10.1007/s00604-019-3504-8.

Cho, Suhyung, Bo Rahm Lee, Byung Kwan Cho, June Hyung Kim, and Byung Gee Kim. 2013. "In Vitro Selection of Sialic Acid Specific RNA Aptamer and Its Application to the Rapid Sensing of Sialic Acid Modified Sugars." *Biotechnology and Bioengineering* 110 (3): 905–13. https://doi.org/10.1002/bit.24737.

Chung, Eunsu, Jinhyeok Jeon, Jimin Yu, Chankil Lee, and Jaebum Choo. 2015. "Surface-Enhanced Raman Scattering Aptasensor for Ultrasensitive Trace Analysis of Bisphenol A." *Biosensors and Bioelectronics* 64: 560–65. https://doi.org/10.1016/j.bios.2014.09.087.

Costantini, Francesca, Nicola Lovecchio, Albert Ruggi, Cesare Manetti, Augusto Nascetti, Massimo Reverberi, Giampiero De Cesare, and Domenico Caputo. 2019. "Fluorescent Label-Free Aptasensor Integrated in a Lab-on-Chip System for the Detection of Ochratoxin A in Beer and Wheat." *ACS Applied Bio Materials* 2 (12): 5880–87. https://doi.org/10.1021/acsabm.9b00831.

Dalirirad, Shima, Daewoo Han, and Andrew J. Steckl. 2020. "Aptamer-Based Lateral Flow Biosensor for Rapid Detection of Salivary Cortisol." *ACS Omega* 5 (51): 32890–98. https://doi.org/10.1021/acsomega.0c03223.

Dolgosheina, Elena V., Sunny C. Y. Jeng, Shanker Shyam S. Panchapakesan, Razvan Cojocaru, Patrick S. K. Chen, Peter D. Wilson, Nancy Hawkins, Paul A. Wiggins, and Peter J. Unrau. 2014. "RNA Mango Aptamer-Fluorophore: A Bright, High-Affinity Complex for RNA Labeling and Tracking." *ACS Chemical Biology* 9 (10): 2412–20. https://doi.org/10.1021/cb500499x.

Ebrahimi, Mohsen, Mohammad Johari-Ahar, Hossein Hamzciy, Jaleh Barar, Omid Mashinchian, and Yadollah Omidi. 2012. "Electrochemical Impedance Spectroscopic Sensing of Methamphetamine by a Specific Aptamer." *BioImpacts : BI* 2 (2): 91. https://doi.org/10.5681/BI.2012.013.

Eissa, Shimaa, Mohamed Siaj, and Mohammed Zourob. 2015. "Aptamer-Based Competitive Electrochemical Biosensor for Brevetoxin-2." *Biosensors and Bioelectronics* 69: 148–54. https://doi.org/10.1016/j.bios.2015.01.055.

Elshafey, Reda, Mohamed Siaj, and Mohammed Zourob. 2014. "In Vitro Selection, Characterization, and Biosensing Application of High-Affinity Cylindrospermopsin-Targeting Aptamers." *Analytical Chemistry* 86 (18): 9196–203. https://doi.org/10.1021/ac502157g.

———. 2015. "DNA Aptamers Selection and Characterization for Development of Label-Free Impedimetric Aptasensor for Neurotoxin Anatoxin-A." *Biosensors and Bioelectronics* 68: 295–302. https://doi.org/10.1016/j.bios.2015.01.002.

Famulok, Michael, and Günter Mayer. 2014. "Aptamers and SELEX in Chemistry & Biology." *Chemistry and Biology* 21 (9): 1055–58. https://doi.org/10.1016/j.chembiol.2014.08.003.

Feng, Xiaobin, Ning Gan, Huairong Zhang, Qing Yan, Tianhua Li, Yuting Cao, Futao Hu, Hongwei Yu, and Qianli Jiang. 2015. "A Novel 'Dual-Potential' Electrochemiluminescence Aptasensor Array Using CdS Quantum Dots and Luminol-Gold Nanoparticles as Labels for Simultaneous Detection of Malachite Green and Chloramphenicol." *Biosensors and Bioelectronics* 74: 587–93. https://doi.org/10.1016/j.bios.2015.06.048.

Fu, Cuicui, Yi Wang, Gang Chen, Liyuan Yang, Shuping Xu, and Weiqing Xu. 2015. "Aptamer-Based Surface-Enhanced Raman Scattering-Microfluidic Sensor for Sensitive and Selective Polychlorinated Biphenyls Detection." *Analytical Chemistry* 87 (19): 9555–58. https://doi.org/10.1021/acs.analchem.5b02508.

Fu, Jie, Yunjuan Sun, Shaoyou Xia, Lihou Dong, Qingqing Wang, Lun Ou, Xianfei Shen, Zhiguo Lv, and Haifeng Song. 2011. "Enzyme-Linked Bridging Assay Method for the Quantification of Oligonucleotide-Based Drugs in Biological Matrices." *Nucleic Acid Therapeutics* 21 (6): 403–13. https://doi.org/10.1089/nat.2011.0319.

Ge, Huanhuan, and Mario Andrea Marchisio. 2021. "Aptamers, Riboswitches and Ribozymes in s. Cerevisiae Synthetic Biology." *Life* 11 (3): na. https://doi.org/10.3390/life11030248.

Gong, Sheng, Hong Lin Ren, Rui Yun Tian, Chao Lin, Pan Hu, Yan Song Li, Zeng Shan Liu, et al. 2013. "A Novel Analytical Probe Binding to a Potential Carcinogenic Factor of N-Glycolylneuraminic Acid by SELEX." *Biosensors and Bioelectronics* 49: 547–54. https://doi.org/10.1016/j.bios.2013.05.024.

Guan, Weihua, Liben Chen, Tushar D. Rane, and Tza Huei Wang. 2015. "Droplet Digital Enzyme-Linked Oligonucleotide Hybridization Assay for Absolute RNA Quantification." *Scientific Reports* 5 (April): 1–9. https://doi.org/10.1038/srep13795.

Guo, Jiajia, Ying Li, Luokai Wang, Jingyue Xu, Yanjun Huang, Yeli Luo, Fei Shen, Chunyan Sun, and Rizeng Meng. 2016. "Aptamer-Based Fluorescent Screening Assay for Acetamiprid via Inner Filter Effect of Gold Nanoparticles on the Fluorescence of CdTe Quantum Dots." *Analytical and Bioanalytical Chemistry* 408 (2): 557–66. https://doi.org/10.1007/s00216-015-9132-1.

Guo, Xiaodong, Fang Wen, Nan Zheng, Qiujiang Luo, Haiwei Wang, Hui Wang, Songli Li, and Jiaqi Wang. 2014. "Development of an Ultrasensitive Aptasensor for the Detection of Aflatoxin B1." *Biosensors and Bioelectronics* 56: 340–44. https://doi.org/10.1016/j.bios.2014.01.045.

Ha, Tai Hwan. 2015. "Recent Advances for the Detection of Ochratoxin A." *Toxins* 7 (12): 5276–300. https://doi.org/10.3390/toxins7124882.

Han, Seung Ryul, Jaehoon Yu, and Seong Wook Lee. 2014. "In Vitro Selection of RNA Aptamers That Selectively Bind Danofloxacin." *Biochemical and Biophysical Research Communications* 448 (4): 397–402. https://doi.org/10.1016/j.bbrc.2014.04.103.

Handy, Sara M., Betsy Jean Yakes, Jeffrey A. DeGrasse, Katrina Campbell, Christopher T. Elliott, Kelsey M. Kanyuck, and Stacey L. DeGrasse. 2013. "First Report of the Use of a Saxitoxin-Protein Conjugate to Develop a DNA Aptamer to a Small Molecule Toxin." *Toxicon* 61 (1): 30–37. https://doi.org/10.1016/j.toxicon.2012.10.015.

Heredia, Frances L., Abiel Roche-Lima, and Elsie I. Parés-Matos. 2021. "A Novel Artificial Intelligence-Based Approach for Identification of Deoxynucleotide Aptamers." *PLOS Computational Biology* 17 (8): e1009247. https://doi.org/10.1371/JOURNAL. PCBI.1009247.

Hianik, T. 2018. *Aptamer-Based Biosensors. Encyclopedia of Interfacial Chemistry: Surface Science and Electrochemistry*. Elsevier. https://doi.org/10.1016/B978-0-12-409547-2.13492-4.

Hosseinzadeh, Laleh, and Mohammad Mazloum-Ardakani. 2020. *Advances in Aptasensor Technology. Advances in Clinical Chemistry*. 1st ed. Vol. 99. Elsevier Inc. https://doi. org/10.1016/bs.acc.2020.02.010.

Huang, Lei, Shulin Tian, Wenhao Zhao, Ke Liu, Xing Ma, and Jinhong Guo. 2021. "Aptamer-Based Lateral Flow Assay on-Site Biosensors." *Biosensors and Bioelectronics* 186 (March): 113279. https://doi.org/10.1016/j.bios.2021.113279.

Jauset-Rubio, Miriam, Mohammad S. El-Shahawi, Abdulaziz S. Bashammakh, Abdulrahman O. Alyoubi, and Ciara K. O'Sullivan. 2017. "Advances in Aptamers-Based Lateral Flow Assays." *TrAC - Trends in Analytical Chemistry* 97: 385–98. https://doi.org/10.1016/j. trac.2017.10.010.

Jepsen, Mette D.E., Steffen M. Sparvath, Thorbjørn B. Nielsen, Ane H. Langvad, Guido Grossi, Kurt V. Gothelf, and Ebbe S. Andersen. 2018. "Development of a Genetically Encodable FRET System Using Fluorescent RNA Aptamers." *Nature Communications* 9 (1): 1–10. https://doi.org/10.1038/s41467-017-02435-x.

Joeng, Choon Bok, Javed H. Niazi, Su Jin Lee, and Man Bock Gu. 2009. "SsDNA Aptamers That Recognize Diclofenac and 2-Anilinophenylacetic Acid." *Bioorganic and Medicinal Chemistry* 17 (15): 5380–87. https://doi.org/10.1016/J.BMC.2009.06.044.

Kaiser, Lars, Julia Weisser, Matthias Kohl, and Hans Peter Deigner. 2018. "Small Molecule Detection with Aptamer Based Lateral Flow Assays: Applying Aptamer-C-Reactive Protein Cross-Recognition for Ampicillin Detection." *Scientific Reports* 8 (1): 6–15. https://doi.org/10.1038/s41598-018-23963-6.

Kalra, Priya, Abhijeet Dhiman, William C. Cho, John G. Bruno, and Tarun K. Sharma. 2018. "Simple Methods and Rational Design for Enhancing Aptamer Sensitivity and Specificity." *Frontiers in Molecular Biosciences* 5 (May): 1–16. https://doi.org/10.3389/ fmolb.2018.00041.

Khosravi, Solmaz, Patrick Schindele, Evgeny Gladilin, Frank Dunemann, Twan Rutten, Holger Puchta, and Andreas Houben. 2020. "Application of Aptamers Improves CRISPR-Based Live Imaging of Plant Telomeres." *Frontiers in Plant Science* 11 (August): 1–13. https:// doi.org/10.3389/fpls.2020.01254.

Li, Xiaoxia, Lihua Shen, Dongdong Zhang, Honglan Qi, Qiang Gao, Fen Ma, and Chengxiao Zhang. 2008. "Electrochemical Impedance Spectroscopy for Study of Aptamer–Thrombin Interfacial Interactions." *Biosensors and Bioelectronics* 23 (11): 1624–30. https://doi.org/10.1016/J.BIOS.2008.01.029.

Liu, Jiaxin, Qiwei Qin, Xinyue Zhang, Chen Li, Yepin Yu, Xiaohong Huang, Omar Mukama, Lingwen Zeng, and Shaowen Wang. 2020. "Development of a Novel Lateral Flow Biosensor Combined With Aptamer-Based Isolation: Application for Rapid Detection of Grouper Nervous Necrosis Virus." *Frontiers in Microbiology* 11 (May): 1–10. https:// doi.org/10.3389/fmicb.2020.00886.

Liu, Lan hua, Xiao hong Zhou, and Han chang Shi. 2015. "Portable Optical Aptasensor for Rapid Detection of Mycotoxin with a Reversible Ligand-Grafted Biosensing Surface." *Biosensors and Bioelectronics* 72: 300–05. https://doi.org/10.1016/j.bios.2015.05.033.

Liu, Lubin, Zeyu Han, Fei An, Xuening Gong, Chenguang Zhao, Weiping Zheng, Li Mei, and Qihui Zhou. 2021. "Aptamer-Based Biosensors for the Diagnosis of Sepsis." *Journal of Nanobiotechnology* 19 (1): 1–22. https://doi.org/10.1186/s12951-021-00959-5.

Lyu, Yifan, Guang Chen, Dihua Shangguan, Liqin Zhang, Shuo Wan, Yuan Wu, Hui Zhang, et al. 2016. "Generating Cell Targeting Aptamers for Nanotheranostics Using Cell-SELEX." *Theranostics* 6 (9): 1440–52. https://doi.org/10.7150/thno.15666.

Maduraiveeran, Govindhan, and Wei Jin. 2019. *Functional Nanomaterial-Derived Electrochemical Sensor and Biosensor Platforms for Biomedical Applications. Handbook of Nanomaterials in Analytical Chemistry: Modern Trends in Analysis.* Elsevier Inc. https://doi.org/10.1016/B978-0-12-816699-4.00012-8.

McKeague, Maureen, and Maria C. Derosa. 2012. "Challenges and Opportunities for Small Molecule Aptamer Development." *Journal of Nucleic Acids*: 1–20. https://doi.org/10.1155/2012/748913.

Moreno, Miguel. 2019. "Sensors | Aptasensors." *Encyclopedia of Analytical Science* 3: 150–53. https://doi.org/10.1016/B978-0-12-409547-2.13934-4.

Ohuchi, Shoji. 2012. "Cell-Selex Technology." *BioResearch Open Access* 1 (6): 265–72. https://doi.org/10.1089/biores.2012.0253.

Paul, Anamika, Ganesh Thapa, Adreeja Basu, Purabi Mazumdar, Mohan Chandra Kalita, and Lingaraj Sahoo. 2010. "Rapid Plant Regeneration, Analysis of Genetic Fidelity and Essential Aromatic Oil Content of Micropropagated Plants of Patchouli, Pogostemon Cablin (Blanco) Benth. - An Industrially Important Aromatic Plant." *Industrial Crops and Products* 32 (3): 366–74. https://doi.org/10.1016/j.indcrop.2010.05.020.

Pehlivan, Zeki Semih, Milad Torabfam, Hasan Kurt, Cleva Ow-Yang, Niko Hildebrandt, and Meral Yüce. 2019. "Aptamer and Nanomaterial Based FRET Biosensors: A Review on Recent Advances (2014–2019)." *Microchimica Acta* 186 (8): 1–22. https://doi.org/10.1007/s00604-019-3659-3.

Pfeiffer, Franziska, and Günter Mayer. 2016. "Selection and Biosensor Application of Aptamers for Small Molecules." *Frontiers in Chemistry* 4 (June): 25. https://doi.org/10.3389/FCHEM.2016.00025/BIBTEX.

Phung, Ngoc Linh, Johanna G. Walter, Rebecca Jonczyk, Lisa K. Seiler, Thomas Scheper, and Cornelia Blume. 2020. "Development of an Aptamer-Based Lateral Flow Assay for the Detection of C-Reactive Protein Using Microarray Technology as a Prescreening Platform." *ACS Combinatorial Science* 22 (11): 617–29. https://doi.org/10.1021/acscombsci.0c00080.

Pietrantonio, Fabio Di, Domenico Cannatà, and Massimiliano Benetti. 2019. *Biosensor Technologies Based on Nanomaterials. Functional Nanostructured Interfaces for Environmental and Biomedical Applications.* Elsevier Inc. https://doi.org/10.1016/B978-0-12-814401-5.00008-6.

Pleiko, Karlis, Liga Saulite, Vadims Parfejevs, Karlis Miculis, Egils Vjaters, and Una Riekstina. 2019. "Differential Binding Cell-SELEX Method to Identify Cell-Specific Aptamers Using High-Throughput Sequencing." *Scientific Reports* 9 (1): 1–12. https://doi.org/10.1038/s41598-019-44654-w.

Pothoulakis, Georgios, and Tom Ellis. 2015. *Using Spinach Aptamer to Correlate MRNA and Protein Levels in Escherichia Coli. Methods in Enzymology.* 1st ed. Vol. 550. Elsevier Inc. https://doi.org/10.1016/bs.mie.2014.10.047.

Purkayastha, J., T. Sugla, A. Paul, S. K. Solleti, P. Mazumdar, A. Basu, A. Mohommad, Z. Ahmed, and L. Sahoo. 2010. "Efficient in Vitro Plant Regeneration from Shoot Apices and Gene Transfer by Particle Bombardment in Jatropha Curcas." *Biologia Plantarum* 54 (1): 13–20. https://doi.org/10.1007/S10535-010-0003-5.

Ray, Partha, and Rebekah R. White. 2017. "Cell-SELEX Identifies a 'Sticky' RNA Aptamer Sequence." *Journal of Nucleic Acids* 2017: 1–9. https://doi.org/10.1155/2017/4943072.

Rodriguez, Marcela C., Abdel Nasser Kawde, and Joseph Wang. 2005. "Aptamer Biosensor for Label-Free Impedance Spectroscopy Detection of Proteins Based on Recognition-Induced Switching of the Surface Charge." *Chemical Communications* 34: 4267–69. https://doi.org/10.1039/b506571b.

Rogers, Tucker A., Grant E. Andrews, Luc Jaeger, and Wade W. Grabow. 2015. "Fluorescent Monitoring of RNA Assembly and Processing Using the Split-Spinach Aptamer." *ACS Synthetic Biology* 4 (2): 162–66. https://doi.org/10.1021/sb5000725.

Rozenblum, Guido T., Ivan G. Pollitzer, and Martin Radrizzani. 2019. "Challenges in Electrochemical Aptasensors and Current Sensing Architectures Using Flat Gold Surfaces." *Chemosensors* 7 (4): 57. https://doi.org/10.3390/chemosensors7040057.

Schneider, Christopher, and Beatrix Suess. 2016. "Identification of RNA Aptamers with Riboswitching Properties." *Methods* 97 (December 2015): 44–50. https://doi.org/10.1016/j.ymeth.2015.12.001.

Schoukroun-Barnes, Lauren R., Samuillah Wagan, and Ryan J. White. 2014. "Enhancing the Analytical Performance of Electrochemical RNA Aptamer-Based Sensors for Sensitive Detection of Aminoglycoside Antibiotics." *Analytical Chemistry* 86 (2): 1131–37. https://doi.org/10.1021/AC4029054/SUPPL_FILE/AC4029054_SI_001.PDF.

Shoeib, Sabry Abdallah, Alaa Efat Abd Elhamid, Enas Sobhi Zahran, and Ahmed Abd Elmottlep Elkalashy. 2020. "Aptamers from Biology to Find Solutions in Immunohematology." *Internal Medicine: Open Access* 10 (3): 10–14. https://doi.org/10.35248/2165-8048.20.10.315.

Singh, Pankhi, Vibhas Chugh, Antara Banerjee, Surajit Pathak, Sudeep Bose, and Ranu Nayak. 2023. "Nanomaterials: Compatibility Towards Biological Interactions. Practical Approach to Mammalian Cell and Organ Culture." In *Practical Approach to Mammalian Cell and Organ Culture*. Springer Nature Singapore: 1059–89. https://doi.org/10.1007/978-981-19-1731-8_19-1.

Sypabekova, Marzhan, Aliya Bekmurzayeva, Ronghui Wang, Yanbin Li, Claude Nogues, and Damira Kanayeva. 2017. "Selection, Characterization, and Application of DNA Aptamers for Detection of Mycobacterium Tuberculosis Secreted Protein MPT64." *Tuberculosis* 104: 70–78. https://doi.org/10.1016/j.tube.2017.03.004.

Tabrizi, Mahmoud Amouzadeh, and Pablo Acedo. 2022. "An Electrochemical Impedance Spectroscopy-Based Aptasensor for the Determination of SARS-CoV-2-RBD Using a Carbon Nanofiber-Gold Nanocomposite Modified Screen-Printed Electrode." *Biosensors* 12(3): 142. https://doi.org/10.3390/bios12030142.

Torrini, Francesca, Pasquale Palladino, Alvaro Brittoli, Veronica Baldoneschi, Maria Minunni, and Simona Scarano. 2019. "Characterization of Troponin T Binding Aptamers for an Innovative Enzyme-Linked Oligonucleotide Assay (ELONA)." *Analytical and Bioanalytical Chemistry* 411 (29): 7709–16. https://doi.org/10.1007/s00216-019-02014-7.

Torshizi, Ramin, Mahdi Zeinoddini, Ali-Asghar Deldar, and Seyed-Mortaza Robatjazi. 2016. "Design of Aptamer-Based Detector for Trinitrotoluene (TNT) and Review of Its Performance." *International Journal of Advanced Biotechnology and Research* 7 (April 2017): 976–2612.

Wang, Yuan Kai, Qi Zou, Jian He Sun, Heng An Wang, Xingmin Sun, Zhi Fei Chen, and Ya Xian Yan. 2015. "Screening of Single-Stranded DNA (SsDNA) Aptamers against a Zearalenone Monoclonal Antibody and Development of a SsDNA-Based Enzyme-Linked Oligonucleotide Assay for Determination of Zearalenone in Corn." *Journal of Agricultural and Food Chemistry* 63 (1): 136–41. https://doi.org/10.1021/jf503733g.

Wolter, Olga, and Günter Mayer. 2017. "Aptamers as Valuable Molecular Tools in Neurosciences." *Journal of Neuroscience* 37 (10): 2517–23. https://doi.org/10.1523/JNEUROSCI.1969-16.2017.

Xu, Jiayao, Wenting Chen, Ming Shi, Yong Huang, Lina Fang, Shulin Zhao, Lifang Yao, and Hong Liang. 2019. "An Aptamer-Based Four-Color Fluorometic Method for Simultaneous Determination and Imaging of Alpha-Fetoprotein, Vascular Endothelial Growth Factor-165, Carcinoembryonic Antigen and Human Epidermal Growth Factor Receptor 2 in Living Cells." *Microchimica Acta* 186 (3): 2–11. https://doi.org/10.1007/s00604-019-3312-1.

Yadav, Gulab Singh, Abhishek Parashar, and Neeraj K. Aggarwal. 2019. "Aptamer: A next Generation Tool for Application in Agricultural Industry for Food Safety." *Aptamers: Biotechnological Applications of a Next Generation Tool* (January): 175–86. https://doi.org/10.1007/978-981-13-8836-1_12/COVER.

Yokobayashi, Yohei. 2019. "Aptamer-Based and Aptazyme-Based Riboswitches in Mammalian Cells." *Current Opinion in Chemical Biology* 52: 72–78. https://doi.org/10.1016/j.cbpa.2019.05.018.

Yoo, Hyebin, Hyesung Jo, and Seung Soo Oh. 2020. "Detection and beyond: Challenges and Advances in Aptamer-Based Biosensors." *Materials Advances* 1 (8): 2663–87. https://doi.org/10.1039/d0ma00639d.

Yue, Fengling, Falan Li, Qianqian Kong, Yemin Guo, and Xia Sun. 2021. *Recent Advances in Aptamer-Based Sensors for Aminoglycoside Antibiotics Detection and Their Applications. Science of the Total Environment.* Vol. 762. Elsevier B.V. https://doi.org/10.1016/j.scitotenv.2020.143129.

Zhong, Yi, Jiayao Zhao, Jiazhao Li, Xin Liao, and Fengling Chen. 2020. "Advances of Aptamers Screened by Cell-SELEX in Selection Procedure, Cancer Diagnostics and Therapeutics." *Analytical Biochemistry* 598 (February): 113620. https://doi.org/10.1016/j.ab.2020.113620.

Zhou, Wenhu, Po Jung Jimmy Huang, Jinsong Ding, and Juewen Liu. 2014. "Aptamer-Based Biosensors for Biomedical Diagnostics." *Analyst* 139 (11): 2627–40. https://doi.org/10.1039/c4an00132j.

Zhou, Wenhu, Yanbin Zhou, Jianping Wu, Zhenbao Liu, Huanzhe Zhao, Juewen Liu, and Jinsong Ding. 2014. "Aptamer-Nanoparticle Bioconjugates Enhance Intracellular Delivery of Vinorelbine to Breast Cancer Cells." *Journal of Drug Targeting* 22 (1): 57–66. https://doi.org/10.3109/1061186X.2013.839683.

Zhuo, Zhenjian, Yuanyuan Yu, Maolin Wang, Jie Li, Zongkang Zhang, Jin Liu, Xiaohao Wu, Aiping Lu, Ge Zhang, and Baoting Zhang. 2017. "Recent Advances in SELEX Technology and Aptamer Applications in Biomedicine." *International Journal of Molecular Sciences* 18 (10): 1–19. https://doi.org/10.3390/ijms18102142.

Zou, Yi, Sophie Griveau, Armelle Ringuedé, Fethi Bedioui, Cyrille Richard, and Cyrine Slim. 2022. "Functionalized Multi-Walled Carbon Nanotube–Based Aptasensors for Diclofenac Detection." *Frontiers in Chemistry* 9 (January): 1237. https://doi.org/10.3389/FCHEM.2021.812909/BIBTEX.

12 Challenges of implementing impedance spectroscopy

Sagnik Sarma Choudhury, Kapil Manoharan,
Nitish Katiyar, and Ranamay Saha

CONTENTS

12.1 INTRODUCTION

Impedance is the resistance to the flow of current, which causes overpotential. Degradation of materials by aging and imperfect electrical contacts, biomolecular binding events, and corrosion results in an impedance change. The quantitative analysis of this impedance change and correlating it to the event resulting in the change is the scope of electrochemical impedance spectroscopy (EIS). In EIS, either a sinusoidal current can be applied and the resulting voltage is measured (galvanostatic mode) or a sinusoidal voltage can be applied and the resulting current can be measured (potentiostatic mode). In either case, a specific frequency and amplitude of the input signal are applied and the output voltage or current with its characteristic phase shift and amplitude are measured as shown in Figure 12.1 (Feliu 2020; Li et al. 2020). The impedance spectrum is then generated by repeating the process in a range of frequencies varying typically from mHz to kHz. The EIS data are commonly represented as Nyquist plot [Real impedance (Z_{Re}) vs. imaginary impedance (Z_{im})] for a range of excitation frequency and Bode plot [logarithm of impedance modulus (|Z|) and phase angle (θ) vs. applied frequency]. The data obtained from EIS is fitted into an equivalent electrical circuit model and studied. Modeling of the EIS data involves fitting them to an analogous electrical circuit made up of resistors (R), capacitors (C) or constant phase elements, inductors (L), and Warburg impedance (W) linked either in series or parallel (Kirkland, Birbilis, and Staiger 2012; Jia et al. 2019). However,

DOI: 10.1201/9781003358091-15

there are limitations in the method as the obtained data from EIS can fit into several equivalent circuit models. Thus, one should have a comprehensive understanding of the underlying processes that result in the obtained EIS data, so that the choice of an equivalent circuit model is made judiciously (Noori et al. 2019). A few points that need to be considered before performing EIS measurements are:

- The type of the sample (solid state or electrochemical).
- The type of measurement (two-, three-, or four-point/electrode measurement).
- The selection of the appropriate frequency window.
- The selection of the proper current range.
- The selection of the proper DC voltage value.
- The sensitivity of the measurement is required.

A three-electrode geometry is the preferred setup for electrochemical systems like energy storage devices and corrosion. For solid-state devices like photovoltaic cells, a two- or four-point setup is used. The frequency window range comes from the time scales of the underlying relaxation processes to be studied and the sample structure/conductivity determines the range of current. The DC voltage value is selected based on the potential range at which the relaxation processes are to be studied. For the proper execution of the EIS technique, the system under study must satisfy some characteristics, viz. causality, stability, consistency, and linearity while taking the measurements (Orazem 2008; Barsukov 2012). When the response is generated by something other than the input, such as a concentration, current, or potential relaxation after the system is disturbed from equilibrium, deviation from causality occurs.

A system's stability is typically not assured, particularly when dynamic electrochemical processes involving batteries or corrosion are studied (Barsukov 2012). The stability of a system can be defined as the retrieval of the original state when the disturbing factors are removed. In practice, it is not possible to have a perfectly stable

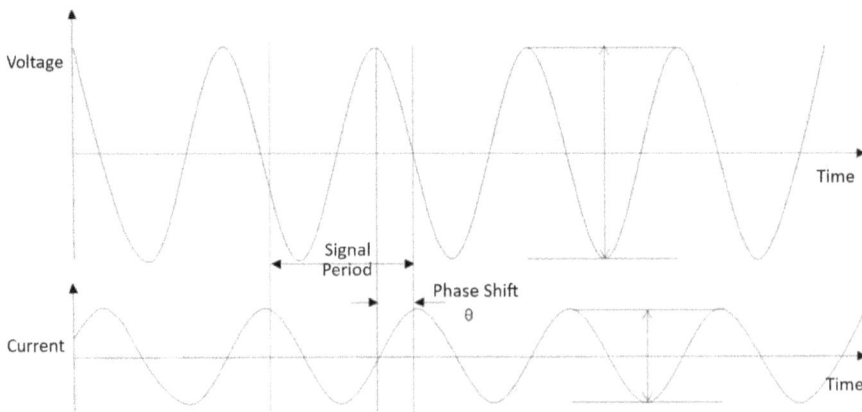

FIGURE 12.1 Representation of input voltage and corresponding output current in EIS. [Reprinted with permission from Feliu (2020). Copyright (2020) MDPI].

system for an infinite time duration. However, for quantitative purpose, the measurement time can be reduced to get a quasi-stable system. For the conditions of causality, stability, consistency, and linearity to be satisfied, a preliminary verification can be done using the Kramers–Kronig relations given by Eq. (12.1).

$$Z_{Im}(\omega) = \frac{2\omega}{\pi} \int_0^\infty \frac{Z_{Re}(x) - Z_{Re}(\omega)}{x^2 - \omega^2} dx$$

$$Z_{Re}(\omega) = Z_{Re}(\infty) + \frac{2}{\pi} \int_0^\infty \frac{xZ_{Im}(x) - \omega Z_{Im}(\omega)}{x^2 - \omega^2} dx \qquad (12.1)$$

where Z_{Re} denotes the real part, Z_{Im} the imaginary part of impedance and x is an independent variable representing the complex frequency. Many researchers use the Kramers–Kronig equations to validate the consistency of impedance data in a certain frequency range. This is done by experimentally measuring the impedance of a system under study and then either the real part or the imaginary part of the experimental data is fed into one of the equations given. The obtained value of the other part of the equation is then compared with the experimental data. Differences in the measured and calculated values are indicative of a failure in fulfillment of one of the conditions of stability, causality, or linearity. Another method to validate the impedance data is by comparing the spectrum of response amplitude with the spectrum of perturbation, where the appearance of strong nonexcited frequencies indicates distortion of data (Van Gheem et al. 2004). A disadvantage of the Kramers–Kronig relation is that often the range of frequency is not adequately large to be integrated over the limits of zero to infinity. To circumvent this problem, impedance data extrapolation is performed, which is useful in cases where there is a single time constant. Hirschorn and Orazem (2009) showed the dependency of the amplitude of AC potential perturbation, bulk material impedance, and the applied frequency range on the sensitivity of Kramers–Kronig relation to nonlinearity. They found that there exists a transition frequency 'f_T'given by Eq. (12.2) above which the Kramers–Kronig relations are not satisfied for high amplitude AC potential perturbations.

$$f_T = \frac{1}{2\pi R_{CT} C_{DL}} \left[1 + \frac{R_{CT}}{R_{SOL}} \right] \qquad (12.2)$$

where R_{SOL} is the bulk material or solution impedance, R_{CT} is the interfacial or charge-transfer impedance, and C_{DL} is the double-layer capacitance. From Eq. (12.3), it can be seen that the transition frequency is dependent on the bulk impedance. This implies that for samples with higher conductivity, the transition frequency will acquire a very high value permitting the Kramers–Kronig relationship to be satisfied in a very broad range of frequencies even for high-magnitude potentials. Thus, for a consistent analysis by EIS, it is needed that the Kramers–Kronig relation be fulfilled, which can be achieved by low-magnitude applied voltage and a high ratio of R_{CT}/R_{SOL}.

12.2   CHALLENGES AND LIMITATIONS OF EIS IN VARIOUS DOMAINS

12.2.1   ENERGY STORAGE DEVICES

EIS finds extensive application in the electrochemical characterization of energy storage devices (ESD) like batteries, supercapacitors, and fuel cells owing to its non-invasive nature. The impedance of an ESD is a critical factor that influences the power density, rate capability, and other important parameters. Generally, an ESD will have electrodes, electrolytes, current collectors, and a separator. Each component will have a characteristic response to the externally applied potential. The time constant of undergoing physiochemical processes that result in energy storage can be analyzed in detail with EIS by tuning the applied frequency.

EIS measurement in batteries is a challenging task as batteries are intrinsically nonlinear systems, that is, the current and voltage do not show a linear relationship while EIS studies are based on linearization. Figure 12.2a shows the deviation from linearity for electrochemical devices at certain voltages. Adjusting the amplitude of the sinusoidal signal to be small enough to get quasilinear conditions can typically solve the nonlinearity issue in the case of AC voltage perturbation. A small AC voltage of 1–10 mV magnitude is often provided to the cell during standard EIS procedures that correspond to a narrow linear region in the current vs. voltage plot of the device. However, this condition will cause a negative impact on the signal–noise ratio if the impedance of the device under study is high. Faradaic processes in batteries tend to show nonlinearity at low frequencies. The kinetics of the charge-transfer reaction in a battery is governed by the Volmer–Butler equation as given in Eq. (12.3).

$$i = i_0 \left[e^{\frac{\alpha z F (V - V_{EQ})}{R_G T}} - e^{-\frac{(1-\alpha) z F (V - V_{EQ})}{R_G T}} \right] \tag{12.3}$$

where V is the applied potential, V_{EQ} is the electrode equilibrium potential, i_0 is the exchange current density, R_G is the gas constant, F is the Faraday constant, z is the number of electrons involved, T is the absolute temperature in K, and α the reaction order. As can be seen from Eq. (12.1), there is an exponential dependence of the reaction current with the interfacial potential. This is because of the diffusion limitations of the ions in the electrolyte, and this effect becomes significant at lower frequencies and DC voltages, which deviates the cell from its linear behavior. A general approach for avoiding this type of nonlinearity could be by carrying out EIS studies with low overpotentials so that the exponential effect is nullified (Vadim F. Lvovich 2012). The electrode typically functions linearly at higher frequencies and nonlinearly at lower frequencies in electrochemical systems (Orjan G. Martinsen 2000).

In nonlinear systems, harmonics of the excitation frequency are generally present in the current response. This phenomenon has been well utilized by some researchers to explore higher harmonic responses in the impedance data and create nonlinear impedance data. Thus the system's linearity can be inferred from the existence or absence of strong harmonic responses as a linear system should not produce harmonics (Jafar, Dawson, and John 1993; Pettit et al. 2002; Giner-Sanz, Ortega, and Pérez-Herranz 2016). An ideal Nyquist plot for a commercial lithium-ion battery is depicted in Figure 12.2b. In the high-frequency region, the plot shows an inductor-like behavior that might be due to interference from the connecting wires between the cell and the device being measured or it might also be due to the cell shape or windings in the cell configuration (Momma et al. 2012; Landinger, Schwarzberger, and Jossen 2019). It is therefore paramount to analyze these scenarios and calibrate the measuring device from such background noises. EIS carried out in a two-electrode setup poses a difficulty in the behavioral study of a single electrode. The EIS spectrum will show curves in the mid-frequency range that will be a combined response from both the electrode's interfacial electrochemical reactions (Vetter et al. 2005). An attempt to tackle this issue is to form a three-electrode setup by introducing a third reference electrode. This will enable one to separate the response from the cathode and anode but will add complexity to data interpretation.

The cell impedance depends on the state of charge (SOC) and state of health (SOH), so it can also give an estimation of these parameters. However, the process is complicated necessitating careful experiment, measurement, and analysis design (Rumpf, Naumann, and Jossen 2017). Additionally, the degree of aging may alter the correlations between impedance and operational parameters, which must be taken into consideration while modeling (Waag, Käbitz, and Sauer 2013). Decoupling the total cell impedance data and obtaining individual cell components or process data is a more reliable method for SOC and SOH estimation (Jungst et al. 2003; Dai, Jiang, and Wei 2018; Zhu et al. 2020).

Supercapacitors are known for their fast charge–discharge rate with higher-energy density than conventional capacitors. EIS is a major electrochemical characterization technique for supercapacitors. Although it is a robust technique to separately study the various physiochemical processes it shows a few discrepancies under certain conditions when compared with other electrochemical characterization techniques like

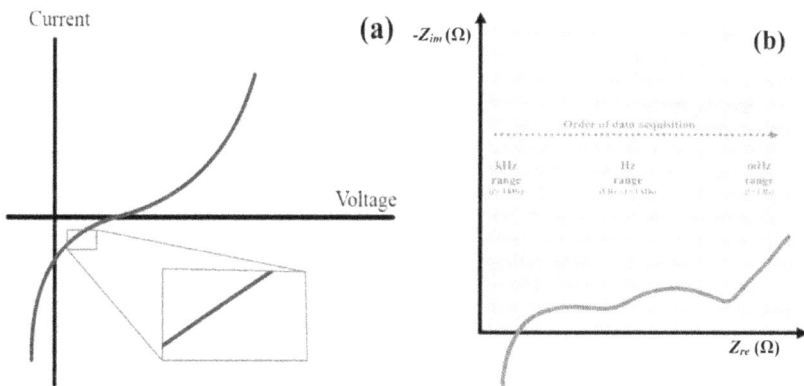

FIGURE 12.2   (a) Pseudolinear behavior in current vs. voltage plot; (b) Nyquist plot of a lithium-ion battery. [Reprinted with permission from Meddings et al. (2020). Copyright (2020) Elsevier].

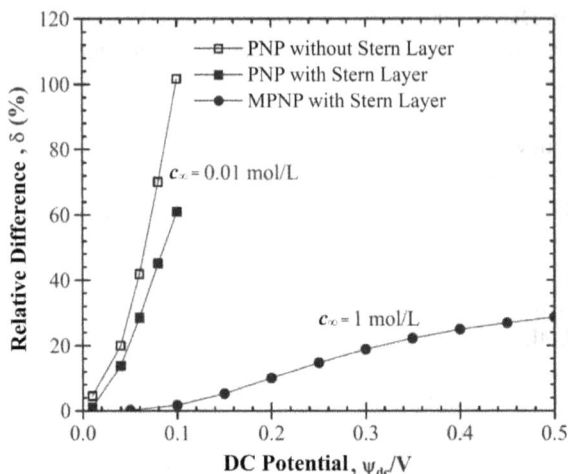

FIGURE 12.3 Relative error between EIS capacitance and electrical double-layer capacitance with increasing DC potential. [Reprinted with permission from Wang and Pilon (2012). Copyright (2012) Elsevier].

cyclic voltammetry and galvanostatic charge–discharge. It has been observed that for electrolytes with dilute concentration, EIS measurement estimated a greater value of electrical double-layer capacitance and a lower value when the electrolyte is concentrated (Ren and Pickup 1992; Kierzek et al. 2004; Fuertes et al. 2005; Zheng et al. 2010; Branzoi and Branzoi 2015). Wang and Pilon (2012) carried out a simulation of EIS measurement on electrical double layers in a planar electrode dipped in an aqueous electrolyte. They used the classical Poisson–Nernst–Planck model and a modified version of the same model for the simulation work. Their results matched the trend in the discrepancy aforementioned. Further, they found that the relative error between capacitance measured by EIS and the double-layer capacitance was increasing with the applied DC potential as shown in Figure 12.3; however, the modified Poisson–Nernst–Planck model gave better results. They concluded that the reason for this discrepancy is the inconsistency of the RC circuit that was used to model the electrical double layer.

12.2.2 Corrosion Study

Measuring the corrosion rate of metals and their alloys is a crucial task for estimating the lifespan of metal-based structures and comparatively assessing their ability to resist corrosion. Among the other techniques like weight loss, H_2 evolution, and polarization curves, EIS is capable of measuring very small rates of corrosion. EIS gives a quantitative analysis and a deeper understanding of the kinetics and mechanics of the corrosion process (Gonzalez-Garcia, Garcia, and Mol 2016; Aparicio and Mosa 2018). Metals like Mg and their alloys that find extensive industrial applications for their high strength-to-weight ratio but are chemically active to corrosion are widely studied using EIS (Esmaily et al. 2017; Choudhury et al. 2020; Feliu

2020; Lakshmi et al. 2020). Reliable estimation of corrosion rates requires measurement during the entire exposure duration as the corrosion rates often tend to vary until a steady state is reached. Corrosion is a slow and complex electrochemical process that involves the formation of adsorbed layers, diffusion, and activation. These slow processes can be located in the low-frequency region of the EIS spectrum (Amirudin and Thieny 1995; Cesiulis et al. 2016). The corrosion resistance measured will be inversely proportional to the rate of corrosion. Additional information such as the dielectric properties of the oxide layer and double-layer formation can also be obtained (Rodrigo de Siqueira Melo et al. 2016).

The major challenge of the EIS technique is the data being equivocal and fitting multiple equivalent circuit models. A comprehensive understanding is therefore much needed for equivalent circuit selection. An alternative way to validate the circuit model could be by comparing the rates of corrosion measured by EIS with the values measured by other techniques. Delgado et al. studied the corrosion behavior of commercial AZ31 alloy by adding various concentrations of aluminum. They validated the EIS data by comparing it with other methods such as hydrogen evolution and weight loss method as shown in Figure 12.4 (Delgado et al. 2017).

Another challenge of EIS in corrosion measurements is the resolution of the technique. As the EIS data is an averaged value over the entire surface of the metal; analysis and study of localized phenomena and micro-defects is difficult (Moreto et al. 2014; Cesiulis et al. 2016). It is also a prerequisite to have the Tafel slope and Stern–Geary coefficient data to be able to transform the polarization resistance and measure the rates of corrosion (Lorenz and Mansfeld 1981). Dispersion in impedance

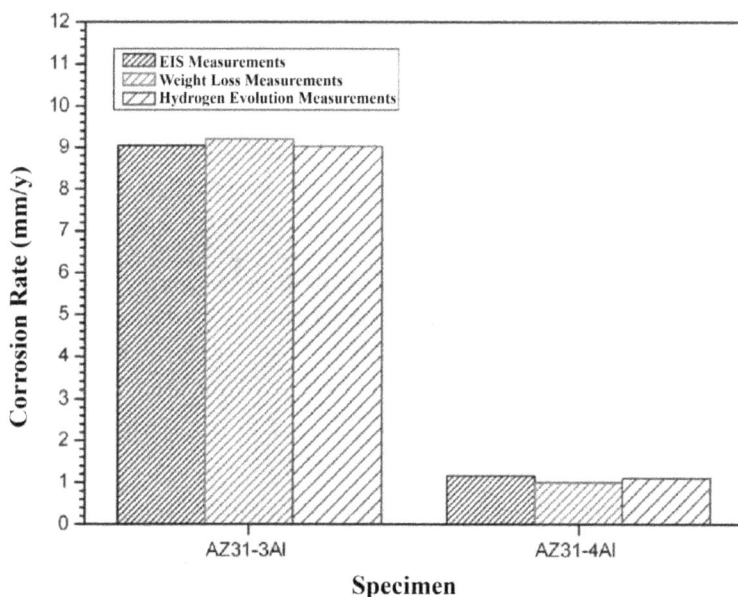

FIGURE 12.4 A comparison of corrosion rates obtained by various methods. [Reprinted with permission from Delgado et al. (2017). Copyright (2017) Elsevier].

values could be seen at the low frequencies when the corrosion process involves instabilities such as pitting corrosion. Furthermore, unlike polarization curves, EIS is unable to identify variations in alloying components, microstructural characteristics (such as secondary phases), or solutions that affect corrosion potentials and relative anodic and cathodic reaction kinetics (Kirkland, Birbilis, and Staiger 2012; Nicholas Travis Kirkland and Birbilis 2014).

Some equivalent electric circuit models that are commonly used for fitting the EIS data are shown in Figure 12.5. The corrosion process dominated by activation control would lead to a semicircular Nyquist plot as shown in Figure 12.5a. A simple Randles circuit with a resistor (R_2) for the charge-transfer resistance in parallel with the oxide layer capacitance (C_1) and another resistor (R_1) in series for solution resistance is used as the model equivalent circuit to fit the EIS data. The thickness of the oxide layer (d) can be determined from the capacitance value using Eq. (12.4).

$$d = \frac{\epsilon A}{C1} \qquad (12.4)$$

where ϵ is the dielectric constant of the oxide layer and A is the exposed area.
In reality, corrosion is a complex process and additional elements need to be taken into consideration while modeling. At the microscopic level, metal surfaces are nonuniform having porosity, roughness, and inhomogeneous surface that does not fit into the assumption of a flat-surfaced capacitor. Thus, a depressed semicircle (Figure 12.5b) is obtained in the Nyquist plot that needs to be treated considering a constant phase element (CPE_1) rather than a capacitor. The impedance of the constant phase element is given by Eq. (12.5).

$$Z_{CPE} = \frac{1}{Y(j\omega)^{\alpha}} \qquad (12.5)$$

where Y is a parameter pertaining to electrode capacitance and α is a coefficient between 0 and 1. The mechanism of the corrosion process is not only limited to the charge-transfer process but also the diffusion of the charged ions through the corrosion layer. This diffusion process manifests itself as a straight line with a 45° magnitude slope in the low-frequency region of the Nyquist plot (Fekry and Fatayerji 2009; Liu, Curioni, and Liu 2018). To model this diffusion phenomenon, an element known as the Warburg diffusional impedance element must be incorporated as shown in Figure 12.5c. The Warburg impedance is given by Eq. (12.6).

$$Z_w = \frac{A_w}{\sqrt{i\omega}} = \frac{A_w}{\sqrt{2\omega}} - i\frac{A_w}{\sqrt{2\omega}} \qquad (12.6)$$

where A_w is the Warburg coefficient, and it can be determined from the diffusion coefficient of ions. An inductive loop may appear at the low-frequency region in the Nyquist plot that is related to the pitting-type corrosion that can be linked with adsorption or desorption of intermediate products on the surface of the metal (Brett et al. 2006; Chang et al. 2007; Coy et al. 2010). The equivalent circuit model for this case is shown in Figure 12.5d. The formation of protective corrosion layers or coatings can result in EIS data with multiple capacitive loops at different frequency

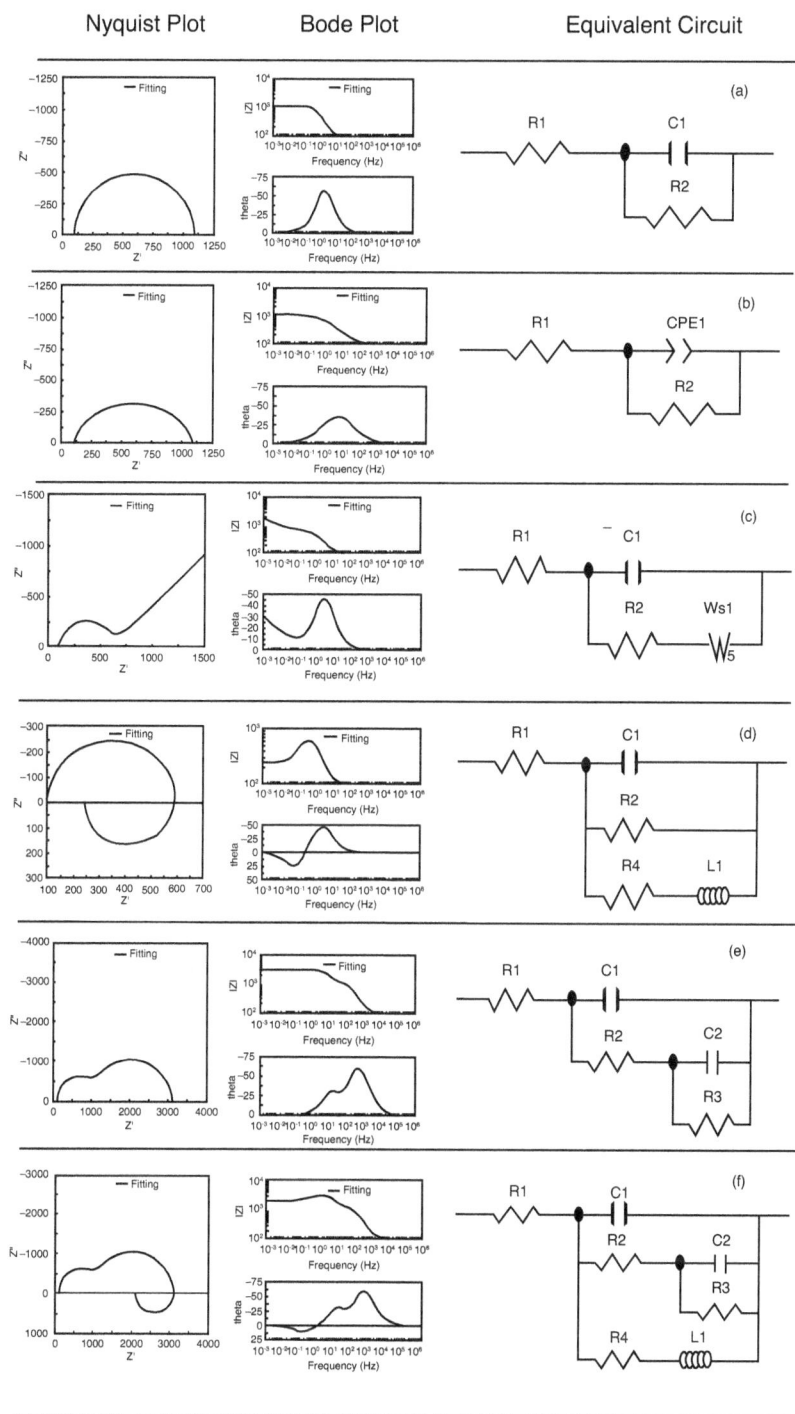

FIGURE 12.5 Nyquist plot and Bode plot with equivalent circuits to model different corrosion phenomena. [Reprinted with permission from Feliu (2020). Copyright (2020) MDPI].

ranges with no inductive behavior (Kirkland, Birbilis, and Staiger 2012). This model is shown in Figure 12.5e where R_1 and C_1 relate to the resistive and capacitive parts of the protective layer, R_2 denotes the charge-transfer resistance and C_2 is the double-layer capacitance. In EIS, complex impedance data as shown in Figure 12.5f are often encountered and multiple interpretations are given by various researchers (Pinto et al. 2011; Cao et al. 2013). A deep understanding of the underlying physiochemical process is necessary to interpret these complex EIS spectra. Thus, we can observe the challenges of modeling corrosion EIS data with various complicated equivalent electrical circuits.

12.2.3 PHOTOVOLTAIC CELL

In the past few decades, impedance spectroscopy has been extensively used to analyze electrochemical systems, such as ESD and corrosion. Nevertheless, using EIS for studying photovoltaic systems offers newer challenges pertaining to the unusual properties of materials and the complicated architecture of the devices. Different phenomena such as lattice distortions, dipole rearrangement, electrode polarization, electronic, and ionic conduction with various relaxation times are being investigated by using EIS. The aim is to find the permittivity (ϵ) of a particular sample. Figure 12.6 shows a comparison of the time for various relaxation processes (von Hauff 2019). In principle, all the possible relaxation processes can be investigated by changing the input signal frequency from zero to a large range. However, the EIS equipment can apply frequencies only in a certain range. The diffusion and recombination of carriers in photovoltaics are mainly analyzed by developing equivalent circuit models from EIS data. The diffusion and recombination are treated as two separate relaxation processes occupying distinct regions in the time scale with the recombination process being nonreversible and of the first order.

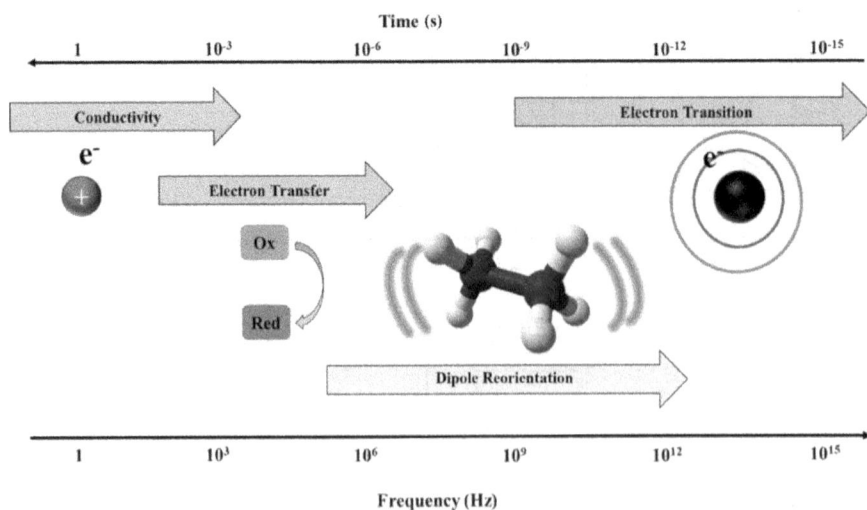

FIGURE 12.6 Time scale and frequencies for various relaxation processes (von Hauff 2019).

Jamnick and Maier (2001) related the phenomenon of carrier transport and recombination in semiconductors to the behavior of resistors and capacitors to model equivalent circuits. In the model, they considered one-dimensional transport of charge and neglected generation and recombination of carriers. There are three distinct components to represent the current density as follows:

$$J_\sigma = q\mu nE \tag{12.7}$$

$$J_{dis} = \epsilon_o \epsilon \frac{dE}{dt} \tag{12.8}$$

$$\nabla J_{\mu A} = -C_\mu \frac{d\mu_A}{dt} \tag{12.9}$$

where E is the electric field, q is the elementary charge, n is the carrier density, ϵ is the dielectric constant, ϵ_o is free space permittivity, $\propto$ is the mobility of the charge carrier, $\propto_A$ is the chemical potential, and C_α is the chemical capacitance. The phase difference between J_σ and the applied ac potential is zero and varies with the semiconductor conductivity proportionally. It is representative of the current flow due to the applied potential that obeys ohm's law. Thus, it can be modeled using an ideal resistor. The displacement current J_{dis} is not in phase with the applied potential by 90° and it signifies the dielectric behavior of the semiconductor. The last term $\nabla J_{\mu A}$ is another expression for the continuity equation considering the chemical capacitance and potential. Figure 12.7 depicts an illustration of electron transfer from a semiconductor's conduction band E_c to a redox system in a single step along with the associated equivalent circuit. C_{DB} represents the capacitive effects of the double layer at the semiconductor–electrolyte interface, C_{SC} represents the depletion layer capacitance and R_{CT} the charge-transfer resistance.

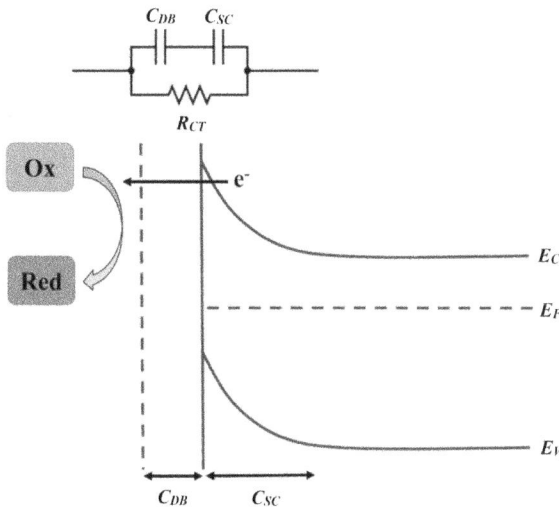

FIGURE 12.7 Electron transfer from the conduction band to a redox system in a semiconductor and the equivalent electrical circuit (von Hauff 2019).

Inorganic crystalline materials like perovskites, and CdTe, which are used in thin film photovoltaics have a very rapid transport and recombination process, which is difficult to be analyzed with EIS. To determine the sample's electronic structure, capacitance vs. frequency spectrum (C_f) is used in this case. The requirement for extracting reliable data with the C_f method is the presence of a depletion region that is well-defined. This becomes difficult when studying the recent photovoltaic devices as the presence of ions and mobile charge disturbs the depletion approximation and results in artifacts in the C_f spectra, as can be seen in the case of metal–halide perovskite cells that are reportedly found to show mixed ionic and electrical conduction (Yuan and Huang 2016). Because of the chemical capacitance and carrier freeze-out, C_f spectra of low-mobility, intrinsic semiconductors frequently exhibit artifacts in the high-frequency range. The signals of mobile ions and charge carriers in the C_f spectra must be carefully identified and interpreted, or in the best-case scenario, eliminated as they can affect the retrieved parameters, such as the built-in field, the dielectric constant, and trap profiles. The analysis processes that are mainly used in this context are summarized in Table 12.1, which shows the extractable parameters, the analysis's underlying assumptions and prerequisites, potential problems when used with the developing photovoltaic devices, and helpful measurement extensions for valid data analysis.

TABLE 12.1
Summary of various analysis procedures along with extractable parameters, assumptions and prerequisites, measurement extensions for valid data analysis, and potential problems for photovoltaics (von Hauff 2019)

Analysis procedure	Equivalent circuit modeling	Capacitance-frequency modeling	Transport analysis
Extractable parameters	Global device response, time scale of various relaxation processes	Concentration of doping	Ion and carrier transport characteristics, knowledge of dispersive transportation
Assumptions and prerequisites	Physical motivation for circuit model, relevant transport processes too slow to give measurable EIS data	Fulfillment of depletion approximation	Relevant transport processes too slow to give measurable EIS data
Measurement extensions for valid data analysis	Control of contact materials, thickness of absorber layer, DC offset, and ambient light conditions	Measurements dependent on temperature to obtain activation energies of trap states	Temperature-dependent measurements, different DC offsets
Potential problems	Lack of circuit model having a unique relation with frequency response	Effect of carrier freeze-out, mobile ions, chemical capacitance, and electrode polarization on impedance spectrum	The difficulty of creating and using transportation models

The requirement of linearity in the EIS technique is crucial. As the current–voltage characteristics are mostly nonlinear in photovoltaic devices, to acquire linear impedance data, one can selectively choose a region where there is a linear variation of current with voltage. This is a pseudolinear region that can be confirmed by varying the input voltage amplitude and checking for no change in the output impedance spectrum. This pseudolinear region depends on the sample, illumination condition, and the applied dc offset voltage. The stability of the system is another essential requirement. Meaningful EIS of photovoltaic systems with unstable nature can be done by sweeping the frequency from high to low range. This is because at low frequencies, the relaxation processes occur and are the ones that are irreversible in nature. Thus sweeping from high to low frequency, one can obtain reliable EIS data as the low-frequency range will be met at the end of the analysis and the irreversible processes would not affect the entire spectrum (von Hauff and Klotz 2022). All the EIS spectra that can be modeled with a single or series of RC circuits satisfy the Kramers–Kronig relations. In solar cells, a negative value of capacitance has been reported in many works (Ehrenfreund et al. 2007; Niu et al. 2018). Therefore, when the Kramers–Kronig compliance of the data is tested, it is a more consistent approach to add a negative value of capacitance rather than an inductor to the circuit design. It is difficult to model higher-order reactions with a single circuit over the entire frequency range. To improve the fit, one can adapt to more complex circuits, but this reduces the ability to interpret and relate to the underlying physical process. At the same time, complex circuits can be simplified mathematically to more understandable circuits, but this often results in the merging of relaxation processes with equivalent time constants. For example, two unique relaxation processes occurring parallelly that can be modeled using two RC elements in parallel say R_1C_1 and R_2C_2 can be equivalently modeled with a single RC circuit with $R = R_1 + R_2$ and $1/C = 1/C_1 + 1/C_2$. This makes it difficult to analyze the two processes separately. Thus, modeling an equivalent circuit that correctly reflects the physical processes of a photovoltaic cell is desirable as it can be effectively used to measure various parameters such as mobility, recombination rates, and diffusion constants at various regions of the current–voltage curve.

12.2.4 BIOSENSING

Sensors are ubiquitous and a crucial element of monitoring in various fields (Bhattacharya et al. 2019; Chauhan, Pandey, and Bhattacharya 2019; Pandey et al. 2019; Choudhury, Pandey, and Bhattacharya 2021; Pandey, Tatiya, and Bhattacharya 2021). The presence or concentration of a chemical or biological target is detected by biosensors. A common element in the biosensor is the recognition element that identifies the specific molecular component of interest in an analyte. The transduction of the recognition event can be done by various methods that involve colorimetry, electrochemistry, and optics. A schematic of the biosensing process is shown in Figure 12.8. Detailed analysis of bio-interaction at the electrode surface, such as enzyme–substrate interaction, antibody–antigen recognition, or cell capture can be done using EIS. Integration of nanomaterials, such as nanocomposites, nanotubes, nanowires, and nanoparticles, has improved the accuracy and reliability in the

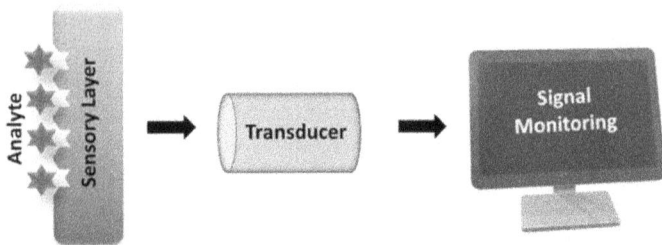

FIGURE 12.8   A schematic of the components of a biosensor.

detection of carcinogenic biomarkers, pathogens, and DNA using EIS. The binding of antibodies and their antigens to the sensor surface in EIS-based sensors results in a change in the output electrical signal. As a result, a variation in the charge transfer is detected, which indicates how many molecules are attached (Kim and Kang 2008).

Despite being a robust noninvasive technique for sensing, impedimetric biosensors have certain limitations and technical challenges. Detection of small entities, adequate stability, and repeatability are some of the issues faced in EIS-based sensors. Out of all other issues, the major one is the selectivity for use in real samples, as they frequently have several unknown nontarget molecules. Antibodies are the mostly used biorecognition element in biosensing. It is crucial that the nonspecific binding event can be reduced as very often the concentration of the target molecule is very less compared with the nontarget molecules in the analyte. Nonspecific binding mainly involves attaching proteins in a sample matrix to the surface of the sensor that gives a false biorecognition signal. The use of blocking agents like ethanolamine, bovine serum albumin, or cysteine can be helpful in reducing nonspecific binding (Magar, Hassan, and Mulchandani 2021). The ionic strength of electrolyte is also an important criterion for highly sensitive assays as the binding event happening within the Debye length is more sensitive and the ionic strength controls the Debye length (Tkac and Davis 2009). The selection of a proper redox probe for an electrode is essential for stable response in EIS, but often redox probes like $[Fe(CN)_6]^{3-/4-}$ have been reported to etch Au electrodes forming an intermediatory complex that affects the electrode's properties, resulting in a shift in the response (Lazar et al. 2016; Hua, Xia, and Long 2019). Redox cycling and extended incubation in the buffer can stabilize the baseline shift in the Au electrode in the presence of $[Fe(CN)_6]^{3-/4-}$ (Kanyong and Davis 2020). Utilizing the redox pair $Ru(NH_3)_6^{2+/3+}$ produced in situ electrochemically from $Ru(NH_3)_6^{3+}$ is another option for a stable EIS output response (Schrattenecker et al. 2019). Bogomolova et al. did an extensive study on various factors that might lead to impedance response from a nonspecific event (Bogomolova et al. 2009). Electrode contamination, the relative change in the position of DNA and protein molecules with respect to the electrode surface, repetitive measurements, additional measurements such as cyclic voltammetry or differential pulse voltammetry, and incubation in buffer, are some of those factors that lead to nonspecific impedance changes. Figure 12.9 shows change in impedance by repeated measurements, which is similar to protein binding (a,b), change in CV curve, and increase in impedance after performing additional CV measurements (c,d), and change in

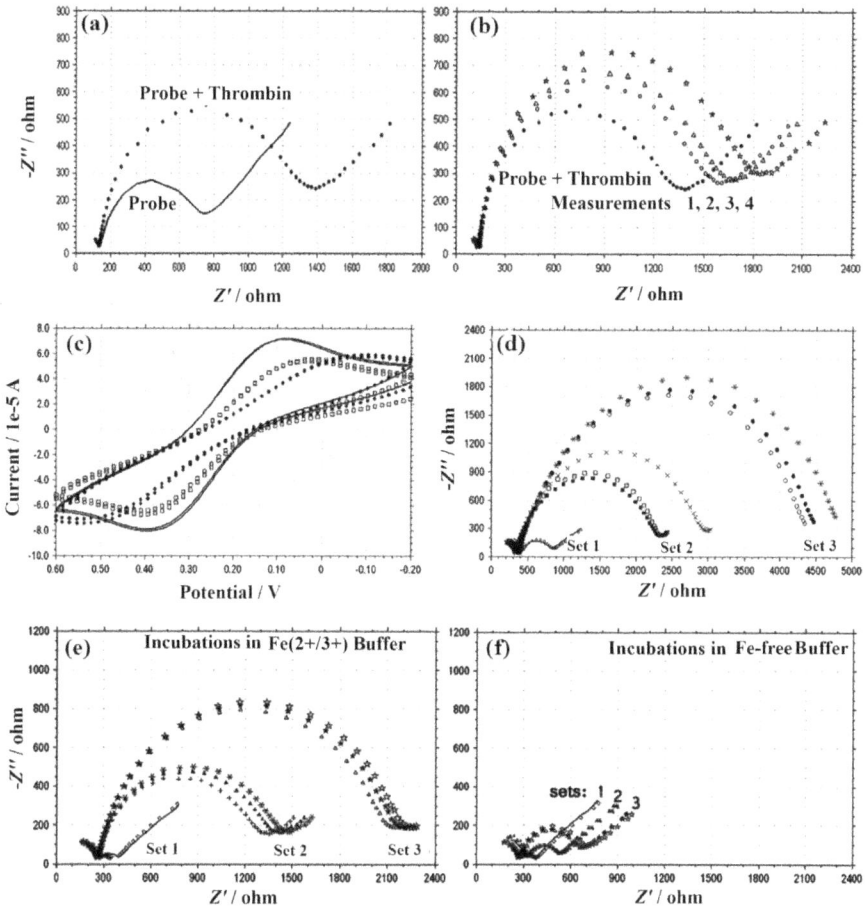

FIGURE 12.9 Change in impedance by various factors other than protein binding. [Reprinted with permission from Bogomolova et al. (2009). Copyright (2009) American Chemical Society].

impedance after incubation in different buffer solution and negligible change when incubated in ferrocyanide free buffer (e,f).

When compared with normal cells, cancer cells have different localized dielectric characteristics, which can be observed as variations in electrical conductivity and capacitance via impedance spectroscopy. EIS has been used in the clinical examination for the diagnosis of breast cancer. An increased conductivity and/or capacitance of a lesion measured through EIS was considered positive for cancer. It is seen that skin alterations, air bubbles, scars, naevi, contact artifacts, interfering bone, and hairs often lead to a false positive result. However, the experienced medical practitioner can identify such inaccuracies (Malich et al. 2003). EIS has also been extensively used in blood glucose monitoring. The blood glucose level has a significant effect on the electric and dielectric properties of human blood (Abdalla, Al-ameer, and Al-Magaishi 2010). The Cole impedance model considering the various layers of skin is shown in Figure 12.10. The main challenge that is faced in skin impedance-based glucose monitoring is the specificity of glucose response compared to other interferences attributable to varying physiological conditions. The presence of an air gap between the skin and the EIS sensor results in a parasitic capacitance that must be considered for accurate measurements. Minimizing the changes to the skin surface due to sensor attachment, optimizing the model for multiparameter prediction and better calibration are some key factors that can reduce interferences and improve EIS-based glucose monitoring. Table 12.2 shows the improvement of EIS-based glucose sensors in the past few years. However, further improvement is much needed for commercial use.

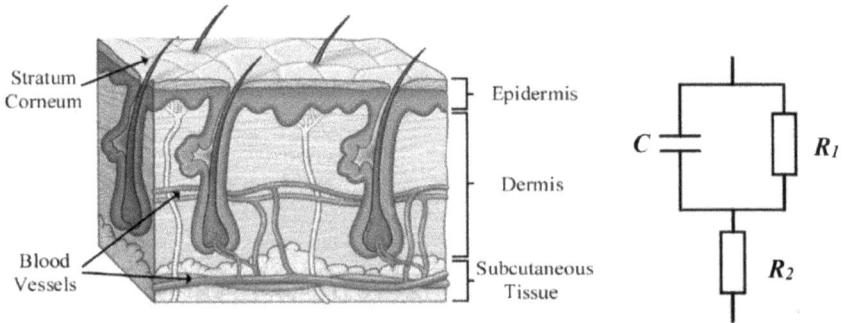

FIGURE 12.10 Cole skin impedance model. [Reprinted with permission from Huang, Zhang, and Wu (2020). Copyright (2020) Elsevier].

TABLE 12.2

Improvement of EIS-based glucose sensors. [Reprinted with permission from Huang, Zhang, and Wu (2020). Copyright (2020) Elsevier]

Year of publication, ref	2006 (Caduff et al.)	2011 (Mueller et al.)	2015 (Andreas Caduff et al.)	2017 (Geng et al.)
Sensors	Impedance sensor; temperature sensor for skin and device.	Long, mid, and short impedance sensor; temperature and humidity sensor; sweat sensor	Long, mid, and short impedance sensor; temperature and humidity sensor; sweat sensor	Low and high-frequency impedance sensor; temperature and humidity sensor
Root mean square error (RMSE)	45	36	NA	15
Mean absolute relative difference (MARD)	NA	21.5%	21.1%	NA

12.3 CONCLUSION AND FUTURE PERSPECTIVES

EIS being a well-established quantitative technique finds vast application in evaluation and diagnostics. Although EIS has several advantages over other electroanalytical techniques like small signal analysis, better surface sensitivity with minimal interaction with the fluids, and lesser detection times, there are several disadvantages of the same.

The major disadvantage of EIS measurement is the cruciality of the requirement of linear behavior between current and voltage characteristics. Although many techniques have been used like small area analysis, especially in the case of photovoltaic cells, there are still issues with the variability in the larger domains. Another issue is the problem of the unknowns and differentiating the same from the targeted analytes or agents. This can be seen in both biosensing as well as corrosion studies of various metallic elements. Several methods have been developed like coating the electrodes with agents or dopants that reduce nonspecific binding or reduce the variability at the boundary surface. A change in physical, chemical, or environmental factors of the base medium during analysis can itself change the evaluations carried out by the EIS and affect its stability due to its high sensitivity.

EIS, a powerful tool in itself, is a very complex electrochemical method, and understanding the intricacies related to the interactions, sensing, and quantification of the various detection parameters and their effects on the current–voltage characteristics with the specific experimental conditions need to be studied properly before implementation of the EIS tool for distinct applications.

REFERENCES

Abdalla, S., S. S. Al-ameer, and S. H. Al-Magaishi. 2010. "Electrical Properties with Relaxation through Human Blood." *Biomicrofluidics* 4 (3): 034101. https://doi.org/10.1063/1.3458908.

Amirudin, A., and D. Thieny. 1995. "Application of Electrochemical Impedance Spectroscopy to Study the Degradation of Polymer-Coated Metals." *Progress in Organic Coatings* 26 (1): 1–28. https://doi.org/10.1016/0300-9440(95)00581-1.

Aparicio, Mario, and Jadra Mosa. 2018. "Electrochemical Characterization of Sol–Gel Coatings for Corrosion Protection of Metal Substrates." *Journal of Sol-Gel Science and Technology* 88 (1): 77–89. https://doi.org/10.1007/s10971-018-4785-9.

Barsukov, Yevgen, and Macdonald, J. Ross and others. 2012. "Electrochemical Impedance Spectroscopy." *Characterization of Materials* 2: 898–913.

Bhattacharya, Shantanu, Avinash Kumar Agarwal, Om Prakash, Shailendra Singh, Mohit Pandey, and Rishi Kant. 2019. "Introduction to Sensors for Aerospace and Automotive Applications." In *Sensors for Automotive and Aerospace Applications*, edited by Shantanu Bhattacharya, Avinash Kumar Agarwal, Om Prakash, and Shailendra Singh, 1–6. Singapore: Springer Singapore. https://doi.org/10.1007/978-981-13-3290-6_1.

Bogomolova, A., E. Komarova, K. Reber, T. Gerasimov, O. Yavuz, S. Bhatt, and M. Aldissi. 2009. "Challenges of Electrochemical Impedance Spectroscopy in Protein Biosensing." *Analytical Chemistry* 81 (10): 3944–49. https://doi.org/10.1021/ac9002358.

Branzoi, Florina, and Viorel Branzoi. 2015. "The Electrochemical Behaviour of PEDOT Film Electrosynthesized in Presence of Some Dopants." *Open Journal of Organic Polymer Materials* 05 (04): 89–102. https://doi.org/10.4236/ojopm.2015.54010.

Brett, Christopher M. A., Lidia Dias, Bruno Trindade, Robert Fischer, and Sascha Mies. 2006. "Characterisation by EIS of Ternary Mg Alloys Synthesised by Mechanical Alloying." *Electrochimica Acta* 51 (8–9): 1752–60. https://doi.org/10.1016/j.electacta.2005.02.124.

Caduff, A., F. Dewarrat, M. Talary, G. Stalder, L. Heinemann, and Yu. Feldman. 2006. "Non-Invasive Glucose Monitoring in Patients with Diabetes: A Novel System Based on Impedance Spectroscopy." *Biosensors and Bioelectronics* 22 (5): 598–604. https://doi.org/10.1016/j.bios.2006.01.031.

Caduff, Andreas, Mattia Zanon, Martin Mueller, Pavel Zakharov, Yuri Feldman, Oscar De Feo, Marc Donath, Werner A. Stahel, and Mark S. Talary. 2015. "The Effect of a Global, Subject, and Device-Specific Model on a Noninvasive Glucose Monitoring Multisensor System." *Journal of Diabetes Science and Technology* 9 (4): 865–72. https://doi.org/10.1177/1932296815579459.

Cao, Fuyong, Zhiming Shi, Joelle Hofstetter, Peter J. Uggowitzer, Guangling Song, Ming Liu, and Andrej Atrens. 2013. "Corrosion of Ultra-High-Purity Mg in 3.5% NaCl Solution Saturated with Mg(OH)2." *Corrosion Science* 75 (October): 78–99. https://doi.org/10.1016/j.corsci.2013.05.018.

Cesiulis, H., N. Tsyntsaru, A. Ramanavicius, and G. Ragoisha. 2016. "The Study of Thin Films by Electrochemical Impedance Spectroscopy." In *Nanostructures and thin films for multifunctional applications: technology, properties and devices*: 3–42. https://doi.org/10.1007/978-3-319-30198-3_1.

Chang, Jian-Wei, Xing-Wu Guo, Peng-Huai Fu, Li-Ming Peng, and Wen-Jiang Ding. 2007. "Effect of Heat Treatment on Corrosion and Electrochemical Behaviour of Mg–3Nd–0.2Zn–0.4Zr (Wt.%) Alloy." *Electrochimica Acta* 52 (9): 3160–67. https://doi.org/10.1016/j.electacta.2006.09.069.

Chauhan, Pankaj Singh, Mohit Pandey, and Shantanu Bhattacharya. 2019. "Paper Based Sensors for Environmental Monitoring." In *Paper Microfluidics: Theory and Applications*: 165–81. Springer. https://doi.org/10.1007/978-981-15-0489-1_10.

Choudhury, Sagnik Sarma, Mohit Pandey, and Shantanu Bhattacharya. 2021. "Recent Developments in Surface Modification of PEEK Polymer for Industrial Applications: A Critical Review." *Reviews of Adhesion and Adhesives* 9 (3): 401–33. https://doi.org/10.47750/RAA/9.3.03.

Choudhury, Sagnik Sarma, Neelabh Jyoti Saharia, Suvan Dev Choudhury, and B. Surekha. 2020. "Influences and Applications of Aluminum Addition on the Mechanical Properties of Pure Magnesium." In *Innovative Product Design and Intelligent Manufacturing Systems: Select Proceedings of ICIPDIMS 2019*: 349–57. Springer Singapore. https://doi.org/10.1007/978-981-15-2696-1_34.

Coy, A. E., F. Viejo, F. J. Garcia-Garcia, Z. Liu, P. Skeldon, and G. E. Thompson. 2010. "Effect of Excimer Laser Surface Melting on the Microstructure and Corrosion Performance of the Die Cast AZ91D Magnesium Alloy." *Corrosion Science* 52 (2): 387–97. https://doi.org/10.1016/j.corsci.2009.09.025.

Dai, Haifeng, Bo Jiang, and Xuezhe Wei. 2018. "Impedance Characterization and Modeling of Lithium-Ion Batteries Considering the Internal Temperature Gradient." *Energies* 11 (1): 220. https://doi.org/10.3390/en11010220.

Delgado, M. C., F. R. García-Galvan, I. Llorente, P. Pérez, P. Adeva, and S. Feliu. 2017. "Influence of Aluminium Enrichment in the Near-Surface Region of Commercial Twin-Roll Cast AZ31 Alloys on Their Corrosion Behaviour." *Corrosion Science* 123 (July): 182–96. https://doi.org/10.1016/j.corsci.2017.04.027.

Ehrenfreund, E., C. Lungenschmied, G. Dennler, H. Neugebauer, and N. S. Sariciftci. 2007. "Negative Capacitance in Organic Semiconductor Devices: Bipolar Injection and Charge Recombination Mechanism." *Applied Physics Letters* 91 (1): 012112. https://doi.org/10.1063/1.2752024.

Esmaily, M., J. E. Svensson, S. Fajardo, N. Birbilis, G. S. Frankel, S. Virtanen, R. Arrabal, S. Thomas, and L. G. Johansson. 2017. "Fundamentals and Advances in Magnesium Alloy Corrosion." *Progress in Materials Science* 89 (August): 92–193. https://doi.org/10.1016/j.pmatsci.2017.04.011.

Fekry, A. M., and M. Z. Fatayerji. 2009. "Electrochemical Corrosion Behavior of AZ91D Alloy in Ethylene Glycol." *Electrochimica Acta* 54 (26): 6522–28. https://doi.org/10.1016/j.electacta.2009.06.025.

Feliu, Sebastián. 2020. "Electrochemical Impedance Spectroscopy for the Measurement of the Corrosion Rate of Magnesium Alloys: Brief Review and Challenges." *Metals* 10 (6): 775. https://doi.org/10.3390/met10060775.

Fuertes, A. B., G. Lota, T. A. Centeno, and E. Frackowiak. 2005. "Templated Mesoporous Carbons for Supercapacitor Application." *Electrochimica Acta* 50 (14): 2799–805. https://doi.org/10.1016/j.electacta.2004.11.027.

Geng, Zhanxiao, Fei Tang, Yadong Ding, Shuzhe Li, and Xiaohao Wang. 2017. "Noninvasive Continuous Glucose Monitoring Using a Multisensor-Based Glucometer and Time Series Analysis." *Scientific Reports* 7 (1): 12650. https://doi.org/10.1038/s41598-017-13018-7.

Gheem, E. Van, R. Pintelon, J. Vereecken, J. Schoukens, A. Hubin, P. Verboven, and O. Blajiev. 2004. "Electrochemical Impedance Spectroscopy in the Presence of Non-Linear Distortions and Non-Stationary Behaviour." *Electrochimica Acta* 49 (26): 4753–62. https://doi.org/10.1016/j.electacta.2004.05.039.

Giner-Sanz, J.J., E.M. Ortega, and V. Pérez-Herranz. 2016. "Harmonic Analysis Based Method for Linearity Assessment and Noise Quantification in Electrochemical Impedance Spectroscopy Measurements: Theoretical Formulation and Experimental Validation for Tafelian Systems." *Electrochimica Acta* 211 (September): 1076–91. https://doi.org/10.1016/j.electacta.2016.06.133.

Gonzalez-Garcia, Y., S. J. Garcia, and J. M. C. Mol. 2016. "Electrochemical Techniques for the Study of Self Healing Coatings." In *Active Protective Coatings: New-Generation Coatings for Metals*: 203–40. https://doi.org/10.1007/978-94-017-7540-3_9.

Hauff, Elizabeth von. 2019. "Impedance Spectroscopy for Emerging Photovoltaics." *The Journal of Physical Chemistry C* 123 (18): 11329–46. https://doi.org/10.1021/acs.jpcc.9b00892.

Hauff, Elizabeth von, and Dino Klotz. 2022. "Impedance Spectroscopy for Perovskite Solar Cells: Characterisation, Analysis, and Diagnosis." *Journal of Materials Chemistry C* 10 (2): 742–61. https://doi.org/10.1039/D1TC04727B.

Hirschorn, Bryan, and Mark E. Orazem. 2009. "On the Sensitivity of the Kramers–Kronig Relations to Nonlinear Effects in Impedance Measurements." *Journal of The Electrochemical Society* 156 (10): C345. https://doi.org/10.1149/1.3190160.

Hua, Xin, Hai-Lun Xia, and Yi-Tao Long. 2019. "Revisiting a Classical Redox Process on a Gold Electrode by Operando ToF-SIMS: Where Does the Gold Go?" *Chemical Science* 10 (24): 6215–19. https://doi.org/10.1039/C9SC00956F.

Huang, Jiamei, Ying Zhang, and Jayne Wu. 2020. "Review of Non-Invasive Continuous Glucose Monitoring Based on Impedance Spectroscopy." *Sensors and Actuators A: Physical* 311 (August): 112103. https://doi.org/10.1016/j.sna.2020.112103.

Jafar, M. I., J. L. Dawson, and D. G. John. 1993. "Electrochemical Impedance and Harmonic Analysis Measurements on Steel in Concrete." *Electrochemical Impedance: Analysis and Interpretation*, 384-384–20. 100 Barr Harbor Drive, PO Box C700, West Conshohocken, PA 19428-2959: ASTM International. https://doi.org/10.1520/STP18081S.

Jamnik, J., and J. Maier. 2001. "Generalised Equivalent Circuits for Mass and Charge Transport: Chemical Capacitance and Its Implications." *Physical Chemistry Chemical Physics* 3 (9): 1668–78. https://doi.org/10.1039/b100180i.

Jia, Ru, Tuba Unsal, Dake Xu, Yassir Lekbach, and Tingyue Gu. 2019. "Microbiologically Influenced Corrosion and Current Mitigation Strategies: A State of the Art Review." *International Biodeterioration & Biodegradation* 137 (February): 42–58. https://doi.org/10.1016/j.ibiod.2018.11.007.

Jungst, Rudolph G., Ganesan Nagasubramanian, Herbert L. Case, Bor Yann Liaw, Angel Urbina, Thomas L. Paez, and Daniel H. Doughty. 2003. "Accelerated Calendar and Pulse Life Analysis of Lithium-Ion Cells." *Journal of Power Sources* 119–121 (June): 870–73. https://doi.org/10.1016/S0378-7753(03)00193-9.

Kanyong, Prosper, and Jason J. Davis. 2020. "Homogeneous Functional Self-Assembled Monolayers: Faradaic Impedance Baseline Signal Drift Suppression for High-Sensitivity Immunosensing of C-Reactive Protein." *Journal of Electroanalytical Chemistry* 856 (January): 113675. https://doi.org/10.1016/j.jelechem.2019.113675.

Kierzek, K., E. Frackowiak, G. Lota, G. Gryglewicz, and J. Machnikowski. 2004. "Electrochemical Capacitors Based on Highly Porous Carbons Prepared by KOH Activation." *Electrochimica Acta* 49 (4): 515–23. https://doi.org/10.1016/j.electacta.2003.08.026.

Kim, Dong, and Dae Kang. 2008. "Molecular Recognition and Specific Interactions for Biosensing Applications." *Sensors* 8 (10): 6605–41. https://doi.org/10.3390/s8106605.

Kirkland, N. T., N. Birbilis, and M. P. Staiger. 2012. "Assessing the Corrosion of Biodegradable Magnesium Implants: A Critical Review of Current Methodologies and Their Limitations." *Acta Biomaterialia* 8 (3): 925–36. https://doi.org/10.1016/j.actbio.2011.11.014.

Kirkland, Nicholas Travis, and Nick Birbilis. 2014. *Magnesium Biomaterials*. Cham: Springer International Publishing. https://doi.org/10.1007/978-3-319-02123-2.

Lakshmi, T. Sree, Sagnik Sarma Choudhury, K. Gnana Sundari, and B. Surekha. 2020. "Investigation on the Effect of Different Dielectric Fluids During Powder Mixed EDM of Alloy Steel." In *Innovative Product Design and Intelligent Manufacturing Systems: Select Proceedings of ICIPDIMS 2019*: 1067–75. Springer Singapore. https://doi.org/10.1007/978-981-15-2696-1_103.

Landinger, Thomas F., Guenter Schwarzberger, and Andreas Jossen. 2019. "A Novel Method for High Frequency Battery Impedance Measurements." In *2019 IEEE International Symposium on Electromagnetic Compatibility, Signal & Power Integrity (EMC+SIPI)*, 106–10. IEEE. https://doi.org/10.1109/ISEMC.2019.8825315.

Lazar, Jaroslav, Christoph Schnelting, Evelina Slavcheva, and Uwe Schnakenberg. 2016. "Hampering of the Stability of Gold Electrodes by Ferri-/Ferrocyanide Redox Couple Electrolytes during Electrochemical Impedance Spectroscopy." *Analytical Chemistry* 88 (1): 682–87. https://doi.org/10.1021/acs.analchem.5b02367.

Li, Jie, Catia Arbizzani, Signe Kjelstrup, Jie Xiao, Yong-yao Xia, Yan Yu, Yong Yang, et al. 2020. "Good Practice Guide for Papers on Batteries for the Journal of Power Sources." *Journal of Power Sources* 452 (March): 227824. https://doi.org/10.1016/j.jpowsour.2020.227824.

Liu, Yuxiang, Michele Curioni, and Zhu Liu. 2018. "Correlation between Electrochemical Impedance Measurements and Corrosion Rates of Mg-1Ca Alloy in Simulated Body Fluid." *Electrochimica Acta* 264 (February): 101–8. https://doi.org/10.1016/j.electacta.2018.01.121.

Lorenz, W. J., and F. Mansfeld. 1981. "Determination of Corrosion Rates by Electrochemical DC and AC Methods." *Corrosion Science* 21 (9–10): 647–72. https://doi.org/10.1016/0010-938X(81)90015-9.

Magar, Hend S., Rabeay Y. A. Hassan, and Ashok Mulchandani. 2021. "Electrochemical Impedance Spectroscopy (EIS): Principles, Construction, and Biosensing Applications." *Sensors* 21 (19): 6578. https://doi.org/10.3390/s21196578.

Malich, A., T. Böhm, M. Facius, I. Kleinteich, M. Fleck, D. Sauner, R. Anderson, and W. A. Kaiser. 2003. "Electrical Impedance Scanning as a New Imaging Modality in Breast Cancer Detection—a Short Review of Clinical Value on Breast Application, Limitations and Perspectives." *Nuclear Instruments and Methods in Physics Research Section A: Accelerators, Spectrometers, Detectors and Associated Equipment* 497 (1): 75–81. https://doi.org/10.1016/S0168-9002(02)01894-6.

Meddings, Nina, Marco Heinrich, Frédéric Overney, Jong-Sook Lee, Vanesa Ruiz, Emilio Napolitano, Steffen Seitz et al. 2020. "Application of Electrochemical Impedance Spectroscopy to Commercial Li-Ion Cells: A Review." *Journal of Power Sources* 480 (December): 228742. https://doi.org/10.1016/j.jpowsour.2020.228742.

Momma, Toshiyuki, Mariko Matsunaga, Daikichi Mukoyama, and Tetsuya Osaka. 2012. "Ac Impedance Analysis of Lithium Ion Battery under Temperature Control." *Journal of Power Sources* 216 (October): 304–07. https://doi.org/10.1016/j.jpowsour.2012.05.095.

Moreto, J. A., C. E. B. Marino, W. W. Bose Filho, L. A. Rocha, and J. C. S. Fernandes. 2014. "SVET, SKP and EIS Study of the Corrosion Behaviour of High Strength Al and Al–Li Alloys Used in Aircraft Fabrication." *Corrosion Science* 84 (July): 30–41. https://doi.org/10.1016/j.corsci.2014.03.001.

Mueller, Martin, Mark S. Talary, Lisa Falco, Oscar De Feo, Werner A. Stahel, and Andreas Caduff. 2011. "Data Processing for Noninvasive Continuous Glucose Monitoring with a Multisensor Device." *Journal of Diabetes Science and Technology* 5 (3): 694–702. https://doi.org/10.1177/193229681100500324.

Niu, Quan, N. Irina Crăciun, Gert-Jan A. H. Wetzelaer, and Paul W. M. Blom. 2018. "Origin of Negative Capacitance in Bipolar Organic Diodes." *Physical Review Letters* 120 (11): 116602. https://doi.org/10.1103/PhysRevLett.120.116602.

Noori, Abolhassan, Maher F. El-Kady, Mohammad S. Rahmanifar, Richard B. Kaner, and Mir F. Mousavi. 2019. "Towards Establishing Standard Performance Metrics for Batteries, Supercapacitors and Beyond." *Chemical Society Reviews* 48 (5): 1272–1341. https://doi.org/10.1039/C8CS00581H.

Orazem, Mark E. and Tribollet, Bernard. 2008. "Electrochemical Impedance Spectroscopy." Wiley, New Jersey, 1–525.

Orjan G. Martinsen, and Sverre Grimnes. 2000. *Bioimpedance and Bioelectricity Basics*. Academic Press.

Pandey, Mohit, Shreyansh Tatiya, Shantanu Bhattacharya, and Shailendra Singh. 2019. "Sensors in Assembly Shop in Automobile Manufacturing." In *Sensors for Automotive and Aerospace Applications*, edited by Shantanu Bhattacharya, Avinash Kumar Agarwal, Om Prakash, and Shailendra Singh, 193–207. Singapore: Springer Singapore. https://doi.org/10.1007/978-981-13-3290-6_10.

Pandey, Mohit, Shreyansh Tatiya, and Shantanu Bhattacharya. 2021. "Design and Development of MEMS-Based Sensors for Wearable Diagnostic Applications." In *MEMS Applications in Biology and Healthcare*, 1–34. AIP Publishing. https://doi.org/10.1063/9780735423954_010.

Pettit, C. M., J. E. Garland, N. R. Etukudo, K. A. Assiongbon, S. B. Emery, and D. Roy. 2002. "Electrodeposition of Indium on Molybdenum Studied with Optical Second Harmonic Generation and Electrochemical Impedance Spectroscopy." *Applied Surface Science* 202 (1–2): 33–46. https://doi.org/10.1016/S0169-4332(02)00798-5.

Pinto, R., M. G. S. Ferreira, M. J. Carmezim, and M. F. Montemor. 2011. "The Corrosion Behaviour of Rare-Earth Containing Magnesium Alloys in Borate Buffer Solution." *Electrochimica Acta* 56 (3): 1535–45. https://doi.org/10.1016/j.electacta.2010.09.081.

Ren, Xiaoming, and Peter G. Pickup. 1992. "Ionic and Electronic Conductivity of Poly-(3-methylpyrrole-4-carboxylic Acid)." *Journal of The Electrochemical Society* 139 (8): 2097–2105. https://doi.org/10.1149/1.2221185.

Rodrigo de Siqueira Melo, Simone Louise Delarue Cezar Brasil, Ladimir José de Carvalho, Aricelso Maia Limaverde Filho, and Cid Pereira. 2016. "Assessment of the Antifouling Effect of Exopolysaccharides Incorporated into Copper Oxide-Based Organic Paint." *International Journal of Electrochemical Science* 11: 7750–63.

Rumpf, Katharina, Maik Naumann, and Andreas Jossen. 2017. "Experimental Investigation of Parametric Cell-to-Cell Variation and Correlation Based on 1100 Commercial Lithium-Ion Cells." *Journal of Energy Storage* 14 (December): 224–43. https://doi.org/10.1016/j.est.2017.09.010.

Schrattenecker, Julian D., Rudolf Heer, Eva Melnik, Thomas Maier, Günter Fafilek, and Rainer Hainberger. 2019. "Hexaammineruthenium (II)/(III) as Alternative Redox-Probe to Hexacyanoferrat (II)/(III) for Stable Impedimetric Biosensing with Gold Electrodes." *Biosensors and Bioelectronics* 127 (February): 25–30. https://doi.org/10.1016/j.bios.2018.12.007.

Tkac, Jan, and Jason J. Davis. 2009. "Label-Free Field Effect ProteinSensing." In *Engineering the Bioelectronic Interface*, 193–224. Royal Society of Chemistry, Cambridge. https://doi.org/10.1039/9781847559777-00193.

Vadim F. Lvovich. 2012. *Impedance Spectroscopy: Applications to Electrochemical and Dielectric Phenomena*. John Wiley & Sons.

Vetter, J., P. Novák, M. R. Wagner, C. Veit, K.-C. Möller, J. O. Besenhard, M. Winter, M. Wohlfahrt-Mehrens, C. Vogler, and A. Hammouche. 2005. "Ageing Mechanisms in Lithium-Ion Batteries." *Journal of Power Sources* 147 (1–2): 269–81. https://doi.org/10.1016/j.jpowsour.2005.01.006.

Waag, Wladislaw, Stefan Käbitz, and Dirk Uwe Sauer. 2013. "Experimental Investigation of the Lithium-Ion Battery Impedance Characteristic at Various Conditions and Aging States and Its Influence on the Application." *Applied Energy* 102 (February): 885–97. https://doi.org/10.1016/j.apenergy.2012.09.030.

Wang, Hainan, and Laurent Pilon. 2012. "Intrinsic Limitations of Impedance Measurements in Determining Electric Double Layer Capacitances." *Electrochimica Acta* 63 (February): 55–63. https://doi.org/10.1016/j.electacta.2011.12.051.

Yuan, Yongbo, and Jinsong Huang. 2016. "Ion Migration in Organometal Trihalide Perovskite and Its Impact on Photovoltaic Efficiency and Stability." *Accounts of Chemical Research* 49 (2): 286–93. https://doi.org/10.1021/acs.accounts.5b00420.

Zheng, J. P., P. C. Goonetilleke, C. M. Pettit, and D. Roy. 2010. "Probing the Electrochemical Double Layer of an Ionic Liquid Using Voltammetry and Impedance Spectroscopy: A Comparative Study of Carbon Nanotube and Glassy Carbon Electrodes in [EMIM]+[EtSO4]−." *Talanta* 81 (3): 1045–55. https://doi.org/10.1016/j.talanta.2010.01.059.

Zhu, Jiangong, Mariyam Susana Dewi Darma, Michael Knapp, Daniel R. Sørensen, Michael Heere, Qiaohua Fang, Xueyuan Wang, et al. 2020. "Investigation of Lithium-Ion Battery Degradation Mechanisms by Combining Differential Voltage Analysis and Alternating Current Impedance." *Journal of Power Sources* 448 (February): 227575. https://doi.org/10.1016/j.jpowsour.2019.227575.

Index